International Review of  A Survey of
**Cytology** **Cell Biology**

**VOLUME 159**

# International Review of Cytology

## A Survey of Cell Biology

Edited by

**Kwang W. Jeon**
Department of Zoology
The University of Tennessee
Knoxville, Tennessee

**Jonathan Jarvik**
Department of Biological Sciences
Carnegie Mellon University
Pittsburgh, Pennsylvania

VOLUME 159

**ACADEMIC PRESS**
San Diego   New York   Boston   London   Sydney   Tokyo   Toronto

Copyright © 1995 by ACADEMIC PRESS, INC.

Academic Press, Inc.
A Division of Harcourt Brace & Company
525 B Street, Suite 1900, San Diego, California 92101-4495

*United Kingdom Edition published by*
Academic Press Limited
24-28 Oval Road, London NW1 7DX

International Standard Serial Number: 0074-7696

International Standard Book Number: 0-12-364562-X

PRINTED IN THE UNITED STATES OF AMERICA
95  96  97  98  99  00  EB  9  8  7  6  5  4  3  2  1

# CONTENTS

v

# Nuclear Remodeling in Response to Steroid Hormone Action

### Klaus Brasch and Robert L. Ochs

# Effects of Axotomy, Deafferentation, and Reinnervation on Sympathetic Ganglionic Synapses: A Comparative Study

### Jacques Taxi and Daniel Eugene

# Chondrocyte Differentiation

### Ranieri Cancedda, Fiorella Descalzi Cancedda, and Patrizio Castagnola

# CONTRIBUTORS

Numbers in parentheses indicate the pages on which the authors' contributions begin.

Klaus Brasch (101), *Department of Biology, California State University, San Bernardino, California 92407*

Ranieri Cancedda (265), *Centro di Biotecnologie Avanzate, Istituto Nazionale per la Ricerca sul Cancro, and Istituto di Oncologia Clinica e Sperimentale, Università di Genova, 16132 Genova, Italy*

Patrizio Castagnola (265), *Centro di Biotecnologie Avanzate, Istituto Nazionale per la Ricerca sul Cancro, 16132 Genova, Italy*

Gillian W. Cockerill (113), *Hanson Center for Cancer Research, Institute of Medical and Veterinary Research, Adelaide 5000, South Australia, Australia*

Fiorella Descalzi Cancedda (265), *Centro di Biotecnologie Avanzate, Istituto Nazionale per la Ricerca sul Cancro, 16132 Genova, and Istituto Internazionale di Genetica e Biofisica, Consiglio Nazionale delle Ricerche, 80125 Naples, Italy*

Daniel Eugene (195), *Institut des Neurosciences, C.N.R.S., Université Pierre et Marie Curie, 75252 Paris, France*

Jennifer R. Gamble (113), *Hanson Center for Cancer Research, Institute of Medical and Veterinary Research, Adelaide 5000, South Australia, Australia*

Robert L. Ochs (161), *W. M. Keck Autoimmune Disease Center and Department of Molecular and Experimental Medicine, The Scripps Research Institute, La Jolla, California 92093*

Vladimir R. Pantić (1), *Serbian Academy of Sciences and Arts, 11000 Belgrade, Yugoslavia*

Jacques Taxi (195), *Institut des Neurosciences, C.N.R.S., Université Pierre et Marie Curie, 75252 Paris, France*

Mathew A. Vadas (113), *Hanson Center for Cancer Research, Institute of Medical and Veterinary Research, Adelaide 5000, South Australia, Australia*

# Biology of Hypothalamic Neurons and Pituitary Cells

Vladimir R. Pantić
Serbian Academy of Sciences and Arts,
11000 Belgrade, Yugoslavia

Results obtained by examining hypothalamic neurons producing precursors to neurohormones, and pituitary cells synthesizing peptide and glycoprotein families of hormones, and recent advances in comparative endocrinology, have been summarized and considered from the following viewpoints: species specificity in the organization and communication of the hypothalamic neurons with different brain areas lying inside the BBB and with CVOs; sensitivity of hypothalamic neurons and pituitary cells to the environmental stimuli; gonadal steroids as modulators of gene expression needed for neuronal differentiation and synaptogenesis; dose(s)-dependent pituitary cell proliferation and differentiation; an inverse relationship between PRL and GH synthesis and release and also between degree of hyperplasia and hypertrophy of PRL cells and retardation of GTH cell differentiation; and responsiveness of neurons producing CRH, and of neurons and pituitary cells synthesizing POMC hormones, to stress and glucocorticosteroids. These data show that growth of the animals may be stimulated, retarded, or inhibited; reproductive properties and behavior may be under hormonal control; and character of responsiveness in reaction to stress, and ability for adaptation and other related functions, may be controlled.

**KEY WORDS:** Hypothalamic neurons, Pituitary cells, Neurohormones, Peptide/glycoprotein hormones, Ontogenic development.

## I. Introduction

The discovery of neurosecretory cells (NSCs), their significance, their common pathway in neuroendocrine integration, and general principles of their mode of communication were described by pioneers in neuroendo-crinology (E. Scharrer, 1952, 1966; B. Scharrer, 1967, 1974; Bern, 1967, 1970; Blackwell and Guillemin, 1973; Schally *et al.,* 1973, 1977; Flerkó, 1975). Light and electron microscopy methods were used for this early work and NSCs in the hypothalamus of different species were identified. Radioimmunoassay (RIA) and immunocytochemistry (ICC) have been used so that a specific antibody may recognize an identical amino acid sequence in the precursor molecules and neurohormones.

Evolutionarily "old" neurons in coelenterates are strongly peptidergic in nature. Most coelenterate neuropeptides have a long lifetime after release. All neuropeptides are synthesized as precursor molecules. In coelenterates, these molecules are processed by enzymes that are similar to those of higher animals (Grimmelikhuijzen *et al.,* 1992), and the precur-sor proteins have been identified for most neuropeptides isolated from these animals. Precursor molecules for neuropeptides are synthesized in neurons in all animal species from coelenterates to mammals, and their unity and diversity have been reported (Joosse, 1990).

In annelids the neurosecretory cells are developed in the cerebral gan-glion. The main site of neuropeptide release in these animals and in insects is the neurohemal organ (B. Scharrer, 1968; Al-Jousuf, 1992).

The significance of NSCs for the synthesis and release of neuropeptides as biologically active molecules that play a role in evolution, distribution, and immunocytochemical detection of pituitary hormones in fish, amphibi-ans, and reptiles has been described (Li, 1972; Van Oordt, 1968; Doerr-Schott, 1976; Ball, 1981; Bern, 1966, 1983). A similarity in the disposition of different hypophysial cell types in amphibians and reptiles has also been described (Doerr-Schott, 1976). Doerr-Schott showed that adreno-corticotropic hormone (ACTH) and prolactin (PRL) cells are encountered in the rostral zone and growth hormone (GH) cells in the caudal zone, whereas gonadotropic hormone (GTH) and thyroid-stimulating hormone (TSH, or thyrotropic hormone) are present in both lobes.

The cytological properties of the hypothalamic neurons and pituitary cells in several fish species have been described (Pantić, 1974b, 1980, 1981). Neuroendocrine cells in mammals comprise specific brain neurons in nuclei of the supraopticus and paraventricularis. The topography and distribution of neuropeptides in the brain areas of different animal species has become one of the most widely investigated areas in neuroendocrinol-ogy. Rats have been the most extensively investigated species (Elde and Hökfelt, 1978; Palkovits *et al.,* 1978; Palkovits, 1982).

The specificity of adenohypophysial cells, the regulation of their activ-ity, and their response to steroid and thyroid hormones have been exten-sively investigated in fish, birds, mammals, and other animal species (Her-lant, 1964, 1967; Farquhar, 1971; Pantić, 1975; Doerr-Schott, 1976).

The selective topographic affinity of pituitary cells, their interrelationships, and interactions, as well as the nature of cell–cell communication, have been summarized by Denef *et al.* (1984). The effects of peptide hormones on target cells interacting with specific receptors on the plasma membrane; the internalization of the ligand–receptor complex, which occurs at different rates in different cell types; the staining of neuropeptides and bioamines in the same neurons and even in the same granule; cosecretion; and the mechanism by which the hormones exert long-term biological effects, including regulation of receptor number in target cells, are increasingly dominating the discussions of endocrinologists.

## II. Genetically Programmed Neuroendocrine Cell Development and Modulation of Cell Activities

The main focus of comparative experimental investigations has been on the properties of hypothalamic neurons and pituitary cells from the following viewpoints: the specific origin of the adenohypophysial cells and the disruption of their connection with the stomodeal epithelium from which they originate, species specificity between the hypothalamus and adenohypophysis during evolutionary and ontogenic development, and the specificity of gene expression needed for the proliferation and differentiation of these cells.

Specific hypothalamic neurohormones are synthesized as a result of programmed expression of genes. Their neurohormonal pathways of communication with the pituitary and different areas of the central nervous system (CNS) need to be elucidated. It is now known that there are more than 100 different neuropeptide molecules, and that regulatory peptides are derived from common ancestral molecules. *Tetrahymena* cells, being phylogenically low but highly differentiated, have been proposed as a suitable model for receptor research (Csaba, 1985). Hormone receptors and hormone molecules have ancestors in food receptors and similar ancestors also exist at the unicellular level (Csaba, 1992). Nature has provided a method for the coevolution of ligand and receptor-binding sites that is based on complementarity of genetic information and its intrinsic self-correction. Any mutation or changes on one strand would be reflected in the other, and a complementary change in the amino acid sequence would occur (Blalock, 1992).

Anterior hypothalamic, preopticohypothalamic, midhypothalamic (tuberal), and posterior hypothalamic (mammillary) regions as well as magnocellular and parvocellular neurosecretory cells have been reviewed (Mikami, 1986). At birth the rat is neurologically more immature than the guinea pig or chicken. Only during embryogenesis and postnatal development is the CNS characterized by rapid proliferation of neuroblasts and glioblasts, followed by axonal growth, synaptogenesis, and neuronal maturation. It appears that astrocytes are differentiated much earlier than oligodendroglia. As the proliferation of neuroblasts and glioblasts ceases, differentiation of neurons and glial cells proceeds.

In our work gonadal steroids were used as modulators to stimulate, retard, or even to inhibit the proliferation rate of pituitary precursor cells and to some extent to direct their differentiation. Representative animals for distinct animal species were selected.

## A. Pituitary Cells Programmed for Synthesis of Peptide and Glycoprotein Hormones

Experimental investigations have used cytological, ultrastructural, immunocytochemical, and RIA methods to follow common and species-specific properties of neuroendocrine cells and the nature of their reactions to different stimuli, the similarity and specificity of cell-type reactions to stress, and their role in regulating growth or reproduction. On the basis of the observation that the amino acid sequences of pituitary hormones have a certain common overlapping biological activity and specificity (Li, 1972), and keeping in mind that the pituitary cells are genetically programmed and differentiated from chromophobes, I proposed three groups of cell types: (1) ACTH and MSH (melanocyte-stimulating hormone) cells, (2) PRL and GH cells, and (3) glycoprotein hormone-producing cells (Pantić, 1975).

Data obtained by our experimental approach and advances in comparative endocrinology have confirmed this classification. From an evolutionary and ontogenic viewpoint, chromophobes are precursor cells from which cells producing three families of pituitary hormones differentiate: first, cells producing the family of proopiomelanocortin (POMC) hormones; second, cells producing the family of prolactin and growth hormones; and third, cells producing the family of glycoprotein hormones (Pantić, 1990a).

Selected papers, monographs, reviews, and even abstract dealing with these exceptionally important and rapidly developing fields of comparative neuroendocrinology are discussed in this article. This article presents my view of the biological meaning of hypothalamic neurons and pituitary cells during evolution and ontogenic development of animal species, including humans. Pituitary cells, as producers of families of peptide and glycoprotein hormones, are discussed with this view in mind.


## B. Animals Used for Experimental Investigations of Neuroendocrine Cells

Our research program focused on the reactions of hypothalamic neurons and pituitary cells during the sexual cycle, their sensitivity, and the character of their reactions during various stages of ontogenesis, under conditions of sufficiency or deficiency of steroid and thyroid hormones.

Representative animals selected for an examination of specificity in their reactions during the sexual cycle were *Torpedo ocellata* for viviparous fish (Pantić and Sekulić, 1975, 1978b); *Serranus scriba* for hermaphrodites (Pantić and Lovren, 1977); the teleostean fish *Alburnus alburnus* for lake fish; *Alosa falax* for fish that live in the sea but migrate to lakes or rivers during the spawning season (anadromous fish) (Pavlović and Pantić, 1973, 1975); elasmobranchs and acipenserida (Sekulić and Pantić, 1977) and deer for wild animals (Žgurić *et al.,* 1968).

The sensitivity and nature of the reactions of hypothalamic neurons and pituitary cells to a sufficiency or deficiency of steroid and thyroid hormones were examined using chickens, rats, and piglets as experimental animals. Common and species-specific reactions were examined in animals given varing doses of gonadal steroids or thyroid hormones during perinatal, juvenile, peripubertal, adult, and senescent stages (Pantić, 1980, 1990a).

The characteristics of the reactions of hypothalamopituitary cells to a deficiency of steroid hormones have been examined in gonadectomized (Genbačev and Pantić, 1975; Pantić and Šimić, 1977a,b) and/or adrenalectomized adult rats (Hristić and Pantić, 1973) and neonatally gonadectomized piglets. Thyroidectomized rats were used (Stošić and Pantić, 1973) to observe reactions to a thyroid hormone deficiency.

Hypothalamic lesions have been created in newborn rats. Pituitary or target organ cells have been transplanted into ectopic areas of hypophysectomized rats and the nature of reactions and the ultrastructural properties of grafts examined (Martinović *et al.,* 1969, 1982).

Data obtained from these experiments and other work by neuroendocrinologists are reviewed in this article. My aim is to consider them from evolutionary and ontogenic viewpoints, in the hope of contributing to the elucidation of the mechanisms regulating the expression of genes needed for proliferation and differentiation of neuroblasts and pituitary cells, and to modulate the development of the neuroendocrine cells so that reactions to stress, adaptation, growth, and reproduction are controlled.

## III. Properties and Species Specificity of Hypothalamic Neurons and Pituitary Cells

### A. Common Properties

Progress in comparative neuroendocrinology during the past decade has provided new information on the development of neuroendocrine cells and their activities in invertebrates and vertebrates. The development of the "lower" forms of animals may be used as a model for better understanding the evolution of bioamines and peptide hormones, the properties of their receptors, and the mechanisms of ligand–receptor actions.

Neurohormones are synthesized as segments of larger ancestral common precursors, stored in specific vesicles or granules, and cleaved by enzymes. Both cleaved and uncleaved molecules may be coreleased by exocytosis. Neurosecretory cells in the hypothalamus are monopolar, bipolar, or multipolar. They are organized as nuclei composed of parvocellular and/or magnocellular neurons. Their axons terminate on the same neuronal body, on other neurons in the same nucleus, in the circumventricular organs (CVOs), especially in the neurohypophysis (NH) and the median eminence (ME), in the limbic system, and other brain areas. As typical neurons, they are able to conduct action potentials and to release their contents as neurohormones via terminals into hypophysial long or short portal blood vessels, or into the systemic circulation. The cytological heterogeneity of the hypothalamic neurons and adenohypophysial cells is expressed in the intracellular mechanisms involved in regulation of biosynthesis at the levels of transcription, processing, translation, storage, and release of heterogeneous granular content.

### B. Specificity of Hypothalamic Neurons

One of the most important phenomena created by nature during evolution is specificity in the organization and communication of the hypothalamus and pituitary, and their target cells. They also have common and species-specific properties for communication. These properties and their role in regulating the vital activities of neuroendocrine cells during ontogenesis are factors that determine behavior and sexual characteristics.

## 1. Annelids as Animals with Neurosecretory Cells Developed within the Cerebral Ganglion

The nature of neurohormones and the genes encoding their precursors in annelids are of basic interest for understanding the phylogeny of numerous animal groups. Thus, annelids were proposed as a model for investigating neurogenesis and neuronal development (Al-Jousuf, 1992). The CNS in the annelid is composed of a cerebral ganglion that is connected to the subesophageal ganglion by circumpharyngeal connectives (Al-Jousuf, 1992). Their neuropeptides are released into the neurohemal organ (Scharrer, 1968).

## 2. Fish as Animals Containing both Hypothalamic and Caudal Neurosecretory Cells

From an evolutionary viewpoint fish have been examined mainly to obtain information important for understanding the role of neuroendocrine cells in the regulation of the life cycle, growth, reproduction, migration, and other processes in other animal species, including mammals.

Three organized endocrine glands in nonmammalian vertebrates are known: (1) the ultimobranchials, which in fish synthesize calcitonins (these are more potent in humans than are human calcitonins, but their role in fish is still not clear); (2) the Stanius corpuscles, which synthesize hypocalcemic factors in teleostean fishes; and (3) the caudal NSCs. These cells are present in fish, but not in any land-leaving vertebrates. They produce urotensin I and urotensin II as neurohormones. As a partial somatostatin (SS) analog, they are absent in all tetrapods whether tailed or not (Bern, 1984, 1985). Somatostatin and the urotensins, in common with other neurohormones, are synthesized as larger precursors (prepro-hormones) that are converted to mature secreted forms by a specific posttranslational proteolytic cleavage (Conlon, 1990).

The caudal NSCs, as a classic example of cells containing more than one type of granule and small vesicle, are localized in the tail end of the fish. In *Tilapia* these cells are able to regenerate from the ependyma. These caudal NSCs seem to have been mostly examined in the elasmo-branchs and the teleosts (Bern, 1970). The regeneration of caudal NSCs in adult *Tilapia,* after total removal, indicates the importance of the ependyma in replacing extirpated caudal neurons, at least in lower vertebrates, but also shows the capability of *de novo* differentiation of ependymal into neurosecretory cells.

The properties of hypothalamic neurons, especially in lower verte-brates, their dendrite-like processes projecting into the third ventricle, and the secretion of caudal neurosecretory neurons into the cerebrospinal fluid have been studied (Bern, 1970). The bidirectional secretory capacity of these neurons is analogous to that of the thyroid follicular cells and they are capable of secreting apically or basally.


## C. Hypophysiotropic Regulatory Neurohormones

Since the first monographs on hypothalamic regulatory hormones were published (Schally *et al.*, 1973), the number of articles, reviews, and monographs has grown rapidly, but only some of them are mentioned here: immunoreactive neurosecretory pathways in mammals (Barry and Dubois, 1976); brain cells as producers of releasing and inhibiting hormones (Sétáló *et al.*, 1978); aminergic and peptidergic pathways in the nervous system, with special reference to the hypothalamus and distribution of peptide-containing neurons (Elde and Hökfelt, 1978; Hökfelt *et al.*, 1978); and topographical distribution of neuronal peptides in the hypothalamus and mammillary body (Palkovits, 1982).

The effects of neuropeptides and monoamines, which are synthesized in neurosecretory cells, are expressed as neurohormones on target endocrine cells, or as neuromodulators and neurotransmitters on neurons. They are also released from neurosecretory axons, via the hypothalamohypophysial portal blood vessels, to the pituitary cells and/or into the cerebral ventricles.


## IV. Synthesis of Precursors to Hormones and Nature of Their Receptors


## A. Evolution of Hormones and Their Precursors

The messenger molecules involved in cell–cell communication in vertebrates may have their origin in the unicellular organism (Roth *et al.*, 1984). This suggestion is based on the finding that steroid hormones (gonadal hormones, corticosteroids, and ecdysone) and peptide hormones, and messenger molecules of vertebrates, in both plants and invertebrates are identical with or similar to the neuropeptides of vertebrates.

Unicellular eukaryotes probably have an essential role in the evolution of life, as intermediates between prokaryotes and eukaryotes. The concept of evolutionary homology is based on the structural identity of chemicals used as single-molecule second messengers (e.g., ATP, GTP, cyclic AMP, phosphoinositol metabolites, and calcium) in cells requiring shared and conserved recognition modules that evolved by convergent evolution or gene duplication (Palme, 1992).

## B. Common Properties of Precursor Molecules

Neuroendocrine cells synthesizing neuropeptides within the cell body have mainly the characteristics of neurons. It is now accepted that most neuropeptides are synthesized as a part of a common precursor that, as a large polypeptide chain, is cleaved by enzymes inside the granules, transported along the axons, and secreted as active peptides. For example, POMC is a common precursor molecule for peptides of the POMC family of hormones. The genes are expressed in the brain neurons, pars intermedia (PI), and pituitary ACTH cells, but a number of this family of peptide hormones may differ in the structural organization of the genes encoding them. Nakanishi *et al.* (1979) showed that preproenkephalins A and B contain a set of three or more enkephalin (ENK)-containing peptides and differ from POMC in only a single copy of the enkephalin sequence.

The search for precursor molecules to biologically active neuropeptides led to the discovery of an increased number of brain peptides. One of the main problems was that peptide hormones, known as releasing (RH) or inhibiting (IH) hormones (factors), are present *in vivo* only in minute quantities. The messenger ribonucleic acid (mRNA) for a given polyprotein could also be present at low concentrations. For example, somatostatin was found in small amounts, whereas vasopressin (VP), oxytocin (OXY), and neurophysins (NFs) are present in greater concentrations. Lutein-releasing hormone (LRH), a decapeptide that occurs in lower concentrations than SS, may not be easily detected. Somatostatin is identified in pancreatic islets and intestinal epithelium, but also in the urinary tract of amphibians.

On the basis of the existence of amino acid precursor uptake decarboxylation (APUD) cells, paraneurons, the epithelial cells in the intestine, and the cells in skin that produce peptides, the following concepts are now generally accepted: (1) a single neuron or pituitary cell may synthesize and release two or more molecules representing different transcriptional segments of a single gene; (2) a single preprohormone synthesized in one neuron may be enzymatically cleaved into peptide molecules. This is the case with neurons synthesizing POMC or procorticotropin-releasing hormones, hormones involved in stress, as well as preprosomatostatin and prolutein-releasing hormones. There is no doubt that precursor molecules are synthesized on granular endoplasmic reticulum (GER), glycosylated in the cavities of the GER, and processed through a cascade of proteolytic events in specific granules.

The common precursor to ACTH and endorphin (END), and the role of glycosylation in processing of the precursor and secretion of both ACTH and END, were reviewed by Herbert *et al.* (1980). It has been shown that the bovine precursor to arginine vasopressin (AVP), NF II, consists of 166 amino acids containing AVP, NF II, and a glycopeptide of 39 amino acids (Land *et al.*, 1982). The data obtained so far may contribute to a better understanding of the evolutionary pathways related to biosynthesis of neuropeptides and their precursor molecules. Some of the results of the search for precursors of neuropeptides are the identification of the common precursors to corticotropins and endorphins (Mains *et al.*, 1977), and the characterization of proopiocortin as a precursor to opioid peptides and corticotropin. Richter *et al.* (1980) showed that neuropeptides illustrate a new perspective in mammalian protein synthesis, the composite common precursor. The presence of presequence (signal sequence) in the common precursor to ACTH and END, and a role for glycosylation in processing the precursor and secretion of ACTH and END, have been suggested (Herbert *et al.*, (1980). De Loof (1992) showed that a majority of peptide families have members in vertebrates as well as in invertebrates, and expects that the neuropeptides identified in invertebrates might outnumber those found in vertebrates.

Neuropeptide synthesis is initiated in response to signals by ligands bound to receptors and causes differentiation of the target cells. Two neuropeptides may occur in a common precursor, but are encoded on different duplicated exons. Members of a peptide family may differ in genetic relationship. Preproenkephalins A and B, which contain a set of three or more enkephalin-containing peptides, differ from POMC, which produces only a single copy of the ENK sequence (Nakanishi *et al.*, 1979).

## C. Membrane Receptors and Transport of Molecules and Ions

Proteins involved in transport contain mainly oligosaccharide chains attached to those parts of molecules that lie outside the plasma membrane. Some of them have enzymatic activities, as is the case for ATPase, or they have binding affinity for one or more ligands. A highly dynamic cell response to extracellular signals is based on the receptors. As the trigger for a cascade of intracellular secondary messengers, the ligand–receptor complex is involved in the generation of specific cell responses to hormonal, neuromodulatory, and neurotransmitter signals. Ion channels play an essential role in the transfer of neuronal information, and their structural properties, varieties, and function have been reported (Gage, 1993). Gage thought that protein macromolecules, which allow the rapid movement of ions across a cell membrane, appear to be influenced by electrical and hormonal forces. When ion channels in neurons are open, current flow in the form of ions generates an electrical potential across the membrane. The movement of ions through channels, and signal transfer in neurons, are generally common to all ion channels. Ion channels differ in selectivity, conductance, kinetic, and other properties. Gage (1993) stated that many questions need to be answered before we can understand the function of ion channels in terms of conformational changes in their three-dimensional structure.

A variety of second intracellular messengers exist, such as cAMP, which are involved in signal transduction originating from hormones, neurotransmitters, neurohormones, growth factors, and so on. These messengers are involved in qualitative cell responses. On the basis of the importance of cloned genes that encode membrane-bound proteins such as receptors, Shine (1993) grouped these messengers into gene superfamilies. Among the many signaling molecules, Shine mentioned the G protein receptors as probably the largest class that is coupled, via GTP-binding proteins, to a variety of intracellular signaling molecules such as enzymes. Shine pointed out the importance of signal transduction through G protein-coupled receptors, second messenger systems, and ion channels in determining the control of cellular homeostasis and the response of the cell to environmental factors.

The thermodynamic basis of pumps and coupled transporters involved in active and passive membrane transport has also been explored (Frömter, 1993). Frömter proposed an evolutionary tree for the cation-pumping family of ATPases. Viewing $Na^+/K^+$-ATPase as a transmembrane-spanning protein consisting of two subunits, Frömter described the pumping of the $Na^+/K^+$-ATPase as an enzymatic reaction cycle and looked at the secondary structure of $\alpha$ and $\beta$ subunits. The $\alpha$ subunit has a length of 1016 amino acids and contains 8 membrane-spanning domains of $\alpha$-helical structure as well as binding sites for ATP and ADP and for the inhibitors vanadate and ouabain. The $\beta$-subunits is highly glycosylated and consists of 30 amino acids (Frömter, 1993). On the basis of thermodynamic reasoning Frömter distinguished five types of transport phenomena: permeability, conductance, active transport (pumps), cotransport, and countertransport. Depolarization-operated $Ca^{2+}$ channels are those that open in response to membrane depolarization, and receptor-operated $Ca^{2+}$ channels are those that are controlled by receptor activation. $Ca^{2+}$ action is initiated by binding growth factors to specific receptors. It appears that $Ca^{2+}$-sensitive, phospholipid-dependent protein kinase plays a crucial role in cell growth and proliferation. Further understanding of a mechanism of transport will depend on knowledge of the three-dimensional structure of the glycoproteins as transport proteins (Kuchel, 1993). In addition to changes in the genetic code, the distribution of ion pumps and channels over the plasma membrane, changes in ion pump and channel activity, and changes in the symmetry of mitosis may have been important mechanisms in evolution (De Loof, 1986).

## V. Neurogenesis and Hypothalamic Organization

### A. Neurogenesis

### 1. Origin and Proliferative Rate of Neuroblasts

Neuroblasts develop from the special neurogenic region of the ectoderm. These cells, known as "bipolar neuroepithelial cells," may be clearly observed prior to the onset of proliferation in the ectodermal germ layer. The mitotic cycle and/or the detachment from the ventricular epithelium might be considered as signals for the migration of these cells to their final destination. Glioblasts also migrate, after division, from subventricular regions to their final area.

The neuroepithelium of the neural plate proliferates during neurogenesis so that from at most several thousand promordial cells, about 100 billion neurons and several hundred billion glial cells are differentiated. The main events following mitosis are migration, differentiation, synaptogenesis, and neuronal maturation.

### 2. Migration

Cell migration during embryogenesis is characterized by ameboid motility and in some ways is a process similar to the intracellular movement of organelles. The following mechanisms must be coordinated: transduction of chemical to mechanical energy, propagation of mechanical forces throughout the cell body, and transmission through the intercellular matrix between neighboring cells. To transmit forces to the substrate, migratory cells use adhesion, but to fasten to the substrate they use electrical and van der Waals forces (Albrecht-Buehler, 1990). The migration of neuroblasts starts first with the large ones, which are followed by intermediate and finally small neuroblasts.

Cell migration, ligand–receptor interactions, communication between cells, and the nature of their response to hormonal stimuli, antigens, and other substances are dependent on their intracellular and extracellular contents, the nature of these molecules, and their interactions and forces. Molecules present in the plasma membrane migrate toward cytoplasm and vice versa. There is no doubt that microtubules (MTs) and microfilaments (MFs) are the main cytoplasmic organelles involved in the movement of molecules and other cytoplasmic organelles inside the cells. They are also involved in regulation of gene expression. This indicates that the cytoskeleton is the main integrator of cytoplasmic events (Meininger and Binet, 1989).

Neuroblasts migrate from the site of origin to those brain areas in which they later reside. During development, neuroblasts are motile and move through intercellular spaces, being affected by molecules of the intercellular matrix and neighboring cells. Levi et al. (1990) described the mode of cell migration in the vertebrate embryo.

The guidance of neurons through the intercellular matrix is also dependent on the biochemical composition and nature of the molecules in the matrix. The routes of cellular migration are influenced by the balance between fibronectin, acting as an adhesive molecule, and chondroitin sulfate proteoglycan, which acts as an antagonist (Newgreen and Erickson, 1986).

Hypogonadal mutant mice have been proposed as a model for investigating neurotrophic factors involved in axon guidance (Charlton, 1986). The mechanisms regulating the guidance of migrating cells are different and depend on the distance from the hypothalamic regions. A series of events occur, such as cell individualization, directionality, and arrest. The migration of cells is dependent on the dynamic potential provided by the lectins, glycoconjugates, and $Ca^{2+}$ concentration.

At different stages of embryogenesis, especially during the intensive proliferation of neuroblasts and their migration, growth, and development, the genetically programmed degeneration of cells (apoptosis) occurs. This process is of great importance for maintaining the capacity to control proliferation rates and for the specificity of hypothalamic organization in the form of distinct nuclei, separated by numerous neuronal fibers terminating on well-defined neurons.

## 3. Differentiation

The sequential steps of differentiation that occurred during evolution usually take place in the same order during embryogenesis. Differentiation of a particular cell is based on modification of the development of a simpler cell type and is thought to be due to the earlier production of more conserved mRNA. As a result of a quantitative and qualitative restriction of the transcription process a relative increase in less conserved mRNA led to the differentiation of specific cell types such as nerve and endocrine cells. Specific stages of differentiation involve the exposure of different sets of genes. Approximately 30% of the genome in mammalian cells in converted from the sequestered to the exposed stage in a single action governed by effectors such as cAMP, retinoic acid, and nerve growth factors acting on transformed cells. The evolutionary aspects of cell differentiation have been reviewed by Plickinger (1982).

During differentiation, the mammalian cell cytoskeleton becomes part of an information transmission system that extends from the cell membrane and its receptor site through the cytoplasm and terminates in specific points on each chromosome, so that exposure and sequestration occur in specific domains. Genome exposure is necessary, but not sufficient, for gene activation. The differentiation of neuroblasts and growth of axons prior to the growth of dendrites were observed by Meininger and Binet (1989). During the growth of these cytoplasmic extensions, the interconnection of the neurons occurs and the synapses are formed. It has been observed that in the "aging" of "mature" brain, molecular modification occurs, but the maturation of neurons in the CNS is never arrested (Meininger and Binet, 1989). These workers examined the involvement of microtubules in modification of the aging CNS and suggested that a better analysis is necessary for a clearer understanding of the mechanisms involved in CNS development. Membrane oligosaccharides, as biologically active molecules directly involved in the control and maintenance of cellular capacity for differentiation and immunity, seem to be of great importance for cellular senescence (Mann, 1988).

Differentiation occurs at the end of the cell migration and neurons differentiate before glial cells. The differentiation of astrocytes earlier than oligodendroglia is undoubtedly closely connected to their role in blood–brain barrier (BBB) development and myelinization of nerve fibers. As the differentiation of hypothalamic neurons proceeds their body size increases and the growth of both axons and dendrites is extended (Figs. 1 and 2).

## 4. Differentiation of Hypothalamic Neurons in Culture

None of the currently available hypothalamic cell lines follow the complete pattern of neuronal differentiation in culture. They represent a stable interruption of a program that is continuously evolving *in vivo* (Tixier-Vidal and De Vitry, 1979). The simultaneous expression of neuronal and glial features may be used for an analysis of neuronal and glial ontogeny. It seems that magnocellular neurons undergo a developmental stage found *in vivo* in the early fetus, and that neurohormones, such as thyrotropin-releasing hormone (TRH) and VP, are detectable in 14-day mouse fetus, several days before the differentiation of axon terminals. The formation of cellular contact between neurosecretory axons and pituicytes as their

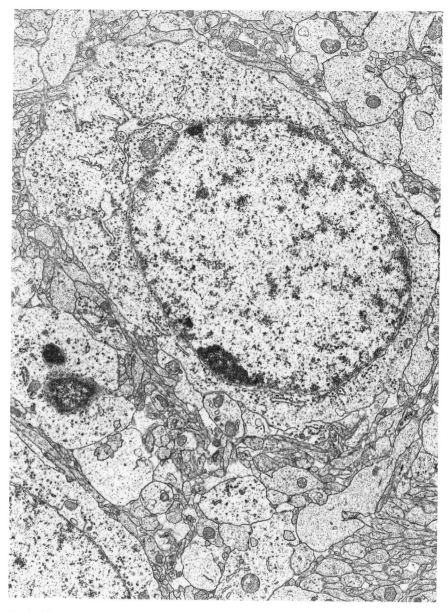

FIG. 1    Neurosecretory cell body in the NSO of a 3-day-old rat. The appearance of polysomes and GER, as well as of rare and small-sized mitrochondria, are clearly signs of differentiation.

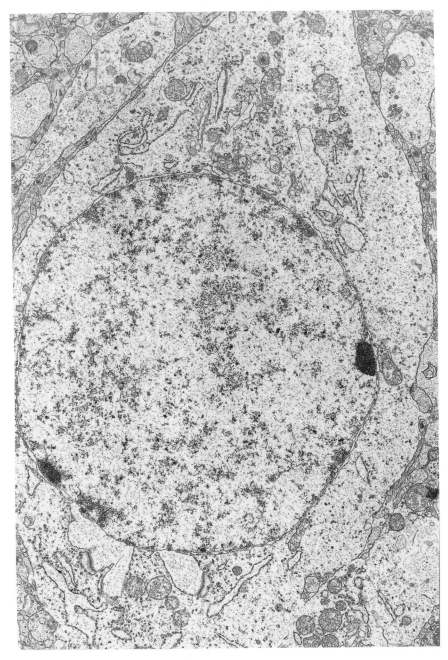

FIG. 2  Neurosecretory cell of a 10-day-old rat, clearly polarized, with an increased amount of ribosomal subunits in the karyoplasm, numerous polysomes, and slightly dilated cavities of GER, shows an advanced stage of differentiation.

target cells might be a necessary step for the completion of their differentiation (Tixier-Vidal and De Vitry, 1979).

The identification of vasotocin, an embryonic form of VP, in primitive ependymal cells of the subcommissural organ (SCO), and SS in tanycyte-like cells, indicates that fetal ependymal cells represent the most primitive cells of the brain (Pavel *et al.,* 1977).

## 5. Synaptogenesis and Maturation

The interconnection of neurons during the growth of neuronal extensions leads to the formation of synapses and the establishment of a specific set of afferent and efferent connections. The formation of synapses implies mechanisms of cell recognition and inhibition of cell movement so that dynamic intercellular junctions can be formed. Each neuron may continually integrate numerous synaptic inputs. The regulation of the number of cells, synapses, and of cell processes is important, because there is an overproduction of neurons forming synapses during development.

The concentration of transferrin in cerebrospinal fluid (CSF) during the fetal development of rats; its transport to the area of intense synaptogenesis; its synthesis in the choroid plexus of the lateral, third, and fourth ventricles of adult rats, which contains as much transferrin mRNA as the liver; and its presence in Schwann cells and oligodendrocytes, undoubtedly indicate an important role for transferrin and iron in the neurogenesis, growth, and metabolism of neuronal tissue (Mescher and Munaim, 1988).

Most if not all neurons may function as multimessenger units (Figs. 3 and 4). The release of neurotransmitters and neuropeptides, which are stored together in small and large dense vesicles (LDVs) within a terminal, may to some extent depend on differences in the molecular organization of the vesicle membranes.

Synapsins I and II are present at high levels in the vesicles of nerve terminals (Thureson-Klein and Klein, 1990). Synaptophysin is a membrane protein associated with small vesicles in neuroendocrine cells. The ability of synaptophysin I to form transmembrane channels may be used to follow the ontogenesis and recycling of small synaptic vesicles.

The absence of significant amounts of synapsin and synaptophysin in LDV membrane may be important for the differential release of neuropeptides and transmitters. Synaptophysin that binds calcium within the binding site, localized on the cytoplasmic side of the small vesicle membrane, could be related to vesicle fusion with the plasma membrane. The supramolecular cytology of coated vesicles was reviewed by Fine and Ockleford (1984).

Many neurons in the CNS, especially in neurosecretory cells, are multimessenger cells containing a variety of storage vesicle or granules. Large,

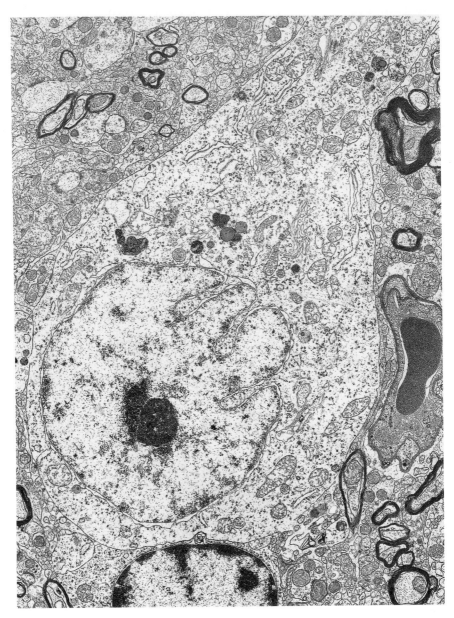

FIG. 3   Part of a neurosecretory cell in the NSO of a young rat, with developed cytoplasmic organelles involved in the synthesis and release of neurohormones. Neuronal fibers surrounding the neuronal body are myelinated.

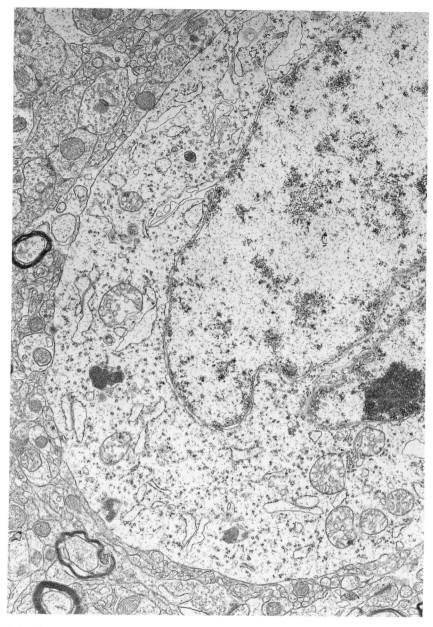

FIG. 4   Neurosecretory cell body in the hypothalamus of a rat, containing developed organelles involved in protein synthesis.

dense-cored vesicles release neuropeptides, and often classic transmitters such as ATP and various protein molecules, across a narrow synaptic gap and interact with receptors on the effector cell (Thureson-Klein and Klein, 1990).

Exocytotic membrane fusion requires very little time, sometimes without any appreciable lag period, and only a minute portion of the total membrane area participates in membrane fusion during secretion (Plattner, 1989).

The pathways for internalization of receptors and their ligands, the fate of the endocytotic contents, and the role of the various intracellular compartments in the sorting signals and their interrelation with the lysosomes and other organelles, need to be further elucidated. Soluble cytoplasmic proteins, which undergo anterograde transport, move at the relatively slow rate of 1–2 mm/day. Large dense vesicles are translocated to the terminals by a fast axonal transport.

Microtubules observed in the presynaptic terminals may be involved in the formation and maintenance of synaptogenesis. Depolarization of the nerve terminals causes an inward transport of $Ca^{2+}$ from the extracellular fluid to the cytoplasm. As a result presynaptic vesicles migrate toward the plasma membrane and fuse with the neurilemma, so that the entire contents of the vesicles are released into the synaptic cleft.

The presynaptic and postsynaptic membrane, as a highly organized structure with specific cytochemical properties and containing receptors and corresponding enzymes, has a crucial role in the direct conduction of bioelectrical information from one neuron to the other or to other cells. It appears that the number, size, and form of presynaptic vesicles are closely related to the kind of mediator or transmitter contents formed during synaptogenesis (Manolov and Ovtscharov, 1982).

## 6. Involvement of Cytoskeleton in Neurogenesis

The cytoskeleton plays an important role during the evolution and ontogenic development of all types of eukaryotic cells, especially neurons. Its role ranging from gene expression, migration of chromosomes, and streaming of cytoplasm to the plasma membrane during exocytosis and endocytosis, is the focus of contemporary research. The cytoskeleton is composed of an actin neurofilament network, microtubules (MTs), and intermediate filaments (IFs). The evolutionarily conserved proteins are actin and tubulin, whereas IF proteins are highly diverse and exhibit cell-type specificity. Gene expression is also closely coordinated with differentiation of cells and development of the entire organism (Albers and Fuchs, 1992).

Amont the three major cytoplasmic polymers, MTs have highly dynamic behavior; they are abundant in neurons and confer a large range of structural and functional capabilities (Meininger and Binet, 1989). They are composed of tubulin and microtubule-associated proteins (MAPs) (Kirschner, 1978; Wiche *et al.*, 1991). The extensive heterogeneity of MAPs (e.g., polypeptides MAP 1a, MAP 1b, and MAP1c), which exhibit different levels of sensitivity to proteolysis, and their distribution in both nerve and glial cells, have been reported. The definition, structure, and other properties of MAPs, and the importance of MTs for many basic biological processes, have been summarized by Wiche *et al.* (1991). The role of tubulin and MAPs is of great importance for neuroblast proliferation, migration, and development especially for the extension of axons and dendrites (Meininger and Binet, 1989).

The major filamentous proteins in nervous tissues are neurofilament proteins, laminin proteins (A, B, and C), and glial fibrillary acidic proteins (NF-L, NF-M, NF-H). Laminin is present at the surface of astrocytes and has a role as a substrate for neuroblast migration and neurite outgrowth. Glial fibrillary acidic protein assembles into filaments as a homopolymer in various glial cell types located throughout the nervous system.

Calmodulin (CaM) is a multifunctional protein. After stimulation of a cell, an increase in the intracellular free $Ca^{2+}$ concentration is due to movement of $Ca^{2+}$ through channels of the plasma membrane, or to its release from intracellular organelles. The CaM molecule can bind as many as four $Ca^{2+}$, producing conformational changes, including a number of cellular activities such as proliferation, motility, transport, metabolic control of cyclic nucleotides, and contraction. The CaM has a role in regulation of the $Ca^{2+}$ complex, it can function directly and rapidly on the ATPase transport system, or indirectly and slowly on protein kinase. Cyclic nucleotides, as regulatory enzymes, also play an important role in cell growth and proliferation (Puck and Krystosek, 1992).

## 7. Proteins as Regulators of Protein Synthesis, Cell Proliferation, Differentiation, and Synaptogenesis

G proteins (guanine nucleotide-binding regulatory proteins) are composed of $\alpha$, $\beta$, and $\gamma$ subunits and are themselves members of a large gene family. G proteins that serve as transducers for membrane receptors, lectins, and laminin, are briefly discussed as regulators of protein synthesis. Three groups of G proteins are known to be involved in regulation of protein biosynthesis through signal transfer, proliferation, and differentiation. These proteins have both GDP/GTP-binding and ATPase activities (Takai *et al.*, 1992).

The first group of G proteins includes those involved in protein synthesis and are initiation factors, elongation factors (EFs), and termination factors. Eukaryotic EF-1 is composed of three different subunits ($\alpha$, $\beta$, and $\gamma$) and $\alpha$ subunits of EF-2 and EF-1 have GDP/GTP-binding and GTPase activities (Takai et al., 1992).

The second group of G proteins comprises a superfamily of the $G_s$, G, $G_0$, and $G_q$ transducin families and is important for signal transfer from membrane receptors to various effectors, such as adenylate cyclase, phosphoinositide-specific phospholipase C, cyclic GMP phosphodiesterase, and the $K^+$ channel. Proteins of this group are composed of three different subunits ($\alpha$, $\beta$, and $\gamma$) and the $\alpha$ subunits have GDP/GTP binding and ATPase activities (Bourne, 1986; Bourne et al., 1991; Taylor et al., 1991).

The third group of G proteins is monomeric and includes the small G proteins. The representatives of this group include the p21ras family and their role is to regulate various cell functions, including proliferation and differentiation (Bourne et al., 1991).

When a membrane-bound receptor acts on a G protein, the GTP-binding or G $\alpha$ subunit dissociates from the G$\beta\gamma$ dimer. It was thought that the G $\alpha$ subunit acts on the enzymes and ion channels controlled by these proteins. Newer evidence indicates that the G$\beta\gamma$ dimer also plays a major part in signal transmission, enhancing the complexity of the possible interactions between the G proteins and their targets. Alteration of the ion permeation pathway contributes significantly to G protein inhibition of N-type calcium channels in cell membrane inhibited by noradrenaline, LRH, $\gamma$-aminobutyric acid (GABA), and glutamate (Kuo and Bean, 1993). Molecular cloning in insects of evolutionarily conserved G protein-coupled receptor cDNAs such as adrenergic receptors, serotonin, and others, which belong to a large "family" of integral membrane proteins, has shown a considerable degree of homology to known vertebrate bioamine and tachykinin peptide receptors (Vanden Broeck et al., 1992). A number of cell membrane receptors for neuropeptide hormones and neurotransmitters are coupled to guanine nucleotide-binding (G) proteins (Martens, 1992).

The function of other molecules, such as lectins, in the differentiation and development of neurons is discussed. Lectins possess binding capacity for endogenous ligands. Various endogenous brain lectins have a role, from viruses to humans, in cell recognition, cell adhesion, intracellular traffic, internalization of external molecules, and transmembrane signaling (Zanetta et al., 1992). The lectin receptor is localized in neurons but is absent from oligodendrocytes and astrocyes. Zanetta et al. showed that endogenous brain lectins have a role in cell recognition, adhesion, and transit, and that the roles could change during ontogenesis. Lectin may play a role in neurite fasciculation. Later (i.e., postnatal days 10–18 in the rat), the antigen is transiently concentrated in Purkinje cell bodies and dendrites, that is, at the period of maximal synaptogenesis. It is thought that recognition of axonal glycoprotein by a dendritic lectin probably constitutes the first stage of synaptogenesis.

It has also been mentioned that concentration of lectin is transiently increased at the time of maximal synaptogenesis, as is the case during late postnatal development of Purkinje cells (Zanetta *et al.*, 1992). Cerebellar soluble lectin (CSL) is a key molecule in the etiopathology of multiple sclerosis (MS) and is the major if not exclusive immunological target in MS. If the level of glycoproteins is reduced, as it is with thyroid hormone deficiency, the number of synapses is reduced.

## B. Hypothalamic Neuronal Organization and Bidirectional Communication

When neuroblasts migrate to the genetically destined hypothalamic regions, differentiation of hypothalamic neurons is initiated. The neuroblasts stop at the base of the diencephalon and become organized into hypothalamic nuclei. About 10,000 neuroblasts migrate to the optic chiasma to form paired suprachiasmatic nuclei. In the rat each nucleus is about 1.5 mm long and fusiform. They consist of magnocellular and parvocellular neurons with short or long cytoplasmic projections. Their axons may be terminated on the same neuronal body, in the same nucleus, or directed toward the median eminence (ME), neurohypophysis (NH), or a different region of the CNS. As the hypothalamic neuronal axons terminate in the above-mentioned regions, bidirectional communication is established between them. This is a result of memory conserved during evolution and the character of this connection is species specific. For example, we examined the pathways of neuronal movement and the nature of their organization in *T. ocelata* and other fish species and the number of neurons per nuclei and the structural properties of developed nuclei in deer and rats. All showed specificity in their properties, especially in their appearance during the sexual cycle and estrous period (Pavlović and Pantić, 1975; Pantić, 1990a).

As small neurons increase in size, they send axons mainly in two directions: one to a different brian area in which the secondary blood–brain barrier is in development, and the second toward the median eminence and neurohypophysis. As differentiation proceeds, synaptic contacts and a flow of information are fully established. The neurons of each nuclei receive information from a different brain area and hypophysis and send it in the opposite direction. The other and very important aspect of hypothalamic neuronal connections is that they communicate bidirectionally, with axons terminating in the neurons in the secondary or primary blood–brain barrier.

## VI. The Blood–Brain Barrier and Circumventricular Organs in Hypothalamic Integration of Neuronal Information and Synthesis and Transfer of Neurohormones

Development, aging, and senescence of the hypothalamic nuclei are closely related to the specificities of their organization, localization, and communication via the primary and secondary BBB. The ultrastructure of blood–brain and blood–CSF barriers in relation to similar noncerebral endothelium and epithelium was reviewed by Van Deurs (1980), who emphasized the importance of the basement lamina and phagocytic peri-cytes in the BBB. Considering that the endothelium of the cerebral mi-crovessels is normally close to proteins and peptides, and that this barrier to lipid-insoluble macromolecules can be ascribed to the tight junctions between the endothelial cells, Van Deurs suggested the need to compare BBB mechanisms in mammals with those of lower vertebrates and inverte-brates.

Owing to the significance of the secondary BBB and the regions in the primary BBB present in circumventricular organs, the main characteris-tics of BBB structural organization and its properties are briefly considered here. Instead of complete (secondary) BBB, the term *BBB* is used.

### A. Circumventricular Organs and the Primary Blood–Brain Barrier

#### 1. General Properties of Circumventricular Organs

The main role of CVOs is to integrate information conducted by monoam-inergic and peptidergic neurons (Bouchaud and Bosler, 1988). Most of the CVO regions are located outside of the BBB and include the organum vasculosum laminae terminalis (OVLT), the median eminence, neurohy-pophysis, the subfornical organ (SFO), the subcommissural organ (SCO), the pineal organ (epiphysis cerebri, or EC), and the area postrema (AP). The choroid plexus, the paraventricular organ, and the collicular recess organ, as specialized ependymal structures, should also be included. The parenchyma of these organs consist of four types of neuroglia: tanycytes and pituicytes, which belong to the hypothalamic regions, and the pinealocytes and ependymocytes, in the pineal gland and SCO, respectively (see Bouchaud and Bosler, 1988). A circumventricular structure, CNS sensors of circulating peptides, and an autonomic control center have been reported (Ferguson *et al.*, 1990).

The typical fenestrations of the endothelium of most capillaries in the CVOs allow the exchange of protein, peptides, bioamines, and other substances as in the endocrine glands. In addition, CVOs have sinusoidal capillaries; the lack of a true (secondary) blood–brain barrier is a characteristic feature. The exception are the SCO and the rostral part of the SFO. Circumventricular organs also receive information that cannot cross the BBB. They are the regions for neurohumoral information and in the brain can be microcenters of integration where multiple transmitter interactions occur (Bouchaud and Bosler, 1988). Bouchaud and Bosler observed that monoamines in some CVOs also participate in neuroendocrine regulation by behaving like true neurotransmitters, acting on proximal and/or distal levels of neuronal targets. Distal interactions occur in the NH, where monoaminergic neurons have a role in modulation of hypophysiotropic and posthypophysial hormonal release through axoaxonic contacts with neurosecretory cell terminals.

Besides their role in regulation of CVO functions, the mode of action of bioamines is also of great importance. A study of the interrelationship between monoaminergic neurons and nonneuronal cells such as tanycytes, pituicytes, pinealocytes, and ependymocytes showed that their terminals establish synaptoid and/or true typical synaptic contact with these cells, providing evidence that 5-hydroxytryptamine (5-HT) and catecholamine play a significant role in regulating their function. It is suggested that monoamines also play a role in modulation and control of the release of hypophysiotropic and posthypophysiotropic hormones, via tanycytes regulating their activities (Bouchaud and Bosler, 1988). The OVLT, SFO, and AP are the important target organs for the action of 5-HT in the control of cyclic GTH secretion, cardiovascular function, the regulation of pituitary cells, and homeostasis. Localization of CVOs plays an important role in integration of information between the nervous and vascular systems and the ventricles.

The primary (incomplete) blood–brain barrier consists of the regions that not only emit, but also receive, neurohumoral information that cannot cross the BBB. The main properties of the primary BBB are semipermeability and reception of neuronal signals. The semipermeable barrier consists mainly of fenestrated endothelium to facilitate the transport of various substances outside of the CSF, as is the case with the choroid plexus. This type of barrier, which is involved in a feedback mechanism, describes the endothelial barrier surrounded by the basal membrane and the pericytes. The structural properties of blood microvessels are similar to those in the endocrine organs.

## 2. Subcommissural Organ

The subcommissural organ (SCO) has an important role in brain evolution. This organ is considered an ancient and persistent glandular structure of the vertebrate brain. During ontogeny the unique ependymal derivatives are among the first secretory cells in the brain to differentiate (Rodriguez *et al.*, 1992). These secretory cells consist of specific ependymocytes and ependyma-derived neuroepithelial (hypendymal) cells.

The SCO represents a highly specialized vascularization within the CNS and in mammals is supplied with serotoninergic and GABA-ergic afferents originating in the raphe nuclei. It is the only organ in the CVO family without fenestration of the capillary endothelium. The capillaries are surrounded by a pervascular space, which because it separates glial end feet from the endothelium, could be responsible for the lack of a BBB within this organ. As a unique structure within the brain, SCO cells are "sequestered" within a double barrier system: the blood–SCO and the cerebrospinal fluid–SCO.

The subcommissural organ consists, in many animal species, of the ependymal and hypendymal cells. During evolution the ependymocytes do not undergo major changes. As evolution proceeds, they are arranged into more layers. Their basal processes come in contact with blood capillaries or the external basement lamina of the brain (Rodriguez *et al.*, 1992). Secretory material forming a film on top of microvilli and cilia, regarded as preReissner's fibers, appears in the form of loosely arranged bundles of thin filaments. It appears that factors from the CSF participate in the formation of Reissner's fibers. As an ancient and one of the first secretory glandular cells to appear during brain differentiation, the SCO is undoubtledly a unique organ from the point of view of innervation and synthesis of various substances, including neuropeptides.

## 3. Subfornical Organ

The proliferation of ependymal cells and their migration into the subfornical organ (SFO) are followed by its vascularization and formation of anastomoses with the blood vessels of the choroid plexus. The development; types of neurons, ependymal and other glial cells, supraependymal cells, and macrophages; their fine structure; as well as vascularity and functional role of this organ have been reviewed by Dellmann and Simpson (1979).

The SFO is a central site in which angiotensin and cholinergic agents cause drinking behavior and increased blood pressure. Their efferent projections terminate within the preoptic area (POA), the OVLT, and within the nucleus supraopticus, and indicate a role for the SFO in the modulation or control of neurohypophysial activities (Dellmann and Simpson, 1979). These authors also mention that the ependymal and subependymal cells of the SFO contain LRH, SS, TRH, dopamine, noradrenaline, serotonin, and histamine. The SFO participates in the central regulation of salt and fluid homeostasis.

### 4. Organum Vasculosum Laminae Terminalis

The organum vasculosum laminae terminalis (OVLT) is known as the "supraoptic crest" and is localized above the optic chiasma, close to the cavity of the third ventricle and rostrally at the level of the preoptic recess. Like the ME, it is essentially a site of distal integration where such neurohormones as corticotropin-releasing hormone (CRH), SS, LRH, and TRH might be released into the blood (Bouchaud and Bosler, 1988). Bouchaud and Bosler summarized data obtained so far and some of them are discussed here. The monaminergic innervation of the mammalian OVLT is predominantly serotoninergic. The OVLT is innervated by afferent fibers from the medial preoptic area (MPA), nucleus ventromedialis, the lateral hypothalamus, and also from the extrahypothalamic brain regions. The axons of the nuclei paraventricularis, dorsomedialis, and periventricularis terminate in the OVLT. Similar to the ME, this organ is also a site where LRH, SS, TRH, and CRH might be released into the blood circulation.

### 5. Epiphysis Cerebri

The epiphysis cerebri (EC) is the outgrowth of the diencephalon roof, known as "pineal" in the lower vertebrates. In these vertebrates, there are neuronal connections between the brain and the epiphysis, whereas in mammals the gland is innervated via its stalk by the sympathetic and rarely by the parasympathetic systems. The gland is composed of pinealocytes, which in mammals have no direct light perception.

### 6. Neurohypophysis

Peptidergic and aminergic hypothalamic neurons terminate in the neurohypophysis (NH) or pars intermedia (Fig. 5). Short portal blood vessels connect the neurohypophysis with the adenohypophysis. It appears that

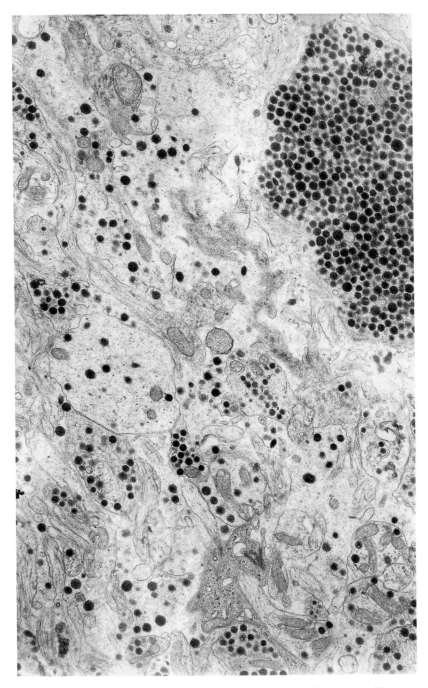

FIG. 5   Nerve terminals in the neurohypophysis of a rat neonatally treated with estrogen. A part of a Herring body is evident.

the NH may store the largest amount of neurohormones and bioamines in the nerve terminals known as Herring bodies. A fine network of CRH-immunopositive fibers terminates around capillary vessels in the neurohypophysis. The terminals of the NSO and NPV and from the other hypothalamic nuclei discharge neurohormones and bioamines into the general circulation. After surgical isolation of the nucleus paraventricularis (NPV), the complete disappearance of CRH-immunoreactive fibers from this lobe was observed (Lengvari et al., 1985).

It appears that both catecholaminergic and serotoninergic terminals make contact with tanycytes and pituicytes and may be involved in the control of neurohormone release.

Changes in the neurohypophysis during spawning of wild carp have been described (Polenov et al., 1986).

A long-term decrease in size, the appearance of pituicytes, and a decreased amount of vesicles and granules in the aminergic terminals of the NH of rats after treatment with gonadal steroids (GSs) are considered by this author to be a result of an inhibitory estrogen effect on the neuronal synthesis of dopamine (DA), and other bioamines, and neurohormones.

Danilova et al. (1982) described the properties of NH and pars intermedia (PI) in rats adrenalectomized at various stages of juvenile and pubertal development.

## 7. Median Eminence

The median eminence (ME) is a link between hypothalamic neurons, the CSF, and pituitary cells. Hypothalamic neurons and pituitary cells are connected via long portal blood vessels, so that blood circulation occurs in both directions. The basic cellular composition of the adult ME, which represents the portion of the tuber cinereum in contact with the pars tuberalis of the anterior lobe, is vascularized by the pituitary portal vessels and is demarcated internally by the wall and floor of the recess of the third ventricle. It is organized as the inner ependymal layer or zona interna, a middle fibrous layer, and an outer layer or zona externa (Kniggc et al., 1976).

A high concentration of hypothalamic neurohormones was detected in the ME, and also in the neural lobe of the hypophysis and the pituitary stalk, which contains a capillary bed closely related to that of long portal blood vessels. A large amount of hypothalamic neurohormones may be accumulated (stored) in both the neurohypophysis and the pituitary stalk and these biologically active substances are released in response to the appropriate stimuli.

The ultrastructure of three distinct zones in the frog ME has been described: an ependymal zone consisting of a single layer of tanycytes;

an internal zone containing predominantly of unmyelinated fibers of the preoptic hypophysial tract; and an external neurovascular zone (Chetverukhin *et al.*, 1986). The morphology and function of tanycytes were reviewed by Flament-Durand and Brion (1985).

The ME of birds has distinct anterior and posterior parts. It is supplied by neurons of preoptic hypothalamic and tuberal axons and consists of a distinct anterior and posterior capillary plexus of portal blood vessels (Mikami, 1986). This suggestion is based on the anatomical relationship between the ME, caphalic, and caudal lobes and the possibility of the presence of a "point-to-point" regulatory system.

The accumulation of neurosecretory substances in the ME of rats adrenalectomized at various stages of ontogenesis, beginning from the early neonatal stage up to 60 days of age, has been examined (Danilova *et al.*, 1980). They have looked at this accumulation of neurosecretion in the external ME during the postoperative period, which was observed in rats adrenalectomized during the first month of life, from the viewpoint of the undeveloped transport system between the hypothalamus and portal blood circulation during the first month of life. The pituitary stalk–median eminence, as the subdivision of the neural lobe, plays an important role in transport of the axonal content from the neurons localized in the anterior perventricular hypothalamus; from the regional neurons lying betweem the optic chiasma, anterior commissure, and the septum, and probably partly from the tuberoinfundibular neurons. It appears that dopamine-containing neurons reach both the ME and the NH (Stark and Makara, 1982).

Matrix cells and differentiating tanycytes of fetuses and neonatal rats possess long basal protrusions reaching the ME and connecting the third ventricle and the primary portal plexus throughout the perinatal period. However, in addition, the basal processes terminate on the neurons of the nucleus arcuatus from the end of the fetal stage of development to the adult, and have a basic role in feedback regulation of neurosecretory neurons by neurohormones circulating in the CSF. The permeability of the CSF–blood barrier in the ME, development of the hypophysial neurons, and the pathways for neurohormone transport were reviewed by Ugrumov (1991). The ME and OVLT are two major CVOs at which LRH-containing axons terminate.

## B. Blood–Brain Barrier

The blood–brain barrier (BBB) is mainly developed in the brain outside the circumventricular organs. Their capillary walls are thinner than the endothelium of microvessels of the primary BBB. Tight junctions are continuous and represent interendothelial junctions (Dermietzel and Krause, 1991).

## 1. Endothelium

As a homeostat between blood and brain neurons, the BBB is highly selective and highly specific, with a high degree of restrictive bidirectional exchanges of substances necessary for neuronal activities. The specificity of the BBB is characterized by a nonfenestrated thin endothelium, a dense perivascular basement lamina, an extracellular matrix, pericytes, and astrocytes. This specificity is designed to allow lumenal absorption through endocytosis and to regulate the transport of various biologically active substance, such as ions, nucleotides, amino acids, glucose, peptides, and enzymes (Dermietzel and Krause, 1991). The structural and functional properties of BBB are species specific and evolutionarily and ontogenetically developed to provide a high degree of protection, selective transport of electrolytes and nutrients, and a highly specific integrative activity. The development of BBB endothelial restriction of horseradish peroxidase (HRP) and the role of astrocytic factors in initiating closure of the endothelium are temporal and species specific and considered a clear sign of BBB maturation. The mechanism governing the pattern of BBB maturation is not limited to the interaction between glial and endothelial cells; the effects of neurons on glia, and glial cells on the BBB endothelium, are also important (Dermietzel and Krause, 1991).

## 2. Cells Surrounding Endothelium

The cells surrounding the endothelium are pericytes and astrocytes. Pericytes, as perivascular cells, are localized spirally around the endothelium. As the myoepithelial cells, they are involved in regulation of contractile capacity, and play a crucial role in blood flow. However, they are responsible for maintaining homeostasis between blood and brain neurons, and as active microglia they participate in phagocytosis. The astrocytic processes in the form of astroglial end feet, and "tight" BBB endothelium, play an important role in the selective permeability and homeostasis of the BBB (Dermietzel and Krause, 1991).

## 3. Basement Lamina

The basement lamina (membrane) is composed of collagenous and noncollagenous components. A highly selective basement lamina, synthesized by adjoining cells, is composed of laminin, collagen IV, proteoglycans, fibronectins, nidogen, and entactin (Inoué, 1989; Dermitzel and Krause, 1991).

Laminin is a glycoprotein; collagent IV is a microfibrillar molecule; proteoglycans possess an affinity for collagen IV resembling that of laminin fibronectins, a group of glycoproteins with an adhesive capacity for binding to collagen fibrils. Nidogen and entactin as glycoproteins are found in small amounts in the basement membrane. The basement lamina is synthesized by adjoining cells, which are connected with the basement lamina via fine filaments. At the electron microscopy level, three main areas may be distinguished: an inner, electron-dense layer, the lamina densa; the lamina lucida (rara), a less electron-dense layer; and the lamina (pars) fibroreticularis (Inoué, 1989). The molecular content and organization are of great importance for a protective mechanical role and highly selective transport of molecules, but also for neuronal and glial migration, differentiation, and activity. Dermitzel and Krause (1991) summarized the role of the BBB as a specific transporter of glucose, specific enzymes, and ions, as well as in peptide and protein transcitosis, the crucial role of the endothelium in separating the blood from the interstitial cerebral compartment and the cerebrospinal fluid, the dense distribution of cells spirally wrapped around the endothelium, and the composition and position of the basement lamina.

## VII. Hypothalamic Neurons Synthesizing Precursors to Neurohormones

Berlind (1977) reviewed the definition of neurosecretory cells in invertebrates, the isolation and characterization of their neurohormones, control of synthesis, and the transport release and feedback control of neurosecretory cells.

Hökfelt et al. (1978) focused on the hypothalamus in summarizing findings on the distribution and pathways of aminergic and peptidergic neurons in the nervous system. The importance of the hypothalamus and amygdala in receiving information from the external environmentals, by way of sensory, visceral, and somatic cells reacting specifically and mostly with coordinated sequences of events, is the subject of many discussions and papers. The nature of peptidic neurons (Hökfelt et al., 1980), and their localization, hypophysiotropic neurohormones (Krulich et al., 1977), and the role of brain peptides in controlling anterior pituitary hormone secretion (McCann et al., 1986) have been examined.

The localization and distribution of CRH, LRH, and TRH neurons and their projections in the CNS of birds were reviewed by Mess and Józsa

(1989). The role of hypothalamic hormones in mediation of distinct events, that is, pituitary behavioral and neuronal outflow, were considered as multisignal integrators by Moss (1979).

The magnocellular hypothalamic neurosecretory system in birds is different from that in the lower vertebrates, but resembles that of reptiles and mammals, and the original preoptic nucleus has become divided into magnocellular preoptic, supraoptic, and paraventricular nuclei. Light and electron microscopic examinations of peptidergic neurons and ME are presented (Mikami, 1986).

The paraventricular nucleus acts as a major relay for the neuroendocrine responses to multiple environmental stimuli and plays a role for hippocampal inputs in setting the basal activity of the hypothalamo–pituitary–thyroid axis in addition to their previously described effects on the hypothalamo–pituitary–adrenal axis (Lightman, 1994).

## A. Precursors to Neurophysins and Associated Neurohormones

### 1. Vasopressin, Oxytocin, and Neurophysins

*a. Vasopressin and Oxytocin* Vasopressin and oxytocin and the associated neurophysins were among the first neuropeptides and proteins discovered. They were identified in the n. supraopticus (NSO) and n. paraventricularis (NPV). As a result of gene duplication, oxytocin and vasopressin-related genes coexist within individual neurons, but it seems that only one of these genes is expressed (Lincoln and Russell, 1986). Lincoln and Russell suggested the coexistence and secretion of oxytocin and vasopressin with corticotropin-releasing hormone (CRH), cholecystokinin (CCK), and opioids. These neuropeptides are transported via neurosecretory granules to the neurohypophysis, costored in the axonal terminals, and synaptically released. Their axons also terminate in the internal and/or external layer of the ME (Vandesande and Dierickx, 1975). Their projections to extrahypothalamic regions, such as the amygdala, septum, and lateral ventricles, and also to the spinal cord, telencephalic choroid plexus, and cerebrospinal fluid, have been described (Buijs et al., 1978; Brownfeld and Kozlowski, 1977; Sofroniev and Weindl, 1978a).

Oxytocin and vasopressin were also in the first peptide hormones to be sequenced and synthesized. These peptides serve as precursors for smaller peptides with entirely different activities. Once the amino acid sequence of a biologically active peptide is known, synthesis of the natural peptide and analogs can be undertaken. Stewart (1982) showed that the C-terminal tripeptide of oxytocin functions as an MSH release-inhibiting factor in the anterior pituitary, and the N-terminal octapeptide of vasopressin serves as an ACTH-releasing factor in the pituitary and in the CNS. In the CNS these neurohormones may have important roles in learning, memory, opiate effects, and behavior. Hypothalamic neurons in the rat producing oxytocin and vasopressin were identified by Dierickx and Vandesande (1975).

*b. Neurophysins*   Cells containing neurophysins were found in the rat supraoptic nucleus, the nucleus paraventricularis, and in the anterior commissural nucleus. Only the NPV has an appreciable number of cells that concentrate estradiol and contain neurophysin. Within the NPV, these cells are found predominantly in the posterior subnucleus (Rhodes *et al.,* 1981).

The distribution and localization of vertebrate oxytocin and vasopressins, and of neurophysins in the hypothalamic magnocellular neurosecretory neuorons, the ME, PI, and extrahypothalamic areas were reviewed by Dierickx (1980), who also discussed the absence of immunocytochemical staining for vasopressin in the nucleus suprachiasmaticus as well as in the hypothalamic magnocellular neurons in homozygous Brattleboro rats with diabetes insipidus. Simultaneous localization of oxytocin and neurophysin I or arginine vasopressin (AVP) in the hypothalamic neurons was demonstrated by Sar and Stumpf (1980).

The biosynthetic origin of neurophysins, of neurophysin–neurophysial peptide complexes (Hough *et al.,* 1980), and of the nucleotide sequence of cloned cDNA encoding the bovine arginine vasopressin–neurophysin II precursor have been determined (Land *et al.,* 1982). The synthesis of vasopressin and its associated neurophysin is stimulated by adrenalectomy, and their content and pathways to the portal capillaries are regulated by glucocorticoids (Stillman *et al.,* 1977).

*c. Preproneurophysins*   Neurophysins are derivatives of an originally biosynthesized form of and their incorporation into noncovalent complexes with hormones derives in some way from postsynthetic processing events (Chaiken *et al.,* 1982). Chaiken *et al.* found that neurophysins do not themselves contain sufficient amino acid sequence information to code for spontaneous folding to active form. The translation products of the neurophysins were defined as preproneurophysins. The evidence that neurophysin and associated neuropeptides originate from a simple polypeptide precursor (Chaiken *et al.,* 1982) and other data undoubtedly led to a clearer elucidation of the links that occur intracellularly during precursor synthesis and storage, proteolytic splitting, transport of the mature proteins and neuropeptides, and, finally, their release from the terminals in the neurohypophysis.

Synthesis and processing of the preproneurophysins I and II have been reported by Schmale *et al.* (1979). Precursors to neurophysin II–arginine vasopressin and to neurophysin I–oxytocin have been immunologically identified by Richter *et al.* (1980). The storage of oxytocin with neurophysin I and of vasopressin with neurophysin II in separate neurosecretory granules was observed by Dean *et al.* (1968).

Strong chemical support has been given to the view that preproneurophysin indeed contains hormone sequences (Schmale and Richter, 1981). The way in which noncovalent complexes of processed neurophysins and associated neurohormones are derived has been described by Land *et al.* (1982) and Chaiken *et al.* (1982), on the basis of the common precursor structure and similarity with the neurophysin I–oxytocin precursor structure.

## B. Corticotropin-Releasing Hormone

Corticotropin-releasing hormone is the predominant regulator of stress-induced ACTH release. As evolved from the precursor molecules, CRH was characterized as a 41-residue ovine hypothalamic hormone that stimulates release of ACTH and β-endorphin (Vale et al., 1981). The evolutionary aspects of CRH were considered by Lederis *et al.* (1990). Allen *et al.* (1984) described the biosynthesis and processing of proopiomelanocortin during fetal pituitary development.

An increased immunoreactivity in the paraventricular terminals of adrenalectomized animals was suppressed by glucocorticoid replacement (Stillman *et al.*, 1977). It is now accepted that the parvocellular neurons of the nucleus paraventricularis (NPV) have a predominant role in producing CRH. Most of these neurons send fibers terminating in the neurohemal zone of the ME by embryonic day 17 (E-17). There are data that the concentration of CRH increases progressively during perinatal development.

The simultaneous expression of glucocorticoid receptors and CRH immunoreactivity in parvocellular neurons supports the concept of a direct feedback action of glucocorticosteroids on CRH-synthesizing neurons (Liposits *et al.*, 1987). An increase in vasopressin and immunoreactive messenger CRH neurons of the nucleus paraventricularis (Wolfson *et al.*, 1985), and a significant increase in hypothalamic mRNAs encoding both CRH and vasopressin precursors, indicate that parvocellular neurons of the NPV may contain both CRH and vasopressin and their precursor molecules (Jingami *et al.*, 1985; Davis *et al.*, 1986).

A significant increase in hypothalamic mRNAs encoding CRH and vasopressin precursors has been demonstrated in adrenalectomized animals by RNA hybridization (Davis *et al.*, 1986). However, these hypothalamic CRH neurons, in which transcription of the prepro-CRH gene occurs, might receive information from monaminergic neurons.

CRH-41, arginine vasopressin, and possibly oxytocin and epinephrins are released from nerve terminals in the zona externa of the ME into the pituitary portal vessels. The neurons of the NPV receive signals from most endogenous and exogenous stressors via multiple pathways. Corticotropin-releasing hormone and VP stimulate ACTH release from the pituitary ACTH cells. However, their neuornal axons terminate on or near the ACTH-producing neurons in the nucleus arcuatus (Sofroniev and Weindl, 1978b). Their role as transmitters of stimuli to other brain ACTH neurons might also suggest their participation in regulation of brain neurons producing families of POMC hormones. A role for vasopressin as a modulator of CRH release has been suggested (Gillies and Loury, 1970). The role of vasopressin in potentiation of cAMP activity and ACTH release, induced by CRH in the anterior pituitary cells *in vivo* and in culture, has been studied (Giguere and Labrie, 1982; Yates et al., 1971; Vincent and Labrie, 1982).

In addition to CRH, AVP plays a major role in regulating ACTH secretion. It interacts with highly specific receptors in the pituitary that are clearly distinct from those that combine CRH-41 and other CRH-like peptides (Koch and Lutz-Bucher, 1986).

Catecholamines and AVP are also involved in mediating stress-induced ACTH release as CRH-potentiating agents (Bruhn *et al.*, 1984). Catecholaminergic modulation of corticotropin-releasing factor and adrenocorticotropin secretion was reviewed by Plotsky *et al.* (1989).

## C. Precursors to Growth Hormone Release-Inhibiting and Growth Hormone-Releasing Hormones

### 1. Preprosomatostatin

The precursors to somatostatin is known as preprosomatostatin. A single gene encodes the 92-amino acid precursor of somatostatin. Somatostatin, as a tetradecapeptide, is derived from this precursor protein by posttranslational processing after enzymatic cleavage (Benoit *et al.*, 1987).

Preprosomatostatin-synthesizing neurons originate in the neuroepithelium of the third ventricle and migrate caudally to the nucleus arcuatus (NA) and nucleus ventricularis on E 18 in the anterior hypothalamus, and to the periventricular region on postnatal days 3–6. Their increase in number and differentiation seem to continue, at least up to the end of the third postnatal week (Ugrumov, 1991). The biosynthesis of somatostatin as the primary product of the translation of SS messenger RNA as a polypeptide (prepro-SS), and the role of two different processing enzymes responsible for generation of SS-28 and SS-14 from the proprecursor, were described by Harmar et al. (1986).

Preprosomatostatin as a precursor of somatostatin-14 and somatostatin-28 was proposed as a model for studying the processing of neuropeptides. Somatostatin(25–34) is highly conserved from catfish to humans. These peptides were isolated from the endocrine cells of the rat stomach, the gastric antrum, and other cells (Benoit et al., 1987).

Several workers have described the regional distribution of brain neurons (Brownstein et al., 1975) and identification of hypothalamic hormones and other peptides in the cell bodies of the nucleus medialis preopticus and nucleus periventricularis, the nerve terminals innervating the nuclei paraventricularis, ventromedialis, arcuatus, and suprachiasmaticus, and in the median eminence and different brain areas outside of the hypothalamus (Hökfelt et al., 1975; Elde and Parsons, 1975; Elde and Hökfelt, 1978, 1979).

Somatostatin fibers projecting to the lateral parts of the ME during the first 2 weeks of postnatal development show the characteristics of adults by the fourth postnatal week. The biosynthesis of SS and vasoactive intestinal polypeptide (VIP) and TRH as brain and gut peptides was described by Goodman et al. (1986).

The relationships of both somatostatin and catecholamine fibers are similar to those in adults from postnatal day 21, when the onset of dopamine control of somatostatin release became evident. The onset of a somaostatin inhibitory effect on the release of GH, beginning from postnatal day 4, and of TSH from embryonic day 20 were reviewed by Ugrumov (1991).

Somatostatins containing nerve terminals occur around the magnocellular neurons of the NPV and NSO, and are present in the pituitary stalk and the neural lobe (Hökfelt et al., 1975). Somatostatin-like immunoreactivity in some peripheral sympathetic noradrenergic neurons was observed by Hökfelt et al. (1977a).

The number of somatostatin-producing neurons per tissue block varied (from 3360 to 4910) and an average of 72.4% of these cells were double labeled (Merchenthaler et al., 1989a). These workers showed that peptidergic neuronal axons terminating around portal capillaries in the external zone of the ME and scattered in the hypothalamus and surrounding brain area are intermixed with other neurons containing the same peptide, but do not terminate in the ME (Merchenthaler et al., 1989b). They also estimated that approximately 70% of the LRH in the anterior hypothalamus and septum, and about the same percentage of SS in the median preoptic area, the anterior periventricular region, and the paraventricular nucleus are projected to the ME. A subcellular description of radioimmunoassayable SS in rat brain was published by Epelbaum *et al.* (1977).

Using a combination of retrograde tracing and immunohistochemical techniques, Epelbaum (1994) observed that SS fibers innervating the arcuate GRH neurons do not originate in the periventricular hypothalamus cell group, but appear intrinsic to the accurate. Epelbaum proposed that the complex anatomical network must be taken into account to interprete the intrahypothalamic interactions between SS and GRH neurons, such as those taking place in GH negative feedback mechanisms.

The other LRH or SS neurons, which do not terminate in the ME, may have axon collaterals reacting with multiple targets within the CNS. The neurons terminating in the other brain areas may function as neuromodulators and/or neurotransmitters. In addition to some differences between the percentage of neurons terminating in the ME and in the extrahypothalamic area, it appears that most of the LRH and SS neurons have a hypophysiotropic role.

## 2. Growth Hormone-Releasing Hormone

The growth hormone-releasing hormone (GHRH)(1–29) amides of rats and humans have been characterized and their stimulative effect on GH release demonstrated (Guillemin *et al.*, 1982). Growth hormone-releasing hormone isolated from a pancreatic islet tumor was charcterized by Rivier *et al.* (1982).

Clark and Robinson (1985) established a "male" type of GH secretory pattern in normal female rats by long-term, pulsing intravenous (iv) infusions of the active human GHRH fragment, GHRH(1–29)-NH$_2$. They reported that this treatment accelerates growth and increases pituitary GH content.

Growth hormone-releasing hormone stimulates GH gene transcription independently of GH release and, conversely, other agents can stimulate GH release without affecting transcription of the GH gene (Barinaga *et al.*, 1985). These researchers showed that the synthesis and release of GH by the GH pituitary cells is under complex hormonal regulation.

Analogs of GHRH(1–30) amide have been developed (Mező *et al.*, 1993) and their *in vivo* and *in vitro* biological activity tested (Kovács *et al.*, 1993; Mező *et al.*, 1993). The potency of the analogs was found to be 1.2 to 2 times stronger than that of the GHRH(1–29) amide.

A galanin-like peptide has been detected in the brain of cartilaginous fish (Vallarino *et al.*, 1990) and in the brain and pituitary of a number of teleostean species (Holmquist *et al.*, 1992). The presence of such widely distributed neurotransmitter neurons in the brains of fish, their biological significance, and their structural similarity to the mammalian galanin molecules indicate a role in regulation of growth, directly and indirectly.

### 3. Prothyrotropin-Releasing Hormone

*a. Prothyrotropin-Releasing Hormone* Prothyrotropin-releasing hormone (PRO-TRH) is widely distributed in tissues containing TRH (Mitsuma *et al.*, 1990). The sequence of PRO-TRH has been identified and its distribution in rat hypothalamus, other brain areas, stomach, and eye demonstrated (Lechan *et al.*, 1986; Wu *et al.*, 1987).

*b. Thyrotropin-Releasing Hormone* Cells containing TRH were first observed in the nucleus dorsomedialis on E 16. As sparse neurons, they appear in the nucleus paraventricularis and the perifornical area on E 18, and in the preoptic area and the lateral hypothalamic nucleus on E 19. A high concentration of TRH was observed in the nucleus paraventricularis on E 21. The most prominent network of neurons containing TRH infiltrates the nucleus dorsomedialis in neonatal rats. The start of TRH control of TSH release has been demonstrated to begin at that point (see Ugrumov, 1991).

Significant quantities of TRH are present in the extrahypothalamic regions of the brain, as well as in the hypothalamic regions. The total quantity of TRH in the rat brain was 18.2 ng. Of this amount, 33.6% was present in the thalamus, 26.7% in the cerebrum, 18.8% in the brainstem, 1.6% in the cerebellum, 0.42% in the posterior pituitary gland, and 0.09% in the anterior pituitary gland (Oliver *et al.*, 1974).

Thyrotropin-releasing hormone is found in the hypothalamic nuclei and also in extrahypothalamic-specific nuclei throughout the CNS (Hökfelt *et al.*, 1975). Hökfelt *et al.* suggested that huge amounts of TRH-like substances in frog skin may be involved in the control of pigmentation in lower vertebrates. Short axons from the nucleus arcuatus extend to the ME.

The relationship between the conformation of this molecule and biological activity was examined and the receptor affinity and conformation of the primary structure reported (Lintner *et al.*, 1982). The highest level of TRH was found in the whole hypothalamus or in the medial hypothalamus (Palkovits, 1982). Palkovits showed that the extrahypothalamic area of the brain contains moderate or low concentrations of this tripeptide. The high plasma level of TRH and its decreased content in the median eminence in hypothyroid animals are undoubtedly the result of increased TRH secretion. The TRH consists of only three amino acids, which are sensitive to substitution and modification. It stimulates TSH, PRL, and GH release, and has a direct effect on target neurons.

Prolactin and GH synthesis are induced by TRH or estrogens. Thyrotropin-releasing hormone is transported via portal vessels to receptors on the pituitary prolactin, GH, and TSH cells. As a neurotransmitter, TRH may also be released from nerve terminals into a narrow cleft and bound to the postsynaptic receptors on the plasma membrane of target cells.

Tonon *et al.* (1980) described TRH as an MSH-releasing hormone, and found a specific TRH receptor in frog PI cells. The receptor affinity of peptides such as TRH, the side chain conformation, and the dependence of both affinity and conformation of the primary structure were reported by Lintner *et al.* (1982).

In the period since native TRH was isolated and the sequence of its amino acids determined, a great number of synthetic analogs have been prepared and other amino acids substituted for specific residues. Their biological stimulatory or inhibitory effects and application were summarized by Schally *et al.* (1976).

## D.  Precursors to Lutein-Releasing Hormone and Gonadotropin-Releasing Hormone-Associated Peptide

### 1.  Genesis and General Properties

Studies of LRH-immunoreactive (LRHir) neurons during ontogenetic development indicate that LRHir neuroblasts differentiating on one side in the olfactory placode are able to inhabit both sides of the forebrain in the course of their migration (Sétáló, 1994). Ontogeny and migration of LRH neurons in human embryos have been reported by Schwanzel-Fukuda (1994).

Lutein-releasing hormone neurons scattered in the preoptic area are first observed on E 17 in the rat hypothalamus (Ugrumov, 1991). From postnatal day 2 until day 20 spinelike processes participating in synaptogenesis appear, increasing from 30 to 60% and demonstrating that undifferentiated smooth cells differentiate into pedunculated ones. At least two populations of LRH neurons in the preoptic septal region and two populations of somatostatin neurons in the medial preoptic and anterior periventricular areas and the paraventricular nucleus of the rat brain have been described: one with access of the portal capillaries of the ME and another that is functionally related to intracerebral neurotransmission or modulation (Merchenthaler *et al.,* 1989a,b). Neurons containing LRH and gonadotropin-releasing hormone-associated peptide (GAP) were classified into two major categories: smooth and fusiform or "spiny." There are also axons containing LRH and GAP that terminate in the amygdala, the bulbus olfactorius, and the region of the central gray matter and that are involved in regulation of sexual behavior. Isolated and incubated granules containing LRH, the role of MgATP, and the mechanism of cooper-stimulated LRH were examined by Barnea (1984).

The GAP and LRH portions of the LRH precursor were localized by immunocytochemistry in prepubertal and adult female rats at different stages of the estrous cycle, and in ovariectomized rats (Merchenthaler *et al.*, 1988). These authors suggested that ovariectomy may result in decreased synthesis and/or processing of the LRH precursor.

The internal segment of the LRH molecule as the site of amino acid substitution, changes in LRH primary structure during evolution, the phylogeny of partially characterized forms of the molecule, and the existence of more than one member of the LRH family have been studied (Sherwood, 1986). Sherwood also discussed the phylogeny of the LRH receptors and structural links of LRH family to older molecules. The structural and functional evolution of LRH has also been reported (Millar and King, 1987).

The characterization and topography of LRH-producing neurons in cats and dogs and in primates, including the human brain, were studied (Barry, 1976; Barry and Carette, 1975; Barry and Dubois, 1975, 1976). The LRH of vertebrates, which is older in origin than that of the amphibians, may have a different chemical structure than that of synthetic LRH, but such a structural difference still allows synthetic LRH to exert biological activity in some lower vertebrates (Deery, 1974). A comparative study of LRH neurons and their projections in human and other mammalian species as well as LRH projections to the ME and OVLT has been made (King *et al.*, 1982; King and Anthony, 1984).

LRH receptor (LRH-R) is a 327 amino acid protein with 7 transmembrane regions, consistent with its coupling to G protein-mediated activation of phospholipase C. Activation of these neuronal LRH-Rs elicits dose-related ($Ca^{2+}$) responses that are dependent on calcium mobilization and entry. The ability of the LRH-R to exert autocrine actions on the secretory neurons could reflect the operation of the endogenous pulse generator and the genesis of the preovulatory LRH surge *in vivo* (Catt *et al.*, 1994).

Lutein-releasing hormone immunoreactive neuronal bodies are localized in the preoptic area and nucleus arcuatus as well as their axons terminating in the ME and OVLT (Knigge *et al.*, 1976; Elde and Hökfelt, 1979). It has been shown that both LRH and GAP are stored in the same neurons (Sar *et al.*, 1987). The failure of LRH to affect the differentiation of LH cells in the rat hypophysis primordium in serum-free culture was observed (Watanabe, 1987).

Lutein-releasing hormone genes have been obtained from the human, rat, and mouse, seven teleostean species, and the chicken. The genes encoding the LRH decapeptide variants have been characterized (Andersen et al., 1992). Andersen et al., observed 77–85% similarity between crucial regions of the salmonids and those of chickens and mammals.

The structures of cDNA encoding common human and rat precursors for LRH and GAP have been determined. A homology of approximately 70% was established between the human and rat GAP. The common distribution of LRH and GAP immunoreactivity in the rat brain strongly supports the theory that GAP is not an LRH precursor (Sar et al., 1987). The cDNA for the precursor of human LRH was also characterized (Seeburg and Adelman, 1984). The isolation of the gene and hypothalamic cDNA for a common precursor of LRH and prolactin release-inhibiting hormone in humans and rats has been reported by Adelman et al. (1986). All the neurons that terminate outside of the blood–brain barrier, that is, in the primary barrier comprising the CVOs were identified.

The mechanisms involved in the regulation of homeostasis and the biodegeneration of LRH in the CNS and pituitary have also been examined (Nikolics et al., 1982; Bauer and Horsthemke, 1984; Koch et al., 1984).

The arcuate dopaminergic neuronal bodies have estrogen receptors. There is evidence that dopaminergic neurons can interact with LRH terminals in the median eminence and that hypothalamic dopaminergic networks have a role in controlling hypothalamic neurons and secretion of pituitary cells.

Subsequent to the isolation and purification of LRH and elucidation of the amino acid sequence of this decapeptide, numerous analogs have been synthesized that are more potent than LRH in releasing LH and FSH. By studying their role in mating behavior, ovulation, and pregnancy, Banik and Givner (1981) showed that, given in an adequate single dose at an appropriate time, LRH and its agonists may be used as contraceptive agents and to avoid abnormal embryonic development.

It is supposed that each pulse of LH in the circulation of the ovariectomized rat reflects a pulse of LRH release from the hypothalamic neurons and that LRH is released in response to the activity of a "pulse generator" located in the nucleus arcuatus. The amplitude of release might represent the amount of LRH release and/or the pituitary GTH cell responsiveness. There is no doubt that the LH pulse starts to increase after ovariectomy and reaches a maximal amplitude by about 2–3 weeks after the operation.

Studies on a mutant hypogonadal mouse hypothalamus, which is deficient in LRH, showed that LRH is essential for the long-term synthesis of GTH. The effect of daily single and multiple injections of LRH on the pituitary and gonads in a hypogonadal mouse was also investigated (Fink et al., 1982; Davey-Smith and Fink, 1983; Charlton, 1986).

Barnea (1984) suggested that there is a molecular aspect to the role of plasma membrane-associated copper in regulating LRH release from hypothalamic neuronal terminals. The mechanism of action of LRH binding with the receptor requires the presence of calcium (McCann et al., 1984). McCann et al. suggested that releasing factors interact at the hypothalamic level to modulate their own release. Somatostatin inhibits the release of both FSH and LH. McCann et al. also demonstrated the probability of the existence of a separate FSH-releasing hormone.

$Ca^{2+}$ depends on the release of neuropeptides and the presence of $Ca^{2+}$ and $Na^{2+}$ voltage-dependent channels on the plasma membrane of LRH- and SS-containing nerve terminals. Their activation leads to membrane depolarization, subsequent $Ca^{2+}$ voltage-dependent channels, and consequently neuropeptide release (Drouva et al., 1982). Drouva et al. also mentioned a role for calmodulin in synaptic transmission.

## 2. Preoptic Area and Sexual Dimorphism

The medial preoptic area (MPA) is undoubtedly of central importance in the regulation of reproduction, but it also has a crucial role in control of the tonic and cyclic pattern of GTH release from gonadotropic cells. These neurons are involved in the control of properties that characterize male and female animals and their behavior. In addition to the MPA, the medial basal hypothalamus (MBH) is also involved in regulation of the tonic or cyclic pattern of GTH release in sexually mature male and female animals. Sexual differentiation of the brain (Dörner, 1980), and sexual dimorphism in the human brain (Hofman and Swaab, 1991), have been reported.

Lutein-releasing hormone neurons first appear in the hypothalamus and in other brain areas during the prenatal period. During differentiation, their axons grow toward the OVLT and the ME. The release of LRH to general and portal circulation is highest during neonatal development in both of these circumventricular organs (Ugrumov, 1991).

As is known, androgen and estrogen of endogenous origin from gonads, or exogenously administered during the perinatal period, may permanently masculinize and defeminize neurons involved in controlling sexually dimorphic brain neurons. Male rats that are castrated perinatally are demasculinized in adulthood after treatment with testosterone. The estrogen may act indirectly via the estrogen receptor-containing neurons and/or directly by neurons of the medial preoptic area that incorporate estrogen.

Masculinization and defeminization of the hypothalamus and limbic structure result from an altered afferent input in these regions. The dynamic interaction among environmental, behavioral, neurotransmitter, and endocrine factors that influence afferent input to the neuronal substratum that underlies reproductive functions has been studied (Beyer and

Feder, 1987). By assuming that the primary sites of gonadal steroids are the medial preoptic area (mPOA), amygdala, septum, and some scattered cells in the brainstem, Pfaff and Keiner (1973) and Beyer and Feder (1987) suggested three distinct types of response: rapid, intermediate, and persistent changes in neuronal function, probably resulting from changes in membrane excitability, protein synthesis, and alteration of chromatin structure. They suggested that defeminization may be related to an increase in the number of cell deaths in the MBH, and masculinization may involve stimulation of dendrite proliferation and reduction of neuronal death in the mPOA.

It has been shown that the rat hypothalamic cDNA encodes amino acid sequences that have over 70% homogeneity with human GAP. This non-LRH portion of the molecule and LRH are also detected in the neurons located within the preoptic area. It appears that prohormone synthesized in the perikarya is transported to terminals and released into the pericapillary space of the ME and OVLT. There is no doubt that cleavage of the precursor molecule may occur anywhere in the neurons. Merchenthaler *et al.* (1984) proposed, using antisera with high titer and specificity, to examine the pro-LRH gene products (LRH and GAP).

Modulation of LRH action by gonadal steroid has been observed (Spona, 1974). The multiplicity of LRH action is expressed in stimulation of reproductive behavior by the CNS and the pituitary GTH cells on steroidogenesis in the gonads. This molecule may have an ancient evolutionary origin as a regulator of reproduction. High-affinity receptors for LRH, localized on the plasma membrane of GTH cells, testicular Leydig cells, and ovarian cells, were stimulated with phosphatidylinositol 4,5-biphosphate and translocated pituitary protein kinase C from the cytoplasm to the membrane (Naor and Childs, 1986).

Sexual dimorphism in the neuropil of the preoptic area and its dependence on neonatal androgen have been studied (Reisman and Field, 1973). Sexual differentiation of LRH neurons is clearly evident as a reaction to high doses of estrogen. There is a reduction in the number of LRH neurons in males and an increased number of LRH neurons and immunoreaction in females.

A neural control that consists of endogenous opioid peptides and steroid-accumulating neurons maintains the hypothalamic adrenergic environment to allow episodic and cyclic secretion of LRH (Kalra and Kalra, 1984). These authors suggested that GS may modulate hypothalamic LRH levels and pituitary LH release.

The efferent neuronal pathway from the brain to the ovary supplements the brain–pituitary hormonal mechanism in the regulation of ovarian steroidogenesis. This neuronal connection seems to be involved in adjusting ovarian responsiveness to gonadotropin stimulation rather than playing a role in maintenance of ovarian steroid secretion. Moreover, neuronal feedback from the ovary exists in addition to the hormonal feedback and plays a role in the regulation of both LRH and pituitary GTH secretion. The complex interaction among neuronal and hormonal control systems in the normal reproductive processes needs to be elucidated (Kawakami et al., 1987a,b).

The brain may modulate hypophysial receptors. After destruction of the MBH, rats have reduced [$^3$H]estradiol binding and uptake in the anterior pituitary, which is closely related to hypersecretion of PRL and hyposecretion of the other pituitary hormones.

The major groups of LRH neurons projecting to the ME have a role in regulation of gonadotropin release, whereas the neurons terminating in other brain areas may function as neuromodulators and/or neurotransmitters. The distribution of LRH-producing neurons and their role in the control of pituitary GTH cell activities were also studied (Barry, 1979).

Culler et al. (1988) suggested that long-term orchidectomy results in decreased biosynthesis of the prohormone and/or in partial increases in LRH and GAP degradation or transport. They suggested that LRH and its prohormone synthesis and processing are affected by testicular factors. They also observed that in both the hypothalamus and the ME, GAP and LRH gradually declined in parallel through 14 days after orchidectomy.

The incubation of synthetic LRH with pituitary plasma membrane resulted in the partial degradation of this neuropeptide. The N-terminal tripeptide was the main degradative product. This tripeptide has a primary structure that is similar to that of TRH, and its possible role as one of the factors involved in the regulation of PRL release was suggested (Koch et al., 1986).

## E. Proopiomelanocortin Hormones

The main sources of the different peptides that belong to families of POMC hormones are brain neurons, pituitary ACTH cells, and PI cells. Brain neurons are producers of a family of POMC hormones and contain different neurohormones. As the cells characterized by specific reactions to various types of stressful stimuli, brain neurons are common target cells for neurohormones, corticosteroids, and bioamines. Their releasing hormones and other peptides are released from the hypothalamic magnocellular and parvocellular neurons. These neurons are regulated by feedback from the adrenocortical cells producing corticosteroids, predominantly glucocorticosteroids, and also by short and autofeedback mechanisms.

Adrenocorticotropic hormone is released during acute stress from the neurons located in the nucleus arcuatus (the main source of ACTH). Neuronal axons containing ACTH, MSH, oxytocin, and vasopressin, originating from the hypothalamus, are also found in the medulla oblongata. Thyrotropin-releasing hormone, ACTH, oxytocin, vasopressin, enkephalin, and other neuropeptides have been found in the spinal cord (Palkovits, 1982). The neuropeptide-containing axons are often seen together with aminergic, cholinergic, and amino acid-containing fibers, but they reach their target as solitary fibers as well. The axons of hypothalamic neurons containing peptides, which belong to the family of POMC hormones, also terminate in areas throughout the CNS.

Different types of neurons containing families of POMC hormones have been identified through advances in methodology. In these cell types, a family of POMC hormones is synthesized, the prohormones and peptides split from them are localized in their cytoplasmic granules, and their content is released by exocytosis in different regions of the CNS, via neuronal axons terminating in the NH and PI, or via portal blood vessels into the anterior pituitary. Owing to the differences in their site of location and connections with the target cells, their pathway can be summarized as follows: short or long axons of the smaller neuronal cell bodies differentiate and terminate in various brain areas or use the spinal cord to reach the target cells. Their granular content is transported via axons and split or unsplit prohormones and peptides are released into the synaptic cleft.

Adrenocorticotropic hormone has been found in the single-cell eukaryote *Tetrahymena* (Le Roth *et al.,* 1982). Because the peptides are important in sexual behavior Stewart (1984) suggested that one of them is LRH, which appears to have evolved from an "active core." The number and types of mutation must have taken place early in evolution, with the order of evolution being ACTH–LRH–$\alpha$-factor, or the reverse.

Two main biological actions of ACTH have been suggested: (1) a direct effect on the adrenal cortex, including adrenal steroidogenesis, ascorbic acid depletion, and growth processes; and (2) indirect effects mediated by the adrenal cortex, such as thymus involution, erythropoiesis, and galactopoiesis (Li, 1972).

Proopiomelanocortin is a common precursor molecule for different peptides such as ACTH, MSHs, opioids, substance P (SP), and others. These peptides, except endorphins, are evolved from a common heptapeptide (Li, 1972). The common precursor molecules for ACTH and/or opioids are also known as "pro-ACTH-endorphin" or proopiocortin. Processed products of POMC such as ACTH, $\alpha$-, $\beta$-, and $\gamma$-MSH, and $\beta$-endorphin contain the Met-enkephalin sequence at the N terminals (Al-Jousuf, 1992).

The distribution of enkephalin-immunoreactive cell bodies in the rat CNS was described (Hökfelt *et al.,* 1977b).

Proopiomelanocortin mRNA levels increase in adrenalectomized rats treated with glucocorticosteroids. The changes in POMC mRNA levels observed after treatment with estrogen might be a result of this hormonal action on transcription of genes in the periarcuate neurons of the hypothalamus. These neurons send their axons into the preoptic area, laterally to the amygdala, and posteriorly down to the brainstem. The amount of POMC mRNA in neurons that project to the preoptic area of the rat hypothalamus has been identified and quantitated (Roberts *et al.,* 1986). The feedback control of ACTH secretion by glucocorticoids at a neural level was suggested by Feldman and Saphier (1984). Stressors or disturbance of corticosteroid equilibrium during perinatal development produce a long-lasting alteration in the brain adrenergic mechanism that participates in the control of the hypothalamo–pituitary–adrenocortic axis (Naumenko, 1984).

Less than 10% of the POMC cells in the nucleus arcuatus accumulate radioactive estrogen. It has been shown that estrogen inhibits POMC gene transcription within 60 min after hormone administration and eventually reduces the synthesis of POMC-derived peptides.

The opioid peptides have been isolated and their amino acid sequences determined (Guillemin *et al.,* 1977). It has been shown that $\beta$-endorphin, as the most potent opioid, has a higher analgesic potency than morphine. This new field of scientific research on the biological and chemical properties of opioids has resulted in an intensive program to synthesize them and more than 100 analogs have been produced.

Enkephalin neurons occur in many different areas in the CNS and mainly have short axons. In general, they have local modulatory functions, principally inhibitory in nature (Stewart, 1984). Stewart showed that the brain ACTH–endorphin system consists of a single set of perikarya located in the nucleus arcuatus and that their long axons terminate in several limbic areas, the midbrain, and reticular formation nuclei. A role for ACTH, vasopressin, substance P, and other peptides in learning and memory has been suggested (Stewart, 1984).

It has been reported by De Wied (1994) that many receptors have been found for ACTH/MSH neuropeptides, and for neurohypophysial and other peptide hormones, in the brain. Numerous studies of the action of these hormones on brain function indicated effects on learning and memory, on social behavior, maintenance, maternal and social behavior, and on temperature and drug tolerance. Structure–activity studies indicated that the behavioral effect of ACTH is located in the portion 4–13 of the molecule and that such fragments may be generated from the parent hormone in the brain.

## F. Lesions of the Hypothalamus

In this section only data obtained through surgical operations on the hypothalamic region, or after head irradiation, are mentioned.

Hyperluteinization was first observed by Flerkó (1954) in the ovary of rats with lesions dorsal to the nuclei paraventricularis and dorsomedialis, in which the dorsal part of the latter had been destroyed. A transverse cut with a fine steel knife in the posterior hypothalamus of newborn rats caused the accumulation of corpora lutea far above their usual number, and this increased number was maintained throughout the life of the animal (Martinović et al., 1968a). Posterior hypothalamic damage and hyperluteinization was also studied (Milovanović et al., 1968b).

A transverse cut through the mammillary body of newborn rats, which bypasses the stalk and median eminence, will produce a disturbance in the normal exchange of information between the hypothalamus and the pituitary gland. As a result, numerous corpora lutea were clearly evident in the ovaries; in some cases this was accompanied by advanced opening of the vagina (Martinović et al., 1970, 1977; Milovanović et al., 1988). Martinović et al. (1968b) also observed hyperluteinization in rats exposed to X rays during intrauterine development. The precocious sexual maturation of female rats with hypothalamic lesions (Bogdanova and Schoen, 1959) and of a female rhesus monkey with posterior hypothalamic lesions (Tarasawa et al., 1984) was observed.

In reexamining CRH-like immunoreactivity and plasma ACTH in rats with bilateral lesions of the NPV, Bruhn et al. (1984) concluded that CRH originating from neurons within the NPV is the predominant regulator of stress-induced ACTH secretion; catecholamines and AVP are involved in mediating stress-induced ACTH secretion, most probably as CRH-potentiating agents; and pituitary hyperresponsiveness to exogenous CRH results from removal of endogenous CRH.

Considerable basal PRL and LH release persists 4 weeks after surgical ablation of the medial basal hypothalamus and this continued secretion may be regulated by neurohormones such as LRH, which arise from areas outside of the MBH (Turpen et al., 1978).

Complete removal of the hypothalamic NPV results in an almost complete disappearance of CRH-immunopositive fibers from the ME, 2 or 3 weeks after the surgery (Liposits et al., 1983). Liposits et al. also observed that practically all CRH terminals of the ME originate exclusively in the NPV. In examining the distribution of CRH-immunoreactive neurons in rats 12 days after inducing various types of lesions within the hypothalamus, Liposits et al. (1983) observed that in all cases, when the mammillothalamic tract was transected, CRH was detected in the neurons of the mammillary body.

The lesions of the rostral hypothalamus or isolation of the MBH by incisions would not be as disruptive to cyclic or tonic gonadotropin secretion in primates as they would be in the rat. In cases in which bilateral lesions in the anterior hypothalamus did disrupt the normal GTH surge, normal resumption of cyclic GTH secretion occurred 4–7 months after surgery (Cogen et al., 1980).

Mechanical lesions in the pituitary stalk of newborn rats caused precocious opening of the vagina, with concomitant or delayed ovulation. The premature opening of the vagina as well as ovulation were obtained by damaging the infundibulum and the distal portion of the third ventricle (Martinović et al., 1977).

Dopamine and norepinephrine levels in the hypothalamus of rats with hypothalamic lesions inflicted on the day of birth and in pubertal controls on the day of vaginal opening were significantly higher than in infantile controls (Ivanišević-Milovanović et al., 1993). The onset of pituitary development in total body X-irradiated rats was accelerated in infant female rats with and without hypothalamic lesions (Martinović et al., 1968b).

## VIII. Adenohypophysial Cells as Producers of Peptide and Glycoprotein Hormones

### A. Chromophobes as Precursor Cells for Differentiation of Adenohypophysial Cells

Adenohypophysial cells differentiate from the oral ectodermal epithelium as invaginations of Rathke's pouch. These undifferentiated chromophobes, as precursor cells, are characterized by a large nuclear-to-cytoplasmic ratio and scarce cytoplasmic organelles. It is also known that the first differentiated pituitary cells are surrounded by vascularized mesoderm. Extensive examination of pituitary stem cells during embryogenesis of different animal species showed that the mitotic rate of chromophobes and differentiation of cell types are governed by many different inductive factors.

The pathways of differentiation of pituitary cells and their specificities in different animals under various experimental conditions were summarized by Pantić (1980, 1982). Adrenocorticotropic hormone cells first appear in Rathke's pouch of rats on day 13 of fetal development. The onset of ACTH synthesis is followed 2 days later by the simultaneous differentiation of PRL, GH, GTH, and TSH cells (Watanabe, 1987; Nemeskeri and Halasz, 1989).

The undifferentiated fetal pituitary does not require hypothalamic neurohormones for proliferation and cytodifferentiation, and its development might be modulated by circulating trophic hormones of the host animals (Nemeskeri *et al.*, 1990).

It appears that the hypothalamus exerts an inductive influence on the differentiation of chick pituitary cells from the epithelial primordium removed at the 17- to 24-somite stage and associated with mesodermal tissue without the presence of hypothalamic neurohormones (Mikami, 1986). Mesenchymal tissue was also necessary for differentiation of pituitary cells after the 25-somite stage.

Chromophobic cells in all vertebrates were described as follicular, folliculostellate cells, as single or grouped cells, but also as the cells forming follicles and involved in the digestion of degraded pituitary cell products (Farquhar, 1971; Pantić, 1975). These cells were considered the most sensitive to estrogen (Pantić and Genbačev, 1969). The proliferative rate of chromophobes and their differentiation into PRL cells were increased in the pituitary of rats treated with a single or repeated dose of gonadal steroids. Polarized cells with microvilli on the apical surface were observed in the pituitary of rats neonatally treated with GS (Pantić, 1980). These findings implicate a phagocytic role for these pituitary cells.

## B. Properties of Adenohypophysial Cells

The specific pituitary cells are synthesizers of hormones and from these cells corresponding families of peptidic or glycoprotein hormones are released. As a source of these hormones these cells interact with the neighboring cells. They are target cells for peptides, steroids, and bioamines, and they also have receptors for molecules present in the intercellular matrix. Besides monoamines, the role of histamine in the regulation of anterior pituitary cell secretion was also described (Weiner and Ganong, 1978).

As methodological approaches have advanced, cytological and molecular biological criteria have been used and different pituitary cell types characterized. The main parameters are the cell size and shape and the nature of nuclear and cytoplasmic organelles. The ultrastructure and immunocytochemistry, the nature of granules and the amount of hormones, and synthetic and secretory capacities in all of these cell types have been studied under various experimental conditions (Pantić, 1974b; Denef *et al.*, 1984). The interaction between GTH and PRL cells during exposure to neuropeptides and after exposure to dopamine in reaggregated cell culture showed that GTH cells appear to be capable of transmitting inhibitory signals to PRL cells (Denef *et al.*, 1984).

The nature of these cells may be altered during each stage of development, in the adult, and in senescence. The character of changes may be expressed in the number of receptors, their binding capacity to ligands, and their gene expression, so that altered properties of cell organelles may be clearly evident.

The adenohypophysial pars distalis and PI of the teleostean fish is the functional equivalent of the adenohypophysis of mammals.

The light and electron microscopic properties of the pituitary cells of some birds, mainly chicken and the Japanese quail, have been documented from an early stage of development, that is, the extension of Rathke's pouch from the stomodeal ectoderm to the base of the infundibulum, throughout embryogenesis. The cytology and immunocytochemistry of the adenohypophysial cells of the intact and castrated birds were discussed by Mikami (1986).

In view of the frequency of close apposition between gonadotropic and prolactin cells, ACTH and GH and the other cells, there is no doubt that communication between neighboring pituitary cells is of great importance, not only for inductive interactions and pathways of specific differentiation, but also for their role in regulation of their synthetic and secretory capacity.

The anterior pituitary, as an endocrine gland, is composed of different cell types, each producing different peptide hormones (Herlant, 1964). Adrenocorticotropic hormone, PRL, GH, TSH, and GTH are secreted by separate adenohypophysial cells (Herlant, 1964; Farquhar, 1971). Nakane (1970) suggested that FSH and LH are produced by a single type of cell. However, in addition to these hormones being produced by specific pituitary cells, these cells also synthesize and release, by paracrine and endocrine secretion, other hormones and molecules into the intercellular matrix. These molecules have an extremely important role in the regulation of neighboring pituitary cells, their differentiation, activities, behavior, aging, senescence, or degeneration.

Anterior pituitary cells producing PRL, GH, GTH, and TSH were separated mainly according to size and density (Denef et al., 1982). Denef et al. proposed the gravity sedimentation method as a fruitful approach to a better understanding of the regulation of pituitary hormone release, the mechanism of action of the hypophysiotropic regulatory hormones, and the interactions among the various pituitary cell types.

## C. Adenohypophysial Cells and Hormones Involved in Reactions to Stress

The PI cells and pituitary ACTH cells are producers of families of POMC hormones. Families of PRL and growth hormones are produced by the corresponding pituitary PRL and GH cell types.

## 1. Genesis and Properties of Pars Intermedia Cells Producing Proopiomelanocortin Hormones

A heptapeptide structure was proposed by Li (1972) as the ancestral molecule for MSH–ACTH–lipotropin (LPH). This molecule processes MSH-like activity and occurs in the structure of those three hormones. Li also suggested that a molecule even smaller than the heptapeptide may be the ancestor.

The common biosynthetic precursors to ACTH, $\beta$-lipotropin and $\beta$-enkephalin, coexist in the cells of the pars intermedia and in the pituitary anterior lobe, but not in the neurohypophysis. Common precursors to corticotropin and endorphins (Mains et al., 1977) and the ontogeny of hypothalamic oxytocin, vasopressin, and somatostatin gene expression were described by Almazan et al. (1989).

In the human fetal pituitary, ACTH-containing cells first appear immunohistochemically at 5 weeks of gestation. The other anterior pituitary hormones appear at 13 weeks of gestation. Pars intermedia cells differentiate as a distinct band predominantly toward ACTH, $\beta$-endorphin, $\gamma$-MSH, and $\alpha$-MSH, but this structure becomes indistinct in adults (Osamura and Watanabe, 1985).

Environmental stimuli such as stress, light intensity, photoperiod, and temperature are conveyed via the hypothalamus to the pars intermedia cells of the amphibian *Xenopus laevis* and are transduced by PI cells into multipeptidergic secretory signals consisting of POMC-derived neuropeptides providing a corresponding response (Roubos et al., 1992).

The genesis and properties of pituitary ACTH-, MSH-, PRL-, and GH-producing cells have been the subject of many articles. It is now accepted that highly specific cells of the PI differentiate later than pars anterior (PA) cells and, as multihormone-producing cells, are able to synthesize prohormones and to cleave them into ACTH, LPH, and endorphins as hormones. After studying data obtained up to that time (Pantić, 1982), I posed the question of what the term "MSH cells" implies; I suggested that MSH, ACTH, and opioid molecules are fragments of prohormone, stored and split in the same granules. This indicates that ACTH and MSH cells are involved in the biosynthesis and secretion of lipotropin, which means that ACTH cells from PA and PI cells are producers of POMC families of hormones. As a result of the release of these substances containing ACTH, MSHs, LPHs, and opioids, corticosteroidogenesis, melanogenesis, and lipogenesis are affected. At that time I also questioned whether we have developed cytochemical methods in light and electron microscopy to a level of precision that would allow us to characterize these cell types.

## 2. Peptides and Bioamines as Regulators of Pars Intermedia Cell Activities

Many articles deal with the role of peptides and bioamines in the regulation of PI cells. Corticotropin-releasing hormone and TRH, diffused from the pars nervosa, stimulate PI cells to release MSH. Corticotropin-releasing hormone occurs only in magnocellular cells of the preoptic nuclei, whereas the secretagogue TRH was found in the preoptic and/or infundibular part of the hypothalamus in *X. laevis*. Corticotropin-releasing hormone and TRH appear to be produced by magnocellular hypothalamic neurons, and released from the terminals in the neuronal lobe influencing the PI cells (melanotrophs) via the circulation (Tainhof *et al.*, 1992). Corticotropin-releasing hormone stimulation of MSH release from the PI lobe seems to be species specific: it stimulates MSH release in rat, but not in bovine PI.

Frog PI cells are richly supplied by nerve fibers containing catecholamine, GABA, and serotonin as neurotransmitters, and TRH, CRH, AVP, and mesotocin as neurohormones. In both hypothalamic neurons (the nucleus arcuatus) and pituitary MSH cells, melanotropin synthesis and release are clearly evident (Vaudry, 1992).

Melanostatin, dopamine, and GABA, as the inhibitory hypothalamic factors of the frog pars intermedia, coexist in the same nerve terminals involved in the negative control of $\alpha$-MSH release, and are coupled to several intracellular pathways (Desrues *et al.*, 1992). Discovery and distribution of melanin-concentrating hormones (MCHs) in invertebrates and vertebrates have been studied (Baker, 1991). Baker showed the significance of the elucidation of the cDNA structure of salmon, rat, and human MCH prohormones, and expects that an understanding of the nonpigmentary role of MCH and its related peptides will be observed more quickly than the acceptance of MCH as a color change hormone.

$\gamma$-Aminobutyric acid, neuropeptide Y (NPY), and dopamine (DA) are detected in magnocellular hypothalamic nuclei and coexist in synaptic contact on MSH cells and glial-like folliculostellate cells producing DA, GABA, and NPY (Roubos *et al.*, 1992). The synthesis and release of MSH by PI cells in *X. laevis* are inhibited by dopamine, GABA, and NPY. The neurons of the suprachiasmatic nucleus terminating in the PI, in which they coexist, play an important role in inhibiting PI cells from releasing the corresponding family of POMC hormones.

Dopamine, GABA, and NPY coexist in the nerve terminals that contact PI cells of *X. laevis* and all inhibit $\alpha$-MSH release in an additive way (Leenders *et al.*, 1992). The GABA fiber system is rich throughout the PI and produces a biphasic effect on MSH secretion, with first a short stimulation followed by a prolonged inhibition.

## 3. Steroid Hormones and Pars Intermedia Cells

We considered the increase in size of ACTH cells observed during the migration and spawning of acipenserida and teleostean fish (Pavlović and Pantić, 1975) to be the result of stress reactions during these periods. The pars intermedia, as the largest part of the pituitary in the lizard, has a role closely related to rapid changes of color. The PI cells were examined in various animal species, but it was difficult to demonstrate ACTH and MSH cells as separate cell types.

Periodic acid–Schiff-positive cells distributed in the neurohypophysis of *Ciprinus carpio* treated with gonadal steroids were smaller after treatment with both estradiol (Oe) and progesterone (Pr), than after treatment with Oe only (Pantić and Šimić, 1977b). Chromophobes are numerous cells in the PI of intact piglets and these cells predominate in the cranial part of this lobe in 1-month-old animals castrated on the first day of life (Pantić and Šimić, 1977a).

After studying the size and cellular organization in the neurohypophysis and pars intermedia of rats treated neonatally with gonadal steroids, I proposed using these lobes as a parameter to show to what extent corresponding hypothalamic neurons are altered, and under what experimental conditions the reaction is temporary and when it is permanently expressed (Pantić, 1982). Both lobes are reduced in size and the cell population is decreased up to puberty. Pituicytes in the neurohypophysis are irregularly distributed and altered. The nerve terminals of aminergic and peptidergic fibers are more devoid of granules and microvesicles (Pantić and Šimić, 1980).

Hyperplasia of chromophobes, hypertrophy of PRL cells, and a reduced size of GH cells were observed in all teleosts examined, kept in aquaria, and treated with repeated doses of estrogen. The response of PRL cells of male carp to female GS was similar to the reaction of this cell type in mammals: the GER was circularly oriented and light and dark GH cells were present (Pantić and Sekulić, 1978c). We thought that both of these hormones might be synthesized in the same cell type and the possibility of GH cells being transformed into PRL cells was suggested.

Cells containing ACTH, but not $\beta$-MSH or $\beta$-LPH, were discovered in the chick pituitary at the periphery of cellular cords and near the capillary of the anterior part of the adenohypophysis. Dupouy (1980) studied corticomelanotropic cells as multihormonal-producing cells in different animal species.

In view of the common and specific properties of ACTH, MSH, and PRL cells from the viewpoint of their abilities to synthesize and release corresponding hormones, I suggested that ACTH and MSH cells are also the source of LPHs (Pantić, 1975). As our knowledge of the origin, nature, and specificity of these cells, known as ACTH and MSH cells, increased, it become clear that they are producers of a common precursor molecule similar to that found in pituitary ACTH cells and pituitary brain neurons. I pointed out that PI cells, as highly specific and multihormonal-producing cells, are able to synthesize prohormone and to cleave them into ACTH, MSHs, and LPHs (Pantić, 1982).

Our experimental conditions clearly indicated that brain neurons, ACTH, and PI cells are involved in synthesis and release of the family of POMC hormones. The reactions of both neurohypophysis and PI in rats are more clearly pronounced if the animals are treated with three repeated doses of Oe during the "critical" period than after 15 days. Both PI and neurohypophysis are strongly diminished. The effect is doubtless dependent on the degree of hypothalamic neuronal development (Pantic and Šimić, 1980). Lutein-releasing hormone administered as repeated doses to sexually immature rats has a stimulating effect on PI cells.

The pars intermedia in 4- to 6-month-old neonatally castrated piglets is nearly fully developed and vacuolated; gonadotropic cells are predominant in it. This lobe is less developed in the animals treated with gonadal steroids than in intact and castrated animals. The degree of regressive changes in PI cells was intensively pronounced (Pantić and Šimić, 1978). The size of both the PI and the NH was greatly reduced for a long time, after GS was administered to neonatal rats as a single or three repeated doses. The degree of change is dose dependent.

The pars intermedia of the deer hypophysis is highly developed during hunting and mating seasons. This development and an increased number of hypertrophic cells containing ACTH and MSH occur as a reaction to both sexual activity and the stressful environmental conditions (Pantić and Šimić, 1974).

## 4. Pituitary Adrenocorticotropic Hormone Cells as Producers of Proopiomelanocortin Hormones

Differentiation of ACTH-, MSH-, endorphin-, and LPH-containing cells during embryogenesis was reviewed by Dupouy (1980). The ultrastructural properties of differentiating pituitary cells are characterized by the appearance of specific cytoplasmic granules in Rathke's pouch of fetuses from day 13 of incubation. In reviewing data published so far, there is no doubt that differentiation of pituitary cells is dependent on an interaction with mesodermal cells and hypothalamic factors. The influence of hypothalamic factors affecting the specific biosynthetic and secretory capacities of target cells is of great importance for feedback mechanisms involved in the development and establishment of a balanced hypothalamus–pituitary relationship. The common and specific properties of ACTH cells and role of neuropeptides, bioamines, and corticosteroids in regulating their activity have been studied (Pantić, 1975).

The ACTH cells in the pituitary anterior lobe constitute only 0.16% of the population of pituitary cells (Siperstein, 1963). Results obtained by quantitative electron microscopy study of ACTH cells following adrenalectomy were reported by Siperstein and Miller (1973). The size, shape, and location of these cells differ clearly between teleostean fish and mammals, and are closely related to the character of communication and the ability to respond to stress. In addition to species specificity, I demonstrated (Pantić, 1975) in the distribution of ACTH cells and the proliferation rate and activity of PI cells some of the cytological properties of the ACTH cells (Fig. 6). I think that a less developed GER and smaller granules (150–200 nm) than in PI cells (500–1000 nm) are signs of ACTH specificity in granular formation and content storage.

The plasma ACTH concentration in female rats exposed to either continuous light or darkness was determined by Ivanišević-Milovanović et al. (1990).

The interactions of ACTH cells with other neighboring cells and even within the corticotrophs, which involve positive or negative signaling and similar responses, are also clearly evident in PRL and GH cell reactions to the molecules from the intercellular matrix. Data obtained so far clearly show that CRH, catecholamine, ACTH, and glucocorticoids are the most important hormones involved in stress reactions.

Adrenocorticotropic hormone cells are highly specific in their reactions to stress. The release of ACTH is modulated by various types of stress, the circadian rhythm, and glucocorticoids. In addition to CRH, catecholamines can affect ACTH release. Receptors are present in ACTH cells as common target cells for peptides, glucocorticoids, and bioamines. The reactions of cells containing the POMC family of hormones occur rapidly.

The ACTH cells may also contain nonspecific receptors that are capable of releasing ACTH in response to many peptides during their reaction to various stressful stimuli. Two types of feedback mechanism involved in regulation of ACTH release are known: fast feedback with a rapid onset and a delayed feedback.

Several kinds of hormones are synthesized in the same developing and differentiating pituitary cells. Corticomelanotropic cells were described earlier as multiple hormone-producing cells. The relation between ACTH, MSH, LPH, endorphins, and corticotropin-like intermediate peptide (CLIP) in different animal species was reviewed by Dupouy (1980). All of these peptides, except CLIP and endorphins, are evolved from a common heptapeptide (Li, 1972). It is now accepted that they have been formed by the duplication of a single gene and that subsequent progressive substitution led to their divergence (Lowry and Scott, 1975).

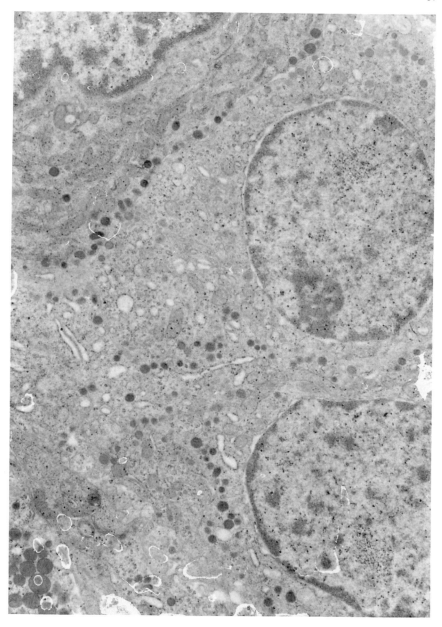

FIG. 6  Hypertrophied ACTH cells, in the pituitary of a rat, with granules localized closely to the plasma membrane, are in contact with GH and GTH cells.

As immunocytochemical methods have advanced, common precursors such as pro-ACTH-endorphin or proopiocortin have been used to locate the corticotropin-like and $\beta$-LPH-like segments in the precursor molecule. Adrenocorticotropic hormone was immunohistochemically localized in the rat brain neurons (Pelletier and Leclerc, 1979).

## D. Pituitary Cells Producing Prolactin and Growth Hormones

### 1. Common and General Properties

The genesis of PRL- and GH-producing cells, their general properties and localization, the nature of their organelles, especially of GER and specific granules and the other constituents that specify their biology, have been summarized (Pantić, 1975, 1982) (Figs. 7 and 8).

Ovine PRL possesses a variety of biological functions in fish, reptiles, amphibians, and mammals. Biologically it is more versatile than GH and therefore it may be assumed that a PRL-like molecule is the ancestor of lactogenic and growth hormone (Li, 1972).

Prolactin and growth hormone belong to a family of hormones similar in structure and evolved from an early gene duplication (Wallis, 1988a), but these hormones have distinct actions. Growth hormone acts via receptors with a high affinity for this peptide, whereas PRL acts via a high binding affinity for PRL but not for GH. However, it seems that human growth hormone has both somatotropic and lactotropic effects and stimulates lactogenesis in mammary explants of different species.

The degree of similarity between PRL and GH and their prohormones in the cells that secrete them was reported by Bern (1983), who stated that these two cell types are concerned with the synthesis, processing, and release of members of a single family of peptides. The basic mechanisms of cell type-specific expression of GH in mammals are inherited from lower vertebrates (Argenton *et al.*, 1992).

Prolactin and GH release increased by various stressful stimuli acting via the hypothalamo–pituitary axis, which converts the stimuli into nervous impulses, and these hormones are bound to receptors. However, only PRL is released in excess when the hypothalamopituitary connection is surgically severed; secretion of all other PA hormones is reduced.

Prolactin and GH cells represent 53.5% of the cell population in the pituitary of the pig. Chromophobes represent 13.1% and are the predominant cell type during lactation. The percentage of GH cells is higher in 17-day-old piglets than in mature animals (Anderson *et al.*, 1972).

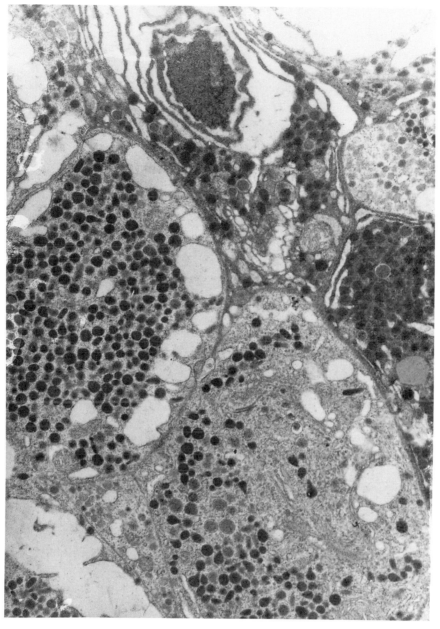

FIG. 7  Pituitary cells of *Torpedo ocellata*, producing the family of PRL and GH hormones.

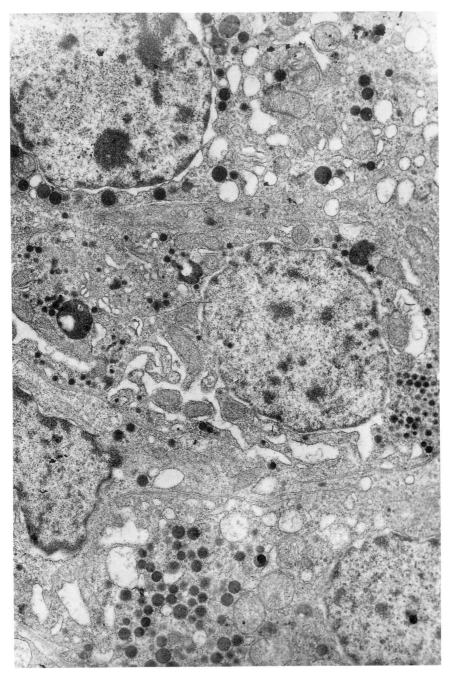

FIG. 8    GH cells surrounding a GTH cell in the pituitary of a rat.

## 2. Common Evolutionary Origin of Prolactin and Growth Hormone

Prolactin is one of the most ancient hormones in vertebrates and shares its molecular form with growth hormone; both of these hormones, as in the prochordates, derive from a common stem cell (Vincent and Liedo, 1992).

Prolactin cells have been proposed as a model for studying neuroendocrine integration (Vincent and Liedo, 1992) and Wallis (1988a,b) summarized the mechanisms of PRL and GH action. Growth hormone cells in culture synthesize and secrete into the culture medium large amounts of PRL and/or GH. These cells respond normally to a variety of physiological signals and appear to be a suitable model for studying the control of PRL synthesis and release. The GH₃ cells were used to examine the effect of TRH on PRL release and the interaction of TRH- and PRL-producing cells (Haug et al., 1982).

On the basis of comparison of amino acid sequences, the concept of a common evolutionary origin for prolactin and growth hormones has been established. The cloning of PRL and GH DNA genes confirmed that these genes are evolved from a common ancestor through several duplications and recombinations before fish and tetrapods diverged evolutionarily. The same general structural differences in five exons and four introns in different animal species, such as some fish, chicken, and mammals, were studied from an evolutionary viewpoint (Belayew et al., 1992). Belayew et al. reported that specific expression of the mammalian GH and PRL genes was induced by the interaction of the transcription factor Pit-1 and specific nucleotide sequences. These data also showed that the genes encoding PRL, GH, and chorionic somatomammotropin are derived from a common ancestor (Belayew et al., 1990).

This family of PRL/GH hormones is produced by the same group of pituitary cells with specific immunocytochemical and cytological properties demonstrating similarity and some specific characteristics. The initial translation product of PRL mRNA is preprolactin. Preprolactin contains an additional 29 amino acids that precede the PRL sequence. Estrogen specifically stimulates PRL synthesis, which appears to be mediated by an increase in the level of pre-PRL mRNA (Maurer et al., 1978).

Localization of PRL-like immunoreactivity in the nerve terminals of rat hypothalamus was demonstrated and the primary structure of PRL was determined by Li et al. (1970). The evolution of hormones and receptors from unicellular organisms to mammalians is closely related to the presence of adenylate cyclase, cAMP, calmodulin, and inositol phosphatase, which as second messengers are able to transmit information to kinases and have an important role in regulation of PRL release.

The role of prolactin in controlling calcium metabolism in the aquatic and terrestrial vertebrate was reviewed by Wenderlaar Bonga and Pang (1991). They studied PRL effects during pregnancy and lactation in mammals, and yolk formation in the lower vertebrates.

## 3. Prolactin Cells as Common Target Cells for Neurohormones, Bioamines, and Gonadal Steroids

The properties of PRL cells in various animal species under different experimental conditions during pregnancy and lactation, and the structural properties of PRL and the role of prolactin-releasing hormone (PRH) and prolactin-inhibiting hormone (PIH) known at that time, have been summarized (Pantić, 1975). Prolactin cells were used as a model for neuroendocrine integration and were separated from rat pituitary by their size and density. It was also mentioned that PRL cells have conserved many common properties with neurons, such as excitability and voltage-sensitive calcium channels.

The secretion and crinophagy of PRL cells were described by Farquhar (1977). These cells increase in number during gestation and lactation. However, it has been shown that acute suckling leads to a rapid release of PRL from PRL cells (Shino et al., 1972).

Work is continuing on efforts to elucidate the complex mechanisms involved in the regulation of proliferation of both chromophobes and PRL cells as well as the differentiation of PRL cells and biosynthesis and release of PRL. Prolactin release is stimulated by PRH, TRH, vasoactive intestinal peptide (VIP), opioid peptide, substance P, neurotensin, and serotonin, and inhibited by dopamine, GABA, and acetylcholine (Rosselin et al., 1982).

*a. Prolactin-Releasing Hormone*  Prolactin-releasing hormone is produced by the hypothalamic neurons containing tyrosine hydroxylase and by pars intermedia cells in which estradiol receptors have been identified (Murai and Ben-Jonathan, 1990; Laudon et al., 1990).

The short portal vessels connect the posterior and anterior pituitary lobes, and carry 20–30% of the total blood flow into the anterior pituitary. Similar to oxytocin, PRH is synthesized in the hypothalamus and transported to the neurohypophysis. Murai and Ben-Jonathan (1990) studied data obtained so far from the following point of view: PRH is located almost exclusively in the pars intermedia; posterior pituitary cells in culture maintain their PRH activity for at least 1 week, whereas OXY is reduced to 5% of its initial value; addition to PRL to cultured pituitary cells reduces their PRH activity by more than 50%. They concluded that these results suggest that PRH is synthesized by pars intermedia cells and its reduction after pituitary stalk lesion is due to elevated PRL.

It is also important to note that estradiol receptors are identified in nucleus arcuatus and PI cells. It seems that Oe has an effect on the hypothalamic nucleus arcuatus from which the tuberohypophysial dopaminergic neurons originate. The fibers of these neurons terminate in both neurohypophysis and pars intermedia. Dopamine receptors are present in both PI and ACTH cells, indicating that DA may be an inhibitor of both PRL and PRH. The neurohypophysis, probably via PRH, is the primary site that mediates the acute effects of estradiol on PRL release (Murai and Ben-Jonathan, 1990).

***b. Thyrotropin-Releasing Hormone*** Thyrotropin-releasing hormone stimulates PRL release. This tripeptide increases the mRNA encoding PRL and it plays a role in the regulation of its own receptors (Gourdji, 1980). An increase in plasma PRL is known to follow stress and suckling. The release from PRL cells depends on the balance between TRH and VIP stimulation and DA inhibition. In examining the role of multiple inhibiting and stimulating factors involved in modulation of PRL secretion at the pituitary cells, Enjalbert *et al.* (1984) also mentioned prolactin-modulating factors. One of the mechanisms involved in modulation of PRL release could be an interaction at the level of the coupling mechanism of the receptor with adenylate cyclase. They recognized three receptors located on PRL cells that appear to be coupled with adenylate cyclase, one positively (VIP) and two negatively (DA and somatostatin).

***c. Vasoactive Intestinal Peptide*** Vasoactive intestinal peptide was isolated from the duodenum as a 28-residue peptide. It acts at different levels on the gastrointestinal, respiratory, cardiovascular, urogenital, and neuronal systems, and on endocrine cells (Rosselin *et al.*, 1982). A high VIP concentration was found in the cerebral cortex, nucleus accumbens, and in the hypothalamic NSO (Palkovits, 1982). The tissue distribution of VIP in the rat as measured by prolactin-inhibiting hormone (PIH) and radioreceptor assay, was discussed by Besson *et al.* (1978).

The maximum VIP content in the hypothalamus was formed in the nucleus suprachiasmaticus. It appears that the nerve terminals containing VIP originate from the neurons present in the mediobasal hypothalamus and the other neurons located outside of the hypothalamus, and that the amygdala may partly control hypothalamic VIP content (Palkovits, 1982). An increased VIP concentration in the adenohypophysis of the adrenalectomized rats returns to normal values when corticosterone or dexamethasone is administered. Like many gut hormones, VIP is present in mammalian and avian endocrine cells that are present throughout the epithelium of the gastrointestinal tract. Prolactin release is stimulated by TRH and VIP in an additive manner from normal and tumoral (GH$_3$) cells (Enjalbert *et al.*, 1984).

Viewing VIP, glucagon, and secretin as structurally related molecules, Rosselin *et al.* (1982) suggested that VIP acts as a neuromodulator, whereas glucagon and secretin act as hormones.

**d. Opioid Molecules** Opioid molecules such as endorphine, Met-enkephalin, and several analogs of Met-enkephalin may stimulate PRL or GH release (Krulich, 1979). However, in most of the experiments, whether the peptides are considered as neurotransmitters or neuromodulators and whether they are stimulators or inhibitors of PRL or GH release, each of them has a dual role: inhibiting PRL and releasing GH. The mode of opioid effects on hypothalamic LRH release and pituitary LH release, as well as site(s) of opioid modulation of LRH secretion, were discussed by Kalra (1982), who also demonstrated the role of catecholamine in mediation of the effects of opioids on LH release.

**e. Catecholamines and Amino Acids** Dopamine and amino acids, such as GABA, glutamate, and glycine have long been known as major neurotransmitters. Dopamine is elevated during stress; the PRL level is reduced and GH release from the pituitary is stimulated. Such an inverse relation between two hormones that belong to the same family of hormones attracted my attention from the point of view of the regulatory mechanism of gene expression and modulation of synthesis of these peptides. Prolactin release is under the tonic inhibition of DA, delivered via the median eminence by the long portal vessels and bound to the receptors on the plasma membrane of the PRL cells.

Catecholamine cell bodies of the neurons are mainly dopaminergic and some neurons of the posterior hypothalamus may be noradrenergic. The distribution and biosynthesis of catecholamines and other putative neurotransmitters such as are acetylcholine, GABA, and other amino acids have been described (Hökfelt *et al.*, 1978).

Dopamine as the major inhibitor of PRL secretion originates from the tuberoinfundibular tract (TIDA), with perikarya in the caudal arcuate nucleus of the medial basal hypothalamus and terminals in the stalk median eminence, and is secreted in the long portal vessels. The second source of DA is the tuberohypophysial tract (THDA) with perikarya in the rostral arcuate nucleus and terminals in the posterior pituitary that reach the anterior pituitary via the short portal vessels (Ben-Jonathan *et al.*, 1977; Ben-Jonathan, 1985). The data indicate that these two dopaminergic systems (TIDA and THDA) regulating PRL release exhibit different responses to estradiol. Estradiol stimulates proliferation of PRL cells and synthesis and release of PRL via receptors located in the PRL and PI cells (Pelletier *et al.*, 1988).

The fibers of dopaminergic neurons passing through the tuberoinfundibular region are responsible for low PRL secretion during the resting state (Krulich, 1979). However, secretion may be increased by stressful stimuli and suckling (Mena *et al.*, 1976).

Cell–cell communication in neuropeptide-stimulated and dopamine-inhibited prolactin release was examined using superfused reaggregated pituitary cells of a mixture of prolactins and gonadotrophs (Denef *et al.*, 1984). Denef *et al.* reported on their own progress in demonstrating that GTH cells can activate PRL cells via paracrine and humoral factors. In studying the interaction between GTH and PRL cells during exposure to DA, they demonstrated the capability of gonadotrophs to transmit inhibitory signals to the PRL cells.

Martin *et al.* (1978) studied the role of central noradrenergic neurons in the regulation of PRL cell activities under different experimental conditions, especially the response of these cells to stress. Significant nonadrenergic prolactin-inhibiting hormone activity is present only in the mediobasal hypothalamus and to a lesser extent in the organum vasculosum laminae terminalis, and GABA in the MBH is able to inhibit PRL release *in vitro* (Enjalbert *et al.*, 1984).

An inhibitory effect of GABA on PRL release was also mentioned by Schally *et al.* (1977), but it seems that its influence is dose dependent; namely, large doses of GABA stimulate PRL release, whereas smaller doses have an inhibitory influence. More data are necessary in order to elucidate the mechanisms of action of these amino acids.

*f. Serotonin, Histamine, and Acetylcholine*   The serotoninergic, histaminergic, and cholinergic neurons have direct or indirect effects on PRL and GH release (Martin *et al.*, 1978). The cell bodies of serotonin-containing neurons is vertebrates as a rule are small, and it is difficult to identify them as individual entities. However, these neurons are larger in invertebrates and retain their functional activity after dissection, sometimes surviving for several days. Thus they have been proposed as models for elucidating the cytology and functional activity of these neurons (Osborne and Neuhoff, 1980).

## 4. Gonadal Steroids as Modulators of Genetically Programmed Differentiation of Cells Synthesizing Prolactin and Growth Hormones

Prolactin cells are atypical cells having a nonpeptidic factor, a fluctuating membrane potential, and a high synthetic and secretory activity. I considered positive and negative feedback action of gonadal steroids on the development of hypothalamic–pituitary cells (Pantić, 1974b, 1980). I had in mind the following data: small doses of estrogen cannot exert an inhibitory feedback action on pituitary FSH secretion until the rat is 15 to 20 days of age; specific plasma protein in neonatal rats has a high affinity for estrogen, the levels of which decrease after 20 days and practically disappear at the end of juvenile period; the receptors for estrogen in a selected brain area are synthesized mainly during the neonatal period of rat development.

An acute effect of $17\beta$-estradiol is suppressed by hypothalamic secretion of dopamine into hypophysial portal blood (Cramer *et al.*, 1979). Progesterone inhibits estrogen-induced PRL synthesis by decreasing the number of receptor sites available (Haug, 1979).

The proliferation rate of both chromophobes and PRL cells may be enormously increased by GS. The responsiveness of these cells is more clearly expressed during the neonatal period and decreases during aging and especially during senescence. These steroids, especially estradiol, are the main regulators of PRL gene transcription. As a result of these stimulative effects on protein synthesis proliferation of the endoplasmic reticulum (GER and smooth ER) appears as concentric whorls in the cytoplasmic areas of PRL cells (Pantić, 1980). An increased number of Golgi complexes and their enlargement were also clearly observed. The differences in the appearance of the GER and Golgi zones are species specific and depend on the dose(s) of GS.

Prolactin and GH are stored in granules that are larger than in the other pituitary cells. Prolactin granules in fish pituitary are mainly ovoid or round in form, whereas in mammals the irregularity of the granules is one of the characteristic storage properties. They are usually larger in PRL than in GH-producing cells.

Mitoses of chromophobes were also observed in chickens treated with estrogen. However, the formation of intracellular membranous whorls in PRL cells in chickens, rats, and piglets, neonatally treated with gonadal steroids, especially with estrogen, is closely related to the blood concentration of estradiol (Pantić, 1974b). The origin and fate of cytomembranes were studied. It seems that, in addition to phospholipids, proteins and other membranous components are stimulated by GS and are incorporated into a preexisting membrane leading to an increased rate of production of endoplasmic reticulum. An increase in $Ca^{2+}$ binding to calmodulin also occurs, so that cytoskeleton activities play a decisive role in the proliferation and character of ER organization (Pantić, 1982).

Specific glandular cell–cell interactions in the pituitary are manifested by increased proliferative rates of one cell type. Stromal–glandular cell interactions play a role in the proliferation or differentiation of cells.

Finally, the intercellular matrix, including molecules of basal lamina, may also be an inductor. Which of these factors will be predominantly involved in these processes also depends on the nature of the receptors and their binding capacity. Localization of gonadal steroid receptors in both brain neurons and pituitary cells, their binding capacity, and other properties are of predominant importance in the responsiveness of these cells to corresponding dose(s) and type of hormone(s) administered to the animals.

It has also been observed that PRL content in the pituitary homogenate is increased in rats neonatally treated with estradiol. As the result of stimulatory effect of PRHs on hypertrophic PRL cells with an increased capacity for PRL release and synthesis, a higher rate of [$^{14}$C]leucine incorporation and an increased band density were more evident in males than in females (Pantić and Genbačev, 1969, 1971, 1972).

Finally, in hypertrophic PRL cells during aging, estrus, and the menstrual cycle, under experimental conditions and after the administration of gonadal steroids (especially estrogens), signs of degradation of the endoplasmic reticulum become more and more clearly expressed. This is due to the lytic activities of the enzymes present in the cavities of the concentrically oriented endoplasmic reticulum, and/or from the lysosomes.

Administration of Oe to ovariectomized rats induced a rapid decline in LH and a delayed, marked increase in PRL. The neurohypophysis, probably via PRH, participates in mediation of the acute effects of Oe on PRL release. Estradiol could act directly on the posterior pituitary, the PI, and the ME, or indirectly via hypothalamic neurons terminating in the posterior pituitary.

## 5. Growth Hormone-Producing Cells

Growth hormone (GH, STH) somatotropin-producing cells are the most numerous cells in the pituitary, in almost all vertebrates. The cells were distinguished as specific cells during the early embryogenesis of deer (Stošić and Pantić, 1966). Growth hormone cells localized in the caudal lobe of ducks and pigeons, as in other avian pituitaries, have been identified after day 12 of incubation. Their secretory granules measure 200–300 nm (Mikami, 1986). These cells in fish, chickens, rats, and piglets, and their reactions to steroid hormones, have been described. The structure of GH has been strongly conserved in evolution (Farmer *et al.*, 1976).

Growth hormone-releasing hormone (GHRH) and somatostatin (SS) as growth hormone release-inhibiting hormones are hypothalamic neurohormones mainly involved in stimulation and inhibition of GH release from GH cells. However, more hormones are involved in the regulation of GH cell activity.

The stress-induced decrease in GH secretion in rats is at least due to an increased release of hypothalamic somatostatin and/or immunologically related GH release-inhibiting hormone (Arimura *et al.*, 1976a).

## E. Pituitary Cells Producing Glycoprotein Hormones

### 1. Common and General Properties

The common general properties of glycoprotein-producing cells, the characteristics of their specific granules, the mechanisms of content release, the nature of gonadal steroid receptors, and their feedback action have been described (Pantić, 1975).

The cytology of teleostean pituitary cells (e.g., their size, shape, and properties of their organelles), as well as their specificity and distribution, have been reviewed (Sage and Bern, 1971; Fontaine and Olivereau, 1975). Gonadotropic hormone (GTH) cells were hypertrophic before and during spawning and in addition attained their maximal size and were vacuolated. We were not able to distinguish the two types of cells and thus it seems to us that only one type of cell was present in the pars distalis posterior (PDP) (Pavlović and Pantić, 1975).

An $\alpha$-FSH-like molecule was proposed as the ancestral molecule of the glycoprotein hormone (Li, 1972). The subunits of vertebrate pituitary glycoprotein hormones are derived from a common ancestral molecule that, early in evolution, gave rise to two types of molecules ($\alpha$ and $\beta$). The partial sequences showed about a 50% isology between fish and mammalian gonadotropins, suggesting conservative amino acid sequences during evolution (Fontaine and Burzawa-Gerard, 1978).

Gonadotropins and TSH share a common subunit of 89–96 amino acid residues and 5 disulfide bonds, designated $\alpha$. The chemically different $\beta$ subunit contains 6 disulfide bonds for 116–147 amino acid residues (Dayhoff, 1976).

It appears that both estradiol and inhibin are involved in FSH release. Secretion of LH appears to be regulated by an interaction of estradiol at the level of the hypothalamic neurons and pituitary cells (Chappel and Spies, 1981).

### 2. Gonadotropic Hormone Cells as Producers of Follicle-Stimulating Hormone and Luteinizing Hormone

The gonadotrophs, round cells usually larger than TSH cells, clearly show signs of changes in size and cytological properties during the sexual cycle. The localization of GTH cells in the pituitary of different animal species, specificity in size, and cytological properties have been described (Pantić, 1975). The two gonadotropic hormones (FSH and LH) were detected in the same pituitary GTH cells of normal male rats and castrated male and female rats (Tougard, 1980).

## 3. Regulation of Gonadotropic Hormone Cell Activity

In addition to the role of LRH in GTH release, there is no doubt that catecholamines, mainly dopamine and norepinephrine, play an important part in regulating the activity of GTH cells (Drouva and Gallo, 1976).

The action of gonadal steroids at the level of brain neurons and pituitary cells, as modulators of neuronal development and activities, was examined. These effects are generally biphasic and expressed early as an inhibitory effect, followed by a stimulatory effect some hours later, or vice versa (Dufy et al., 1982). However, it is difficult to explain the effects of these steroid hormones occurring too rapidly and it is more probable that their action is expressed via different pathways. The hypertrophy of GTH cells continues during migration and before spawning, reaching a maximum size at spawning (Pavlović and Pantić, 1975).

In an attempt to obtain more information on the regulation of hypothalamic–pituitary cells and their target organs, we summarized data related to the sensitivities of hypothalamic neurons and pituitary cells to large doses of gonadal steroids during the early stages of development in fish, chicken, rats, and piglets (Pantić, 1981, 1984a; Pantić and Gledić, 1977a,b). The perinatal and early juvenile periods of differentiation of hypothalamic neurons and pituitary cells were used in order to experimentally control further processes, especially synaptogenesis, neuronal interconnection, and maturation (Pantić, 1990a). The reactions of pituitary GTH cells in immature female rats treated with estrogen and the serum concentration of GTH in old rats treated with GS were reported (Pantić et al., 1980, 1982) (Figs. 9 and 10).

An important role of steroids in the control of cell responses by regulating the expression of specific genes at precise stages during embryogenesis and cell differentiation. It also appears that estrogen receptors belong to a large family of genes encoding transcription and translation regulatory proteins.

Cortisol suppressed LH but not FSH in castrated animals and restored postcastration FSH but not LH secretion 12 hr after combined castration–adrenalectomy. These data were added to the earlier evidence that LH and FSH are regulated by different mechanisms under many experimental conditions, including stress and elevated corticoid levels (Ringstrom and Schwartz, 1984).

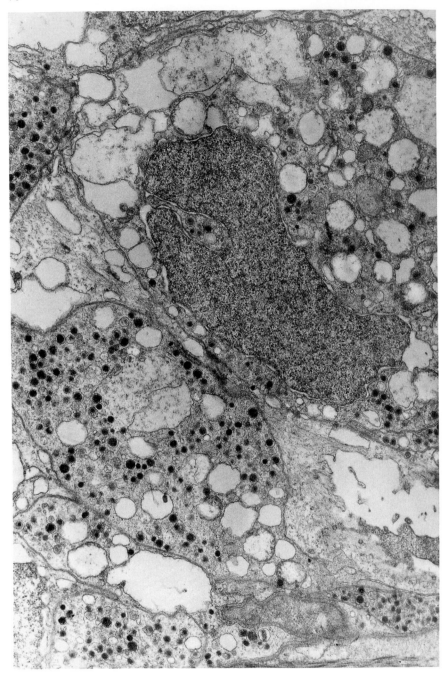

FIG. 9   Pituitary GTH cells of *Torpedo ocellata*. Granules and dilated vesicules predominate in the cytoplasm.

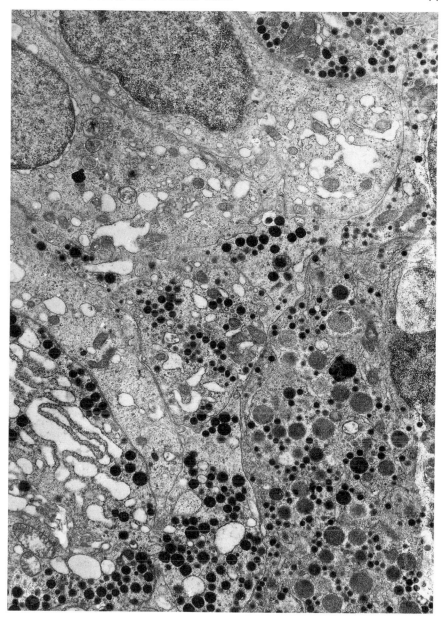

FIG. 10 GTH cells, with numerous specific granules and moderately dense osmiophilic bodies in the cytoplasm, and a portion of GH cell cytoplasm with specific granules, in the pituitary of a rat a long time after treatment with estrogen.

Adrenaline-containing neurons that innervate distinct regions in the brain have been demonstrated for the amygdala, the locus coeruleus, septum, preoptic area, and several distinct nuclei in the hypothalamus. They have a modulatory role in LH release (Elde and Hökfelt, 1978). The action of opioids can be attributed to a link between opioid peptides and noradrenergic neurons that occurs in close vicinity to the LRH/GAP, producing neurons in the preoptic–tuberal pathway (Kalra, 1982). Adrenalectomy or severe stress inhibits LH and possibly FSH secretion by suppressing LRH release. The adrenalectomy-induced suppression of FSH, but not of LH, can be reversed by corticoid pretreatment.

## 4. Thyrotropin Cells

The thyrotrophs, relatively small cells that are polygonal in shape, are characterized by a poorly developed GER, slightly dilated cavities, and the presence of homogeneous electron-dense granules. These cells in the pituitary of animals treated with antithyroid drugs show markedly dilated GER cavities, a well-developed Golgi complex, and degranulation.

Thyrotropin cells are target cells for neuropeptides, thyroid hormones, and bioamines. Several workers have studied the size and shape of these cells in fish and mammals, the specificity of their granules, their sensitivity to thyroid hormone deficiency, and the properties of thyroidectomy cells as well as mechanisms of TRH action *in vivo* and *in vitro* (Blackwell and Guillemin, 1973; Pantić, 1975). Hypothalamic neurons and the reactions of pituitary cells of thyroidectomized rats with intraocular thyroid grafts were also analyzed (Stošić *et al.,* 1969).

We studied mechanisms of TSH stimulation of the subcellular organization of thyroid follicular cells as well as their content of nucleic acids, proteins, iodine, and thyroid hormones after stimulation with TSH, from the viewpoint of the role of neurosecretory and TSH cells in the regulation of thyroid cell activity (Pantić, 1974a).

Single or repeated doses of TSH were followed by an increased [131]I content in the thyroid, the greater part being bound to protein (Pantić and Ekholm, 1963). Thyrotropin induces an increase in DNA in the guinea pig thyroid, so that after 4 days the amount of DNA was about twice that of the controls. The increase in DNA was lower than that of RNA, indicating that a rapid increase in RNA is followed by increased protein synthesis (Ekholm and Pantić, 1963).

Thyrotropin stimulated thyroid follicular cell activity in rats hypophysectomized at the age of 2.5 months and treated 10, 20, and 30 days after the operation (Pantić and Kalušević, 1974). The effect closely depends on the degree in regressive changes of follicular cell organelles expressed at the start of TSH treatment and the doses of this glycoprotein. We reported on the character of ultrastructural changes in TSH cells of *Caras-*

*sius carassius* treated with glycoprotein hormones (Pantić and Sekulić, 1978a).

## F. Inverse Relationship between Prolactin and Growth Hormone and Synthesis and Release of Prolactin and Gonadotropic Hormone Cells

The expression of the genes needed for proliferation and differentiation of both hypothalamic neuroblasts and pituitary cells has an important role in the development and organization of the hypothalamic nuclei and adenohypophysis. During postnatal and juvenile periods of development, proliferation of neuroblasts ceases and hypothalamic neuronal differentiation proceeds. At that stage of development, differentiating specific neurons are sensitive to gonadal steroids. However, both proliferation and differentiation of pituitary cells occurring in this period are also sensitive to GS. An inverse relationship between the activities of PRL and GH cells, as well as between PRL and GTH cells, is expressed for a long time after treatment with single or repeated large dose(s) of these steroids.

The hyperplasia of chromophobes, hypertrophy of PRL cells, the reduced size of GH cells, and signs of reduced activity of these cells in carp and *C. carassius* treated with GS (Pantić and Sekulić, 1978a) indicated an inverse relation between PRL and GH synthesis and release. Transformation of GH cells into PRL cells was observed (Pantić and Sekulić, 1978c). The proliferation of chromophobes during embryogenesis is followed by differentiation of growth hormone-producing cells, one of the most numerous cells found during ontogenesis of mammals.

Pituitary FSH, LH, and PRL are low during the first 2–3 weeks of life in both female and male intact rats. The hormones all increased in females during the third and fourth week, but decreased sharply in the days before vaginal opening. All the data obtained on these hormones showed a distinct sexual dimorphism for pituitary FSH (Döhler *et al.*, 1977).

The prolonged stimulative effect of a single dose of estradiol administered shortly after birth to both male and female rats was expressed as hyperplasia and hypertrophy of PRL cells. An increase in the proliferation rate of granular endoplasmic reticulum in PRL cells of rats neonatally treated with Oe was clearly expressed 15 days after treatment and later. A band of pituitary homogenate corresponding to PRL was clearly pronounced in treated rats and could not be identified in the control (Pantić and Genbačev, 1971). Proliferation of endoplasmic reticulum, an increased density in the corresponding band, and a higher rate of [$^{14}$C]leucine incorporation were less pronounced in females than in males. However, GH cells in the treated male pituitaries were degranulated, the corresponding band density was significantly lowered, and incorporation of [$^{14}$C]leucine was reduced (Pantić and Genbačev, 1972).

When the number of PRL cells is increased in the pituitary of roosters and hens treated with a single dose of estrogen during embryogenesis of after hatching, gonadotrophs are decreased. The differentiation of GH cells is retarded and they are less numerous than in the controls (Pantić and Škaro, 1974; Škaro and Pantić, 1976).

As a result of hyperplasia and hypertrophy of PRL cells in the pituitary of both male and female rats, an increased serum concentration of PRL was clearly evident. The remnants of degenerating cell organelles were observed in the widened intercellular spaces (Pantić, 1984a). Numerous microvilli were also seen on the cell surface. These are clearly signs of altered cell-to-cell communication and of an increased phagocytosis. As the number and activity of PRL cells increased, a decrease in GH cell number and vice versa was evident.

Direct or indirect interaction of gonadal steroids with the receptors for neurotransmitters shows the complexity of neurohormonal transfer of information (Dufy *et al.*, 1982). Dufy *et al.* stated that the rapid effect of steroid hormones can be modulated by the classic steroid receptors, leading to genomic activation and protein synthesis, alteration of the properties of ionic channels, or modification of receptor affinity for peptides and neurotransmitters by changes in the physical state or biochemical composition of membrane lipids.

Prolactin release and synthesis by cells maintained in culture for a long time favor autonomous PRL secretion. A decrease in the release and synthesis of GH *in vitro* favors the idea of a predominant effect of a GH-releasing factor *in vivo* (Cesselin and Peillon, 1980). Prolactin concentration in serum seemed to vary inversely with GH and a negative correlation was found. Prolactin is generally considered to be a stress hormone that responds to a number of different forms of stress such as surgery, exercise, and parachute jumping.

The addition of TRH to cultured GH$_4$ cells results in an increase in concentration of PRL mRNA. The rate of PRL production, as determined by radioimmunoassay for PRL in the medium, increases linearly to a maximum rate of production at 48–72 hr, whereas production of GH and the growth hormone mRNA concentration concurrently decreases approximately 52% at 72 hr after addition of TRH (Evans and Rosenfeld, 1980).

## G. Pituitary Cells as Transplants

Intracerebral, intraventricular, and kidney subcapsular transplants and transplantation into the eye chamber of hypophysectomized animals have been investigated. The main aim was to explore replacement therapy for hypophysial dysfunction and to elucidate the role of hypothalamic and extrahypothalamic brain neurons in regulation of the pituitary cells of intact hypophysectomized animals or after transplantation.

Several clonal strains have been developed using a cell line of pituitary anlage derived from the epithelium of Rathke's pouch: one of these releases ACTH, PRL, and GH, but not glycoprotein hormones, in culture (Hymer *et al.*, 1980).

The capacity of anterior pituitary homotransplants grafted into a hypophysectomized host to influence the growth of the animal, which was fully arrested after hypophysectomy, and to restore the capacity of the animal to reproduce, has been the subject of a research program (Martinović and Pavić, 1960, Martinović *et al.*, 1966; Pavic *et al.*, 1977). The presence of PRL, GH, GTH, and TSH cells many months after transplantation affected the recovery of spermatogenesis in 60–70% of recipients. However, the dislocation of specific granules in homogeneous osmophilic cytoplasm, enlarged intercellular edematous spaces, and altered blood vessels were clearly evident in a direct 4-month-old pituitary eye graft.

Pituitary cells grafted ectopically in an eye chamber or subcapsule of kidney were able to release families of peptide or/and glycoprotein hormones and to stimulate target organs such as gonads, adrenals, and thyroid follicular cells. However, after they released their peptidic and glycoprotein hormones in the amount necessary to stimulate gametogenesis, glucocorticoidogenesis, and thyroid follicular cells, respectively, they lost their characteristic cytoplasmic properties and the subcellular organization of their organelles was altered. The homogeneous cytoplasmic appearance clearly showed signs of proteolysis and no ribosomes or polysomes were seen. The mitochondria were rare and altered, and the capability of these cells for synthesis and secretion was retarded and/or inhibited (Figs. 11 and 12). The regional accumulation of specific granules is clearly a sign that there were no receptors for LRH and no signs of recovery of cells (Martinović *et al.*, 1982). However, in all experiments to date ACTH- and PRL-producing cells survived longer than GH and GTH cells.

## H. Pituitary Mammotropic Tumor Cells as Predominant Producers of Adrenocorticotropic Hormone and Prolactin

Pituitary cells of transplantable mammotropic tumors (METs) are able to synthesize and release ACTH, PRL, and GH. When we examined cells from a transplantable mammotropic pituitary tumor in female rats of the Fischer strain sixty-ninth transfer generation, we found that these tumor

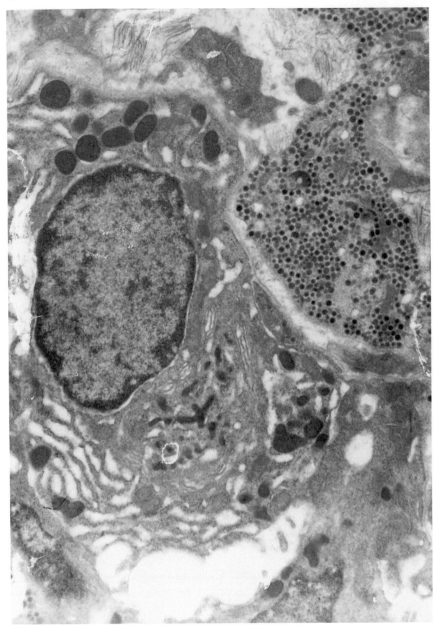

FIG. 11  A PRL cell and part of a TSH cell in a pituitary eye graft four months after transplantation. Only altered specific granules are seen in the homogeneous cytoplasm of TSH cells. In the PRL cells GER and specific granules are altered and enormous dense bodies are present.

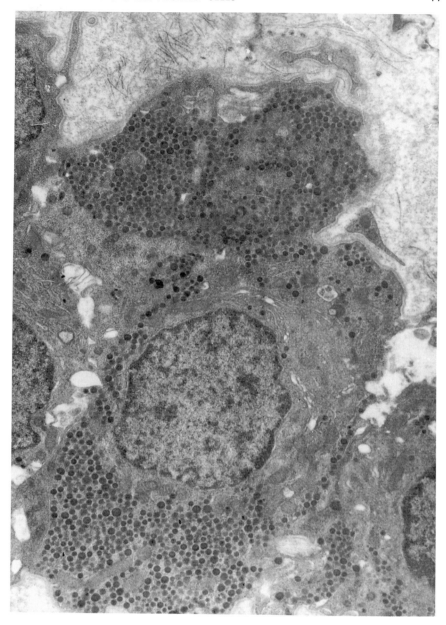

FIG. 12  Glycoprotein-synthesizing cells of the pituitary eye graft four months after transplantation. Synthesis and release are interrupted by proteolysis and only numerous specific granules are present.

cells behave autonomously in secreting large amounts of ACTH and PRL, but little GH or none (Pantić *et al.*, 1971a). Owing to poorly developed GER and rare granules in their cytoplasm, it was not possible to identify specific ACTH, PRL, and GH cells, even by using electron microscopy criteria. As a result of a continuous increase in ACTH and PRL from this cell, stimulation of both adrenals and the mammary gland was clearly evident.

## I. Neuronendocrine and Immunocompetent Cell Communication during Evolution

The neuroendocrine cells that regulate the synthesis and release of immunocompetent cells as are macrophages which are located between hypothalamic neurons, pituitary cells, and other endocrine cells. The focus is often paid to GH cells and those that are involved in regulating their activity.

It is obvious that hypophysectomized animals are more susceptible to infections. More detailed results showed that growth hormone has a stimulative role in production of the superoxide anion and also augments the capability of macrophages to phagocytose microorganisms (Kelley and Dantzer, 1990).

The immunoregulatory involvement of the POMC family of hormones in the immune system of invertebrates seems to have been important in the elucidation and continuity of communication between neurons and immunocompetent cells during the evolution from invertebrates to vertebrates (Stefano, 1990).

A large number of moncytes have been observed to enter the developing brain, especially when the blood–brain barrier is formed. Macrophages from the brain of newborn rats synthesize several peptides that have mitogenic action on astrocytes. Interleukin 1 induces proliferation of astrocytes, and they play a role in both gliogenesis and angiogenesis. The astrocytes have an important role in development, homeostasis, and responses to injury in the CNS (Fedoroff and Vernadakis, 1986). The tight junctions between the endothelium of the blood vessels in the blood–brain barrier are brought about by the interaction of endothelium and astrocytes (Janzer and Raff, 1987). Cytokines, as molecules synthesized by macrophages, are able to act as hormones.

Macrophages enter the circulation as monocytes and then migrate into the central nervous system and into all other tissues. Because these cells also produce nerve growth factor and other growth factors, they play a role in neuronal gliogenesis and angiogenesis. There are more macroglia present in gray than in white matter. The hypothalamus has an average microglial density. In both immature and mature brain microglia play an important role in the removal of degenerating neurons and their processes. They have also been found in pericapillary spaces around the magnocellular terminals of the neurohypophysis. A central role for the microglia in immunologically mediated events in the CNS was suggested by Perry and Gordon (1991).

Age-dependent functional changes, the mechanisms influencing senescence, and the role of immunosenescence in degenerative processes, interactions between neighboring cells, and neoplastic illnesses need to be the focus of research on how aging leads to an immune decline (Miller, 1991).

The thymus, an immunocompetent organ found in all mammalian species from neonatal to peripubertal stages of development, is characterized by higher proliferation rates of both epithelium and lymphocytes than are found in the other lymphopoietic organs. As a target organ for glucocorticoids, in neonatally treated rats, the mitotic rate of these cells is decreased. Proteolysis in lymphocytes is expressed, the immunological properties of this organ are reduced, and the reaction is dose dependent (Miholjčić et al., 1986; Radić and Pantić, 1987). It was suggested that the reaction of the thymus to stress should be included in biomedical research programs, in order to decrease the risk of various diseases, including cancer (Pantić, 1992).

## IX. Responsiveness of Hypothalamic Neurons and Pituitary Cells to Environmental Factors and Steroid Hormones

The responsiveness of hypothalamic neurons to such environmental factors as the changes in daylight hours over the course of the year is of great importance for the adaptation, regulation of reproduction, and other functions of different animal species. In addition to light, temperature also plays a significant role, especially in regulation of growth. The effect is clearly pronounced in fish, but is apparent in other animal species as well (Pantić et al., 1982). Hypothalamic neurons integrate such information transferred from different brain areas and coordinate it with the pituitary and the appropriate target cells. The responses of both hypothalamic and pituitary cells depend on the feedback of the target cells and their own hormones. Some characteristics of the feedback actions of peptide and steroid hormones on their target cells are discussed here.

It is now known that steroid hormones play an important role as regulators of hypothalamic neurons and pituitary cells via long feedback mechanisms. Hormonal action within the hypothalamo–pituitary axis, which is expressed as "short" and "ultrashort" feedback mechanisms, may use

LH, FSH, and LRH. Rodent steroids have profound effects on both hypothalamic neurons and pituitary cells (Kalra, 1976). In the primates, "long" and "short" feedback action is expressed primarily at the level of pituitary cells (Nakai *et al.*, 1978).

A number of experiments with ultrashort feedback circuits suggested that there is a mechanism by which LRH may control its synthesis and/ or release and, as a consequence, GTH is released from the anterior pituitary cells (Flerkó *et al.*, 1987).

The mechanism enabling the female pituitary to release GTH in a cyclic pattern does not function in males and appears to be hormonally induced rather than genetically determined. Intact female rats that are given testosterone, during the first few postnatal days, lose the ability to release GTH in a cyclic manner. This interpretation was that neonatal androgen action may interfere with the normal synthesis of the estrogen receptor and in this way reduce the responsiveness of the neuronal and nonneuronal target tissues to estradiol (Flerkó, 1975).

## A. Responsiveness of Monoestric Animals to Environmental Factors

### 1. Fish

The hypothalamic nucleus preopticus (NPO) and nucleus lateralis tuberis (NLT) of fish were used to estimate the reactions of NSCs to light and other environmental factors in the regulation of the sexual cycle, and some of the data obtained are discussed here.

Lutein-releasing hormone and other neurohormones have an autocrine and paracrine role and play an important part as inductors of neighboring cells. Comparative investigations of the distribution, structure, and action of LRH were reviewed by Chieffi *et al.* (1991). The conserved $NH_2$-terminal sequence of LRH during evolution suggests that this part of the molecule has a vital role in interacting with the receptor. A modulatory role for LRH in the pituitary cells was studied by Fink and Pickering (1980).

At least two forms of LRH are present in teleostean fish, salmon, and/ or chicken II LRH. In vertebrates, six forms of LRH have been identified so far: mammalian, chicken I and II, salmon, lamprey and catfish I LRH. Gonadotropic hormone I (FSH-like) and GTH II (LH-like) are detected in a number of teleostean species. It is known that in mature male catfish, chicken II LRH is about 100-fold more effective in stimulating an increase in GTH II plasma level than catfish I LRH. Lutein-releasing hormone also enhances growth hormone release (Schulz *et al.*, 1992).

The quantity of neurosecretion (NS) in the NPO of both male and female *T. ocellata* increases in sexually mature females caught from February to

May; the highest amount was observed in axons and terminals of pregnant females. The amount is closely related to the activity of GTH cells and gametogenesis in the gonads of both sexes (Pantić and Sekulić, 1975). In the neurosecretory cells of the NPO and the caudal neurosecretory cells in *T. ocellata* and *Acipenser ruthenus,* the amount of neurosecretion in the cell bodies, extensions, and terminals is also closely related to the stages of sexual cycle and pregnancy (Pantić and Sekulić, 1978b).

The involvement of magno- and parvocellular neurons of both the nucleus preopticus and NLT during the sexual cycle of *S. scriba* was evident. The quantity of neurosecretory substances was most clearly seen in the cell bodies of magnocellular neurons of the NPO and in parvocellular neurons in March and May. The terminals of NPO neurons were closely connected to neurons of the NLT. They were mostly filled with neurosecretion during the spawning season, showing that both the NPO and NLT are involved in regulation of the sexual cycle and reproductive activities (Pantić and Lovren, 1977).

The amount of neurosecretion in the NPO, NLT, and neurohypophysis of the teleostean *Alburnus alburnus,* the fish of Skadar Lake, and *A. falax,* which migrate from the Adriatic sea to Skadar Lake before spawning, was also examined. In both species NLT neuronal bodies were hypertrophied before spawning and the amount of NS was highest in the neurons of the medial NLT. It appears that neurons of the NLT are involved more actively before spawning and neurons of the NPO are involved more actively during spawning (Pavlović and Pantić, 1973). Gonadotropic hormone cells in the pituitary of migratory fish were hypertrophic during migration and spawning. They reach maximum size during spawning season. Their degranulated polymorphic nuclei and vacuolated cytoplasm are evidently signs of a high level of secretory activity (Pavlović and Pantić, 1975). However, in carp treated with Oc and progesterone (Pr), there were more granules in the GTH cells in the peripheral part of the mesadenohypophysis than in the controls (Pantić and Lovren, 1978).

The high density of DA and serotonin fibers found in the preoptic and tuberal hypothalamus, and the penetration of both DA and NA fibers into the rostral and proximal pars distalis of teleostean fish (*Poecilia latipinna* and *Dicentrachus labrax*) were studied to see if there is any homology between teleosts and tetrapods in the forebrain area that regulates secretion of pituitary cells (Batten *et al.,* 1990).

## 2. Deer

We studied the ability of the neuroendocrine cells of deer to respond to environmental stimuli, focusing on the role of light in regulating the sexual cycle and the development and shedding of antlers. The hypothalamic neurons, pituitary, and target cells are genetically programmed. However, environmental factors and stressful stimuli may often modify the biology of the NSCs the results of which can be seen in the quality of the antlers.

The amount of neurosecretory granules in the neurosecretory cell bodies and their terminals in the neurohypophysis was increased during the summer, when the growth of antlers is completed and the soft "velvet" dries and is shed, and mating occurs. Decreased neurosecretory cell activity during the winter was clearly evident (Stošić and Pantić, 1966, 1967). As the days become longer, in the adult deer the GH cells degranulate and show signs of degeneration. Growth hormone and TSH cell activity increases during pregnancy.

As a result of the decreased capacity of PRL- and GH-producing cells, the number of hypertrophic GTH cells increases. Thyrotropin cells also increased in number, and both gonads and thyroid glands were stimulated. However, the quality and/or character of the antlers depends not only on the genetic program, but also on the balanced coordination of neuroendocrine cells, environmental conditions, and the capacity to adapt. The responsiveness of neurosecretory cells decreases in most deer older than 10 years, revealing the nature of the cyclic changes; this is evident in the poor quality of the antlers.

## B. Reactions to Gonadal Steroids

Pituitary cell–cell interactions are manifested by an increased proliferative rate for one cell type. Stromal–glandular cell interactions may play a role in cell proliferation and differentiation. The intercellular matrix, including molecules of basal lamina, may also be inductors. Which of these factors will be predominantly involved in these processes depends on the ligands as well as on the receptors and their binding capacity. The localization of gonadal steroid receptors in both brain neurons and pituitary cells, their binding capacity, and other properties are critically important in the responsiveness of these cells to corresponding dose(s) and type of hormone(s) administered to the animals. The topography of estradiol-concentrating neurons in the hypothalamus (Stumpf, 1968) and in the other steroid target cells in the brain was determined by Stumpf and Sar (1976).

Gonadal steroids can significantly alter the responsiveness of the preoptic LRH- and GAP-containing neurons, which seem to be only moderately influenced by the limbic system (Fink et al., 1982). Fink et al. analyzed LRH release into the hypophysial portal blood vessels, the mechanism of their peptide action, and the neuronal mechanism responsible for the spontaneous preovulatory episodic manner and diurnal rhythm of LRH release. They proposed two sites of estrogen action: the first is on the brain neurons of rats, triggering the surge of LRH, and the second is expressed in an increased pituitary cell responsiveness to LRH. The effect of estrogen on the electrical activity of hypothalamic units was discussed by Dufy *et al.* (1976).

The mechanisms of action of hypothalamic neurohormones and their interaction with sex steroids in the anterior pituitary gland were the subject of a discussion published by Labrie *et al.* (1980). Szego (1978) discussed the significance of the plasma membrane in cell function and the dichotomy in the ways steroid and peptide hormones, via their receptors, affect gene expression.

At low plasma concentrations of estradiol, inhibited GTH output is potentiated by progesterone. A relatively sudden increase in plasma estradiol, in the presence of low progesterone, increases GHT responsiveness. Progesterone enhances the responsiveness of the hypothalamopituitary cells sensitized by estrogen (Fink and Pickering, 1980). During the period of spontaneous, preovulatory GTH surge, estrogen and progesterone exert their so-called "positive feedback" action on the hypothalamogonadotropic cells and facilitate GTH release.

The complex role for GS in regulation of the synthesis, release, and degradation of LRH in the peptidergic neurons was summarized by Kalra *et al.* (1981). These authors postulated that estrogen and androgen induce new synthesis of precursor at the level of the LRH perikarya and possibly activate the transformation of precursor molecules into immunoreactive LRH in the terminals. It appears that progesterone initially accelerates this process and facilitates LRH discharge from the terminals into the hypophysial portal blood vessels. Kalra *et al.* supported the hypothesis that the catecholaminergic link plays a modulatory role in cyclic and steroid-induced gonadotropin release in female rats.

In addition to its profound effects on LRH release, GS also modulates the synthesis and release of monoamines and neuropeptides and the responsiveness of LRH-producing neurons (Kalra and Kalra, 1984). Kalra and Kalra reviewed the role of GS adrenergic neurons and endogenous opioid peptides in the control of LRH neuronal activities.

## C. Sensitivity to Gonadal Steroids during Postnatal Development

We have studied the sensitivity and character of the responses of hypothalamic neurons and pituitary cells to gonadal steroids administered during early postnatal development. We used neonatal rats because the hypothalamic–pituitary–gonadal axis in newborns is sexually unidfferentiated, that is, this time of life is known as the "critical" period. We hypothesized that during this period of development in rats, the hypothalamic neurons are in a phase of intensive differentiation and synaptogenesis, and that pituitary cells, especially chromophobes, are characterized by proliferation followed by differentiation of specific cells. The species specificity and sensitivity of these cells were examined during postnatal, juvenile, and peripubertal stages, and in adult and old animals (Pantić, 1981).

The serum concentrations of PRL, FSH, LH, and progesterone in female and male rats from birth to puberty were reported by Döhler and Wütke (1974). The testosterone (T) levels in early fetal testes of domestic pigs was reported by Raeside and Sigman (1975). The similarity in the concentration of T in pig and humans, from the fetal stage to sexual maturity observed by Meusy-Dessolle (1975), encouraged us to investigate the sensitivity of pituitary cells and their target cells in growing piglets, from the neonatal period up to puberty, under an androgen deficit, but also after excess Oe, or Oe and Pr, was administered as a single dose or as repeated doses.

The stimulatory effect of both female gonadal steroids (Oe plus PR) on gonadotropin release is dependent on dosage, age of development, and duration of exposure to the hormones. We have described reactions of GHT cells in prepubertal and gonadectomized animals (Pantić, 1980). The number and size of GTH cells in the pituitaries of rats treated with a single dose of testosterone proprionate (TP) after birth were smaller than in the controls (Pantić and Gledić, 1977a).

In addition, the hypothalamus of newborn rats, of either sex, has the inherent ability to maintain cyclic release of GTHs; noncyclic release in the male is determined by testicular androgen during the first postnatal days, and it is now clear that both male and female rats are under the influence of androgen during the first 15 days of postnatal life, after which they lose the ability to release GTHs in a cyclic manner (Döhler and Haucke, 1978; Vigh et al., 1978). The biphasic effect of estradiol on the response of pituitary cells to LRH was reported by Vilchey-Martinez et al. (1974).

I discussed the specifity of adenohypophysial cells and gonadal steroids (Pantić, 1980) when I posed the following questions: How do cells evolving as producers of the same ancestral molecules react to gonadal steroids? What are the differences in reactions of PRL and GH cells to gonadal steroids? It was clearly evident in all of the animal species examined that for a long time after treatment with large doses of GS, proliferation of PRL cells and the capacity for PRL hormone synthesis and release were stimulated at the expense of GH cell activities.

The interdependence of the reactions of PRL and GTH cells to GS has also been established. Very little GTH is expressed in the pituitary of both male and female rats neonatally treated with a single large dose or two doses of estrogen. The reaction is more clearly expressed in males than in females (Vigh et al., 1978; Török et al., 1982). The activity of PRL and GH cells in the pituitary of rats treated with estrogen on days 10 or 15 after birth was reported by Genbačev et al. (1977). A decreased number of GTH, and to far lesser extent TSH, cells was observed in adult and old rats chronically treated with testosterone propionate (Pantić and Gledić, 1977a,b). Long-term effects of gonadal steroids on pig pituitary cells were found (Pantić and Gledić, 1978). Gonadotropic hormone cells were easily identified by the presence of specific granules. The cavities of granular endoplasmic reticulum in GTH cells in gonadectomized piglets were dilated or cysternoid, and from 30 days on these cells appeared as gonadectomy cells (Pantić and Gledic, 1978).

The stimulatory effect of estrogen on pituitary reaction to LRH is due to its action on the pituitary or the hypothalamus, and an increase in pituitary responsiveness by way of the priming effect of LRH was observed by Fink et al. (1982).

We have summarized the ways in which gonadal steroids effect hypothalamic neurons and pituitary cells under different natural and experimental conditions, and the nature of their reactions (Pantić, 1974b, 1988). The presence and role of estradiol-binding plasma protein were reported by Raynaud et al. (1971). Reconsidering data published so far, I have showed (Pantić, 1990a) that during the early development of fish, rats, chickens, and piglets treated with a large single or repeated doses of Oe, Oe plus Pr, or TP, hypothalamic neurons are sensitive to a sufficiency or deficiency of gonadal steroids, and that both these neurons and pituitary cells may be modulated by GS.

We believe that the density of pituitary LRH receptors is closely dependent on gonadal steroids and that an increase in receptor density after orchidectomy or ovariectormy is correlated with an increase in plasma LH concentration. It is also significant that such an increase may be inhibited by estradiol in the female and by testosterone and estradiol in the male.

Given what we know so far, an important question still needs to be answered: Are the steroid actions expressed on the same cells that concentrate steroids? Kelly et al. (1980) observed that estrogen "inhibits" the activity of 30% of mediobasal hypothalamic neurons and that neurons from other hypothalamic areas might be coupled to steroid-responsive neurons. The ultrastructure of the pituitary GTH cells in castrated male rats was described by Arimura et al. (1976b).

Ovariectomy induced a significant decrease in PRL biosynthesis that was clearly evident 60 days after the operation. Chronic administration of estradiol stimulated PRL synthesis in ovariectomized animals. Our data indicate sex differences in the reaction of PRL cells to gonadectomy; in females these pituitary cells are more sensitive to gonadal steroid deficiency than in males. However, although synthesis of PRL cells was more highly stimulated in the pituitary of gonadectomized females than in males, inhibition of GH synthesis was more pronounced in estrogen-castrated males (Genbačev and Pantić, 1972).

We observed inhibition of PRL cell synthesis in the pituitary of rats bearing MtT during the first 3 weeks of tumor growth. From 3 weeks on, changes in all cell organelles were more clearly pronounced. Ultrastructural properties and PRL content in the pituitary homogenate become nearly undectable 2 months after transplantation. The results indicate that an increased PRL release by tumor cells has a suppressive effect on the PRL cells in the pituitary of tumor-bearing rats (Pantić et al., 1971b).

I have summarized the nature of the response of both neuronal and pituitary cells to gonadal steroid hormones during early development of fish, chickens, and mammals. The following results were emphasized: in addition to the complex mechanisms that are involved in regulating genetically programmed development of neuroendocrine cells in vertebrate species, gonadal steroids could be used as modulators to control growth rate and behavior. The character of NE cell reactions depends on dose(s) and the sensitivity of these cells when the hormones are administered (Pantić, 1988, 1990a). Advances in our understanding of the sequences of differentiation and maturation of specific brain neurons that produce neurohormones and bioamines involved in regulation of the pituitary gonadal axis, sexual dimorphism, and cyclicity show that gonadal steroids may be used as as single dose of estrogen, a combination of estrogen and progesterone, or as testosterone in controlling the reproductive properties of fish and domestic animals (Pantić et al., 1978, 1982; Pantić, 1988, 1990b).

## D. Reactions to Corticosteroids

We studied the decreased number of chromophobes and ACTH cells, their less expressed hypertrophy, and the appearance of adrenalectomy cells in adrenalectomized rats treated with corticosteroids from the viewpoint of their synthetic pathway in animals that had only been adrenalectomized. However, the great number of specific granules in ACTH cells in the animals treated with hydrocortisone undoubtedly is a sign of the inhibitory effect of this steroid. We studied the character of these reactions after adrenalectomy and treatment with corticosteroids or ACTH (Hristić and Pantić, 1973, 1978).

A less developed GER and a Golgi regression indicated a considerably decreased protein synthesis in ACTH cells of adrenalectomized rats 15 and 30 days after head irradiation (Pantić and Hristić, 1975). We also showed that an increased number of granules and granules larger than these in ACTH cells of normal rats were the result of slower storage, transport, and secretion of the products of these cells.

An increased level of CRH mRNA was found by quantitative *in situ* hybridization histochemistry in adrenalectomized rats (Young *et al.*, 1986).

Ultrastructural changes of the adrenals in neonatal rats born of mothers bilaterally adrenalectomized at 7 or 14 days of gestation were studied to determine the loss of maternal glucocorticoids, increased maternal ACTH, and the permeability of placental barrier to maternal ACTH (Nickerson *et al.*, 1978). The ACTH cells of fetuses and newborn rats from mothers adrenalectomized on days 7, 14, or 20 of gestation showed signs of hypertrophy. These were most pronounced in the pars anterior of all fetuses and newborn rats from mothers adrenalectomized during the last week of pregnancy, that is, on days 14 and 20. ACTH-like cells were observed in the PI of newborn rats from mothers adrenalectomized on the fourteenth day of gestation (Hristić *et al.*, 1978). We have described the ultrastructural properties of ACTH cells in intact or adrenalectomized rats treated with corticosteroids or ACTH and sacrificed 15, 30, or 60 days later (Hristić and Pantić, 1973; Hristić *et al.*, 1987).

Halasz *et al.* (1994) studied the effects of maternal glucocorticoids on the steady state levels of hypothalamic CRH and glucocorticoid receptor (GR) mRNA and anterior pituitary POMC mRNA by adrenalectomizing dams prior to the functioning of the fetal pituitary–adrenal axis (on day 8 of gestation). The lack of maternal glucocorticoids influences hypothalamic CRH and GR expression only transiently, but seems to have a prolonged effect on POMC expression in the female offspring.

In view of the fact that CRH activity in the fetal hypothalamus in late pregnancy and fetal pituitary cell responsiveness were established under special experimental conditions (Dupouy, 1975), and in view of our own findings, it appears that maternal ACTH does not cross the placental barrier. I have summarized the characteristic and clearly pronounced changes in hypothalamic neurons and pituitary GTH, ACTH, and TSH cells, after gonadectomy, adrenalectomy, or thyroidectomy of rats (Pantić, 1980). These are the cells primarily (mainly) involved in reactions to deficiency of gonadal steroids, corticosteroids, and thyroid hormones.

We identified ACTH cells after 10 and 20 days in the pituitary of MtT-bearing rats. The heterogeneous density of specific granules; changes in the Golgi zone; irregular form, size, and mitochondrial matrix density;

and the presence of multivesicular bodies were considered the result of inhibition of CRH provoked by an increase in ACTH, which is secreted by tumor cells, and plasma cortisone levels (Pantić et al., 1971a).

Information on the molecular evolution of prohormones as the ancestors of neurohormones and their role in regulating pituitary cells and their targets in mammalian species may contribute to recognition of common features of hypothalamic neurons, pituitary cells, APUD cells, brain–gut–skin, heart, respiratory, and other endocrine cells in the evolution of vertebrates from lower vertebrates and invertebrates. The cleavage of biologically inactive prohormones into biologically active individual hormones that has occurred throughout evolution may provide new insights on the evolutionary history of species specificity in the development of the adrenals in fish and birds, and on the development of the cortex and medulla in the adrenals of mammals.

The development of hypothalamic and extrahypothalamic brain neurons and pituitary cells producing peptide hormones, as well as the adrenocortical cells, as the cells involved in stress reactions is closely dependent on the character and pathways of stressful stimuli. However, pituitary cells, which produce POMC hormones, and PRL- and GH-producing hormones, as the cells involved in stress have an important influence on the cells producing the glycoprotein family of hormones. The character and nature of responses are potentiated by a rapidly increasing number of stressors, medical drugs, steroids, and other hormones (Pantić, 1984b). Brain neurons producing the POMC family of hormones are closely dependent on the character and intensity of stressful stimuli (Pantić, 1992).


## X. Summary and Perspectives

Many data that have been obtained are leading to rapid advances in comparative neuroendocrinology. An attempt has been made to contribute to our understanding of the common general properties and diversity of NSCs from evolutionary and ontogenic viewpoints. Representative animals of distinct species were used to search for genetically programmed development of NSCs, mechanisms of regulation, and modulation of their differentiation and activities.

The specificity of hypothalamic neurons in fish, which are animals containing caudal neurosecretory cells, indicates the capability of lower vertebrates for de novo differentiation of ependymal into neurosecretory cells. Common precursor molecules, such as large polypeptides, and ancestors to neurohormones, may be cleaved by enzymes or not, and released as biologically active molecules or as uncleaved inactive ancestors. The nature of the synthesized molecules is closely dependent on the degree and character of the GER and the other organelles that specify their ability for protein synthesis.

Neurohormone synthesis is initiated in response to the signals of ligand–receptor complexes, whereas the character of cell responses is dependent on the types of ligands for signal transduction and cascades of events, as well as on internalization of the ligand–receptor complex. The fate and mechanism of the responses influence gene expression at the levels of transcription and translation. The role of membrane receptors in the transport of molecules and ions through ion channels for transfer of neuronal information is of great importance for cell functions.

In their origin, proliferation rate, migration, differentiation, and synaptogenesis, neuroblasts show specificity in hypothalamic organization and maintain contact with the areas of their origin. However, species-specific communication of hypothalamic neurons with the extrahypothalamic brain areas lying outside the BBB, and with CVOs, may be altered during each stage of development. The character of the response depends on the types of stressors, hormones, and other chemical substances, as well as on physical and environmental factors.

The biosynthesis of the other main constituents of the cytoskeleton, the major filamentous proteins, G proteins and laminin, may be affected by many factors, so that their major role as the main regulators of protein synthesis, cell proliferation, migration, differentiation, and synaptogenesis may be temporarily or permanently altered. The character of these changes depends on the nature of the chemical substances, such as hormones, and is usually dose dependent. However, the focus should be on the stage of development and the sensitivity of neuroblasts and pituitary cells to these agents. The results are expressed in underdevelopment and paucity or failure of specific cells, which determine the future properties of the NSCs, including vital functions in the organism and the individual.

Data obtained from an examination of the hypothalamic neurons and pituitary cells in different fish species during the sexual cycle, and their responsiveness to GS, are helpful for understanding mechanisms regulating differentiation of NSCs and their activities in all vertebrates, including primates. Fish species demonstrate specificity in receiving signals from light sources, magnetic fields, and temperature changes, and especially in their reactions to stressful stimuli, their strategy for adaptation, and by regulation of growth and reproduction. All of these are of great importance for their protection and survival. Although this search provided some insight into the organization and communication of neuroendocrine cells, further attention was focused on laboratory animals.

Various modulating agents, such as GS, corticosteroids, pituitary peptide, or glycoprotein hormones (especially ACTH, PRL, and TSH), were used and the data obtained were reviewed in the light of many more recently published articles, monographs, and reviews.

The neurons of the NSO and NPV synthesize precursors to neurophysins and associated neurohormones and are also producers of CRH, SS, and other peptides. Signs of these neuronal reactions to either deficiency or sufficiency of gonadal steroids, glucocorticosteroids, and thyroid hormones were clearly evident 2 to 3 weeks after gonadectomy and/or adrenalectomy or thyroidectomy. In some of the experiments, they were visible earlier than in corresponding pituitary cell types.

As the amount of data in the literature and our own experience accumulates, it is clear that the perinatal and juvenile peroids of development have attracted attention from the following viewpoints: (1) proliferation of neuroblasts at that stage has almost ceased; the number of polysomes increases, and the onset of the development of the GER, the Golgi complex, and mitochondria, and the degree of axonal growth toward the areas of their destined termination, clearly show the stage of differentiation; (2) the maturity of the BBB is advancing but is not complete. However, at that period, both cell proliferation and differentiation of pituitary cell types that synthesize protein and glycoprotein hormones are intensive. Moreover, both hypothalamic neurons and pituitary cells at that period of development are sensitive to different agents, especially to GS, and to mechanical injuries and environmental factors.

The reactions to stressful environmental stimuli, such as light intensity, photoperiod, and temperature, are conveyed to the pars intermedia by hypothalamic neurons that produce peptides and bioamines. A species-specific reaction is expressed in some animals, such as *X. laevis,* as a rapidly increased PI cell proliferation rate and/or secretion of POMC hormones. The development of this lobe in wild animals and reduction of their cellular numbers through evolution indicate that these cells are mainly involved in the adaptation and defenses of animals. The innervation specificity of these cells, via peptidergic and aminergic fibers, including those of DA-containing THDA fibers terminating in the NH, and data that PI cells and the NH are a source of PRH as regulators of PRL release, should not be neglected.

The discovery of receptors for GS in the NA as a source of DA and also in the PI, and the reactions of PI cells to GS, which are clearly expressed for a long time after treatment, may be important for further elucidation of the mechanisms regulating these cell activities and the character of their response to gonadal steroids, corticosteroids, and different stressful stimuli. In addition, under our experimental conditions, the reactions of these cells to glucocorticoids are weaker than they are to GS.

Although the signals from different brain areas are conveyed to PI cells mainly via neurotransmitters, the ACTH cells, as the first cells differentiating during embryogenesis, survived as transplants for a longer time than other pituitary cell types. Pituitary ACTH cells are regulated by hormones via long portal blood vessels. The ultrastructural properties of ACTH cells and their long cytoplasmic extensions between other cell types showed their specificity in reactions to stress, but also in their interaction with neighboring cell types.

The reaction of ACTH cells to stress occurs rapidly and the release of members of this family of hormones, as biologically active molecules, is dependent on the effects of CRH, corticosteroids, catecholamine (CA), and other hormones. Through evolution these neurohormones also have and important role in the regulation of GTH and other pituitary cells.

Prolactin and GH genes have evolved from a common ancestor, and as the most ancient hormones in vertebrates are produced from specific PRL and GH cell types. However, estrogen specifically stimulates the gene expression needed for proliferation, differentiation, and hypertrophy of PRL cells. Considering the role of peptides and bioamines involved in stimulation and inhibition of PRL, gonadal steroids are the main modulators of genetically programmed differentiation of cells that synthesize the family of PRL and growth hormones.

The proliferation rate of both chromophobes and PRL cells was enormously increased in all of the animals examined. An inverse relationship between PRL and GH, and between PRL and GTH synthesis and release, was evident in the animals treated with a single large or repeated dose(s) of GS, and was expressed less during aging and senescence.

As the blood concentration of PRL decreased and PRL cells degenerated, GTH synthesis and release increased. However, in the animals neonatally treated with a single dose of GS, the population of differentiated GTH cells in the prepubertal period reached the levels found in intact animals, but the amount of FSH and LH was altered in all of the animals examined.

Knowledge about LRH/GAP neuron topography, regulation of gene expression, the structure of DNA encoding the precursor of LRH/GAP, and the mechanism involved in regulating the activity of these neurons is rapidly advancing. Summarizing some of the more recent data, it appears that androgens and estrogens have a crucial role in the control sexually dimorphic brain neurons. The MPA is of central importance for the regulation of reproduction. Both the MPA and MBH are involved in the regulation of tonic and cyclic patterns of GTH release from the GTH cells. Lesions placed in the mammillary body of newborn rats that bypass the stalk and the ME interrupt the exchange of information between these brain areas and the pituitary. As a result, an inverse relationship between GTH and PRL occurs. The GTH cells lose the ability to release GTH in a cyclic manner, leading to advanced puberty and ovarian hyperluteinization.

The results discussed here show that genetically programmed differentiation and synaptogenesis of hypothalamic neurons involved in regulating the development of specific cells that produce families of POMC, PRL/GH, and glycoprotein hormones are sensitive to steroid hormones.

The results can be summarized as follows:

The neurons producing RH and IH involved in regulation of the PRL/GH family of hormones, the proliferative rate of pituitary chromophobes and PRL cells, the hypertrophy of PRL cells, and the inverse relationship between PRL and GH synthesis and release indicate that growth of animals may be stimulated, retarded, or inhibited.

Reactions of hypothalamic neurons, especially MPA and MBH, as integrators of information received directly from GS, or via peptidergic and aminergic neurons incorporating GS, and the inverse relationship between PRL and GTH release, show that reproductive properties and behavior of the animals may be under hormonal control.

Hypothalamic NPV, as the main source of CRH, VP, and other peptides, as well as neurons, PI, and ACTH cells that produce members of the POMC family of hormones, are also sensitive to hydrocortisone and GS, and the character of their responsiveness in reaction to stress, their ability for adaptation, and other related functions may be controlled.

Summarizing all these data, there is no doubt that memory of neuroendocrine cells, acquired during evolution and expressed in specific, genetically programmed development, may be modulated by steroid hormones during early ontogenesis in order to obtain the appropriate results.

## Acknowledgments

The author thanks Prof. Dr. Olivera Pantić and Sanja Pantić for their encouragement and devoted engagement in the preparation of this manuscript. The author is also grateful to Prof. Dr. Vojislav Vuzevski for supplying the electron micrographs in Figs. 1–4.

## References

Adelman, J. P., Hayflick, J. S., and Seeburg, P. H. (1986). Isolation of the gene and hypothalamic cDNA for common precursor of gonadotropin-releasing hormone and prolactin release-inhibiting factor in human and rat. *Proc. Natl. Acad. Sci. U.S.A.* **83,** 179–183.
Albers, K., and Fuchs, E. (1992). The molecular biology of intermediate filament proteins. *Int. Rev. Cytol.* **134,** 243–279.

Albrecht-Buehler, G. (1990). In defense of "nonmolecular" cell biology. *Int. Rev. Cytol.* **120,** 191–241.

Al-Jousuf, S. A. (1992). Neuropeptides and immunocytochemistry in annelids. *Int. Rev. Cytol.* **133,** 231–308.

Allen, R. G., Pintar, J. E., Stack, J., and Kendall, J. N. (1984). Biosynthesis and processing of proopio-melanocortin derived peptides during fetal pituitary development. *Dev. Biol.* **102,** 43–50.

Almazan, G., Lefevre, D. E. L., and Zingg, H. H. (1989). Ontogeny of hypothalamic vasopressin, oxytocin and somatostatin gene expression. *Dev. Brain Re.* **45,** 69–75.

Andersen, O., Klungland, H., Dunn, I., and Zohar, Y. (1992). Evolutionary aspects of the GTnRH family. *Conf. Eur. Comp. Endocrinol., 16th,* Padova, p. 52.

Anderson, L. L., Peters, J. B., Melampy, P. M., and Cox, D. F. (1972). Changes in adenohypophyseal cells and levels of somatotropin and prolactin at different reproductive stages in the pig. *J. Reprod. Fertil.* **28,** 55–65.

Argenton, F., Bernardini, S. W., Vianello, S., Colombo, L., and Bortolussi, M (1992). Conservation of the growth hormone gene transcription between fish and mammals. *Conf. Eur. Comp. Endocrinol., 16th,* Padova, p. 54.

Arimura, A., Smith, W. D., and Schally, A. V. (1974a). Blockade of the stress-induced decrease in blood GH by anti-somatostatin serum in rats. *Endocrinology (Baltimore)* **98,** (2), 540–543.

Arimura, A., Shino, M., de la Cruz, K. G., Rennnels, E. G., and Schally, A. V. (1976b). Effect of active and passive immunization with luteinizing hormone releasing hormone on serum luteinizing hormone and follicle stimulating hormone levels and the ultrastructure of the pituitary gonadotropes in castrated male rats. *Endocrinology (Baltimore)* **99,** 291–303.

Baker, B. I. (1991). Melanin-concentrating hormone: A general vertebrate neuropeptide. *Int. Rev. Cytol.* **126,** 1–47.

Ball, J. N. (1981). Hypothalamic control of the pars distalis of fishes and reptiles. *Gen. Comp. Endocrinol.* **44,** 135–170.

Banik, U. K., and Givner, M. L. (1981). Fertility and antifertility effects of LH-RH and its agonists. *In* "Reproductive Processes and Contraception" (K. W. McKerns, ed.), pp. 143–160. Plenum, New York.

Barinaga, M., Bilezikjian, L. M., Vale, W. W., Rosenfeld, M. G., and Evans, R. M. (1985). Independent effects of growth hormone releasing factor on growth hormone release and gene transcription. *Nature (London)* **314,** 279–281.

Barnea, A. (1984). Molecular aspects of the release of luteinizing hormone releasing hormone (LHRH) from hypothalamic neurons. *In* "Hormonal Control of the Hypothalamo-Pituitary-Gonadal Axis" (K. W. McKerns and Z. Naor, eds.), pp. 27–38. Plenum, New York.

Barry, J. (1976). Characterization and topography of LH RH neurons in the human brain. *Neurosci. Lett.* **3,** 287–291.

Barry, J. (1979). Immunohistochemistry of luteinizing hormone-releasing hormone-producing neurons of the vertebrates. *Int. Rev. Cytol.* **60,** 179–221.

Barry, J., and Carette, B. (1975). Immunofluorescence study of LRF neurons in primates. *Cell Tissue Res.* **164,** 163–178.

Barry, J., and Dubois, M. P. (1975). Immunofluorescence study of LRH producing neurons in the cat and the dog. *Neuroendocrinology* **18,** 290–298.

Barry, J., and Dubois, M. P. (1976). Immunoreactive LRF neurosecretory pathways in mammals. *Acta Anat.* **94,** 497–503.

Batten, T. F. C., Berry, P. A., Moons, L., and Cambre, M. L. (1990). Immunocytohistochemical study of monoaminergic neurons in the brain and pituitary of two teleosts *Poecilia latipinna* and *Dicentrarchus labrax. Conf. Eur. Comp. Endocrinol., 15th,* Leuven, p. 103.

Bauer, K., and Horsthemke, B. (1984). Degradation of LH-RH. *In* "Hormonal Control of the Hypothalamo-Pituitary-Gonadal Axis" (K. W. McKerns and Z. Naor, eds.), pp. 101–114. Plenum, New York.

Belayew, A., Peers, B., Berwaer, M., Voz, M., Jacquemain, P., Monget, P., Mathy-Hartert, M., Luoette, J., Morin, A., and Martial, A. (1990). Tissue specific expression and hormonal regulation of the human growth hormone/prolactin gene family. *Conf. Eur. Comp. Endocrinol., 15th,* Leuven, p. 29.

Belayew, A., Poncelet, A. C., Sekkali, B., Brim, H., Sovennen, D., and Martial, J. A. (1992). Molecular evolution and transcriptional regulation of the genes encoding growth hormones and prolactins in vertebrates. *Conf. Eur. Comp. Endocrinol., 16th,* Padova, p. 6.

Ben-Jonathan, N. (1985). Dopamine: A prolactin inhibitory hormone. *Endocr. Rev.* **6,** 564–589.

Ben-Jonathan, N., Oliver, C., Winer, H. J., Mical, R. S., and Porter, J. C.. (1977). Dopamine in hypophyseal portal plasma of the rat during the estrous cycle and throughout pregnancy. *Endocrinology (Baltimore)* **100,** 452–458.

Benoit, R., Ling, N., and Esch, F. (1987). A new prosomatostatin-derived peptide reveals a pattern from prohormone cleavage at monabasic sites. *Science* **238,** 1126–1128.

Berlind, A. (1977). Cellular dynamics in invertebrate neurosecretory systems. *Int. Rev. Cytol.* **49,** 171–251.

Bern, H. A. (1966). On the production of hormones by neurons and the role of neurosecretion in neuroendocrine mechanism. *Symp. Soc. Exp. Biol.* **20,** 324–344.

Bern, H. A. (1967). Hormones and endocrine glands in fishes. Studies of fish endocrinology reveal major physiological and evolutionary problems. *Science* **158,** 455–462.

Bern, H. A. (1970). Concluding remarks. *In* "Aspects of Neuroendocrinology" (W. Bargmann and B. Scharrer, eds.), pp. 375–377. Springer-Verlag, Berlin.

Bern, H. A. (1983). Functional evolution of prolactin and growth hormone in lower vertebrates. *Am. Zool.* **23,** 663–671.

Bern, H. A. (1984). The metamorphosis of comparative endocrinology. *Gen. Comp. Endocrinol.* **53,** 428–432.

Bern, H. A. (1985). Elusive urophysis—twenty five years in pursuit of caudal neurohormones. *Am. Zool.* **25,** 763–769.

Besson, J., Laburthe, M., Bataille, D., Dupont, C., and Rosselin, G. (1978). Vasoactive intestinal peptide (VIP). Tissue distribution in the rat as measured by radioimmunoassay and radioreceptorassay. *Acta Endocrinol. (Copenhagen)* **87,** 799.

Beyer, C., and Feder, H. H. (1987). Sex steroid and afferent input: The role in brain sexual differentiation. *Annu. Rev. Physiol.* **49,** 349–364.

Blackwell, R. E., and Guillemin, R. (1973). Hypothalamic control of adenohypophyseal secretion. *Annu. Rev. Physiol.* **35,** 357–390.

Blalock, J. E. (1992). Molecular recognition theory: Sense-antisense peptide binding. *Conf. Eur. Comp. Endocrinol., 16th,* Padova, p. 7.

Bogdanova, E. M., and Shoen, H. C. (1959). Precocious sexual development in female rats with hypothalamic lesions. *Proc. Soc. Exp. Biol. Med.* **100,** 664–669.

Bouchaud, C., and Bosler, O. (1988). The circumventricular organs of the mammalian brain with special reference to monoaminergic innervation. *Int. Rev. Cytol.* **105,** 283–327.

Bourne, H. R. (1986). One molecular machine can transduce diverse signal. *Nature (London)* **321,** 814–816.

Bourne, H. R., Sanders, D. A., and McCormick, F. (1991). The GTPase superfamily: conserved structure and molecular mechanism. *Nature (London)* **349,** 111–127.

Brownfeld, M. S., and Kozlowski, G. P. (1977). The hypothalamo-chorioidal tract. Immuno-histochemical demonstration of neurophysin pathways to the telencephalic choroidal plexus and cerebrospinal fluid. *Cell Tissue Res.* **178**, 111–127.

Brownstein, M, Arimura, A., Sato, H., Schally, A. V., and Kizer, J. S. (1975). The regional distribution of somatostatin in the rat brain. *Endocrinology (Baltimore)* **96**, 1–127.

Bruhn, T. O., Plotsky, P. M., and Vale, W. W. (1984). Effect of paraventricular lesions on corticotropin-releasing factor (CRF)-like immunoreactivity in the stalk-median eminence: Studies on the adrenocorticotropin response to ether stress and exogenous CRF. *Endocrinology (Baltimore)* **114**, (1), 57–62.

Buijs, R. M., Swaab, D. P., Dogterom, J., and Van Lecuwen, F. M. (1978). Intra- and extrahypothalamic vasopressin and oxytocin pathways in the rat. *Cell Tissue Res.* **186**, 423–433.

Catt, K. J., Krsmanović, I., and Stojilković, S. (1994). GnRH receptors and signaling mechanisms in hypothalamic and pituitary cells. *Neuroendocrinology* **60**, Suppl. 1, p. 2.

Cesselin, F., and Peillon, F. (1980). In vitro studies of the secretion of human proclatin and growth hormone. *In* "Synthesis and Release of Adenohypophyseal Hormones" (M. Jutisz and K. W. Mckerns, eds.), pp. 677–721. Plenum, New York.

Chaiken, I. M., Fischer, E. A., Giudice, L. C., and Hough, C. J. (1982). In vitro synthesis of hypothalamic neurophysin precursors. *In* "Hormonally Active Brain Peptides: Structure and Function" (K. W. McKerns and V. Pantić, eds.), pp. 327–347. Plenum, New York.

Chappel, S. C., and Spies, H. G. (1981). Control of gonadotropin and prolactin secretion in rhesus monkeys and rodents. *In* "Reproductive Processes and Contraception" (K. W. McKerns, ed.), pp. 3–25. Plenum, New York.

Charlton, H. M. (1986). The physiological action of LH RH evidence from the hypogonadal (*hpg*) mice. *In* "Neuroendocrine Molecular Biology" (G. Fink, A. J. Harmar, and K. W. McKerns, eds.), pp. 47–56. Plenum, New York.

Chetverukhin, V. K., Belenky, M. A., and Polenov, A. L. (1986). The hypothalamo-hypophyseal system of the frog *Rana temporaria*. Ultrastructure of the median eminence in the adult frog with reference in the distribution of serotoninergic terminals. *Cell Tissue Res.* **243**, 649–654.

Chieffi, G., Pierantoni, R., and Fasano, S. (1991). Immunoreactive GnRH in hypothalamic areas. *Int. Rev. Cytol.* **127**, 1–55.

Clark, R. G., and Robinson, I. C. A. (1985). Growth induced by pulsatile infusion of an amielated fragment of human growth hormone releasing factor in normal and GHRF deficient rats. *Nature (London)* **314**, 281–283.

Cogen, P. H., Antunes, J. L., Louis, K. M., Dyrenfurth, I., and Ferin, M. (1980). The effect of anterior hypothalamic disconnection on gonadotropin secretion in the female rhesus monkey. *Endocrinology (Baltimore)* **107**, 677–681.

Conlon, J. M. (1990). Biosynthesis of somatostatins and urotensins. *Conf. Eur. Comp. Endocrinol., 15th,* Leuven, p. 56.

Cramer, D. M., Parker, C. R., and Porter, J. C. (1979). Estrogen inhibition of dopamine release into hypophyseal blood. *Endocrinology (Baltimore)* **104**, (2), 419–422.

Csaba, G. (1985). The unicellular *Tetrahymena* as a model cell for receptor research. *Int. Rev. Cytol.* **95**, 327–377.

Csaba, G. (1992). Chemical mediators and their receptors in protozoa. *Conf. Eur. Endocrinol., 16th,* Padova, p. 13.

Culler, M. D., Valenca, M. M., Merchenthaler, I., Flerkó, B., and Negro-Vilar, A. (1988). Orchidectomy induces temporal and regional changes in the processing of the luteinizing hormone-releasing hormone prohormone in the rat brain. *Endocrinology (Baltimore)* **122**, 1968–1976.

Danilova, O., Hristić, M., and Pantić, V. (1980). Neurosecretory material in the median eminence of rats adrenalectomized at various ages of ontogenesis. *Acta Vet. (Belgrade)* **30,** (1–2), 13–14.

Danilova, O., Hristić, M., and Pantić, V. (1982). Neurohypophysis and pars intermedia in rats adrenalectomized at various stages of juvenile and pubertal period of development. *Arh. Biol. Nauka* **34,** 1–4.

Davey-Smith, G., and Fink, G. (1983). The effects of daily administration of single and multiple injections of gonadotropin-releasing hormone on pituitary and gonadal function in the hypogonadal (*hpg*) mouse. *Endocrinology (Baltimore)* **133** (2), 535–544.

Davis, L. G., Arentzen R., Reid, J. M., Manning, R. W., Wolfson, B., Laurence, K. L., and Baldino, F. J. (1986). Glucocorticoid sensitivity of vasopressin mRNA levels in the paraventricular nucleus of the rat. *Proc. Natl. Acad. Sci. U.S.A.* **83,** 1145–1149.

Dayhoff, M. O., ed. (1976). "Atlas of Protein Sequence and Structure," Vol. 5, No. 2, pp. 116–119. Natl. Biomed. Res. Found., Georgetown University, Medical Center, Washington, DC.

Dean, C. R., Hope, D. B., and Kazic, T. (1968). Evidence for the storage of oxytocin with neurophysin I and vasopressin with neurophysin II in separate neurosecretory granules. *Proc. Br. Pharmacol. Soc.* **34,** 1928–1936.

Deery, D. J. (1974). Determination of radioimmunoassay of the luteinizing hormone-releasing hormone (LHRH) content of the hypothalamus of the rat and some lower vertebrates. *Gen. Comp. Endocrinol.* **24,** 280–285.

Dellmann, H. D., and Simpson, J. B. (1979). The subfornical organ. *Int. Rev. Cytol.* **58,** 333–421.

De Loof, A. (1986). The electrical dimension of cells: The cell as a miniature electrophoresis chamber. *Int. Rev. Cytol.* **104,** 251–352.

De Loof, A. (1992). Peptide hormone families shared by vertebrates and invertebrates. *Conf. Eur. Comp. Endocrinol. 16th,* Padova, p. 14.

Denef, C., Swennen, L., and Andries, M. (1982). Separated anterior pituitary cells and their response to hypophysiotropic hormones. *Int. Rev. Cytol.* **76,** 225–244.

Denef, C., Baes, M., Schramme, C., and Swennen, L. (1984). The roll of cell–cell communication in neuropeptide-stimulated and dopamine-inhibited prolactin release. *In* "Hormonal Control of the Hypothalamo-Pituitary-Gonadal Axis" (K. W. McKerns and Z. Naor, eds.), pp. 355–366. Plenum, New York.

Dermietzel, R., and Krause, D. (1991). Molecular anatomy of the blood-brain barrier as defined by immunocytochemistry. *Int. Rev. Cytol.* **127,** 57–109.

Desrues, L., Tonon, M. C., Lamacy, M., Stoeckel, M. E., Bosler, O., and Vaudry, H. (1992). Three inhibitory factors of the pars intermedia co-exist in the same nerve terminals and act through three transduction pathways. *Conf. Eur. Comp. Endocrinol. 16th,* Padova, p. 90.

De Wield, D. (1994). Peptide hormones and the brain. *Neuroendocrinology* **60,** Suppl. 1, p. 1.

Dierickx, K. (1980). Immunocytochemical localization of the vertebrate cyclic nonapeptide neurohypophyseal hormones and neurophysins. *Int. Rev. Cytol.* **62,** 120–185.

Dierickx, K., and Vandesande, F. (1975). Identification of the vasopressin and the oxytocin producing neurons in the rat hypothalamus. *Conf. Eur. Comp. Endocrinol. 8th,* Bangor, p. 1.

Doerr-Schott, J. (1976). Immunocytochemical detection, by light and electron microscopy, of pituitary hormones in cold-blooded vertebrates. *Gen. Comp. Endocrinol.* **28,** 487–512.

Döhler, K. D., and Haucke, J. L. (1978). Thoughts on the mechanism of sexual brain differentiation. *In* "Hormones and Brain Development" (G. Dörner and K. Kawakami, eds.), pp. 153–158. North-Holland Publ., Amsterdam.

Döhler, K. D., and Wütke, W. (1974). Serum LH, FSH, prolactin and progesterone from birth to puberty in female and male rats. *Endocrinology (Baltimore)* **94**, 1003–1006.

Döhler, K. D., Muhlen, A., and Döhler, U. (1977). Pituitary luteinizing hormone (LH), follicle stimulating hormone (FSH) and prolactin from birth to puberty of female and male rats. *Acta Endocrinol. (Copenhagen)* **85**, 718–728.

Dörner, G. (1980). Sexual differentiation of the brain. *Vitam. Horm. (N.Y.)* **38**, 325–381.

Drouva, V. S., and Gallo, V. R. (1976). Catecholamine involvement in episodic luteinizing hormone release in adult ovariectomized rats. *Endocrinology (Baltimore)* **99**, 651–658.

Drouva, V. S., Epelbaum, J., and Kordon, C. (1982). Hormonal regulation of and ionic requirements for in vitro release of hypothalamic peptide. *In* "Hormonally Active Brain Peptides: Structure and Function" (K. W. McKerns and V. Pantić, eds.), pp. 99–123. Plenum, New York.

Dufy, B., Partouche C., Poulain, D., Dufy-Barbe, L., and Vincet, J. D. (1976). Effect of estrogen on the electrical activity of identified and unidentified hypothalamic units. *Neuroendocrinology* **22**, 38.

Dufy, B., Dufy-Barbe, L., Arnauld, E., and Vincent, J. D. (1982). Steroids and membrane-associated events in neurons and pituitary cells. *In* "Hormonally Active Brain Peptides: Structure and Function" (K. W. McKerns and V. Pantić, eds.), pp. 235–353. Plenum, New York.

Dupouy, J. P. (1975). CRF activity in fetal rat hypothalamus in late pregnancy. *Neuroendocrinology* **19**, 203–213.

Dupouy, J. P. (1980). Differentiation of MSH-, ACTH-, endorphin-, and LPH-containing cells in the hypophysis during embryonic and fetal development. *Int. Rev. Cytol.* **68**, 197–249.

Ekholm, R., and Pantić, V. (1963). Effect of thyrotropin on nucleic acids and protein contents of the thyroid. *Nature (London)* **199**, 1203–1204.

Elde, R., and Hökfelt, T. (1978). Distribution of hypothalamic hormones and other peptides in the brain. *In* "Frontiers in Neuroendocrinology" (W. Ganong and L. Martini, eds.) Vol. **5**, pp. 1–33. Raven Press, New York.

Elde, R., and Hökfelt, T. (1979). Localization of hypophysiotropic peptides and other biologically active peptides within the brain. *Annu. Rev. Physiol.* **41**, 587–602.

Elde, R., and Parsons, J. A. (1975). Immunocytochemical localization of somatostatin in cell bodies of the rat hypothalamus. *Am. J. Anat.* **144**, 541–548.

Enjalbert, A., Bockaert, J., Epelbaum, J., Moyse, E., and Koredon, C. (1984). Modulation of prolactin secretion at the pituitary level: Involvement of adenylate cyclase. *In* "Hormonal Control of Hypothalamo-Pituitary-Gonadal Axis" (K. W. McKerns and Z. Naor, eds.), pp. 367–383. Plenum, New York.

Epelbaum, J. (1994). Anatomical and physiological interactions between GHRH and SRIH neurons. *Neuroendocrinology* **60**, Suppl. 1, S1.2, p.3.

Epelbaum, J., Brayeau, P., Tsang, D., Brawer, J., and Martin, J. B. (1977). Subcellular description of radioimmunassayable somatostatin in rat brain. *Brain Res.* **126**, 309.

Evans, G. A., and Rosenfeld, M. G. (1980). Hormonal regulation of prolactin mRNA. *In* "Synthesis and Release of Adenohypophyseal Hormones" (M. Jutisz and K. W. McKerns, eds.), pp. 295–309. Plenum, New York.

Farmer, S. W., Papkoff, H., and Hayashida, T. (1976). Purification and properties of reptilian and amphibian growth hormone. *Endocrinology (Baltimore)* **99**, (3), 692–700.

Farquhar, M. C. (1971). *Mem. Soci. Endocrinol.* **19**, 79–124.

Farquhar, M. C. (1977). Secretion and crinophagy of prolactin cells. *In* "Comparative Endocrinology" (H. B. Dellman, J. A. Johnson, and D. M. Klachko, eds.), pp. 37–91. Plenum, New York.

Fedoroff, S., and Vernadakis, A. (1986). *In* "Astrocytes: Cell Biology and Pathology of Astrocytes" (S. Fedoroff and A. Vernadakis, eds.), Vol. 3. Academic Press, Orlando, FL.

Feldman, S., and Saphier, D. (1984). Role of neurotransmitters and electrophysiological changes in the hypothalamus related to central adrenocortical regulation. *In* "Neuroendocrine Correlates of Stress" (K. W. McKerns and V. Pantić, eds.), pp. 36–62. Plenum, New York.

Ferguson, A. V., Donevan, S. D., Papas, S., and Smith, P. M. (1990). Circumventricular structures, CNS sensor of circulating peptides and autonomic control centers. *Endocrinol. Exp.* **24**, (1–2), 19–36.

Fine, R. E., and Ockleford, C. D. (1984). Supramolecular cytology of coated vesicles. *Int. Rev. Cytol.* **91**, 1–43.

Fink, G., and Pickering, A. (1980). Modulation of pituitary responsiveness to gonadotropin-releasing hormone. *In* "Synthesis and Release of Adenohypophysial Hormones" (M. Jutisz and K. W. McKerns, eds.), pp. 617–638. Plenum, New York.

Fink, G., Aiyen, M., Chiappa, S., Henderson, S., Jamieson, M., Levy-Perez, V., Pickering, A., Sarkar, O., Sherwood, N., Spewight, A., and Watta, A. (1982). Gonadotropin-releasing hormones: Release into hypophyseal portal blood and mechanism of action. *In* "Hormonally Active Brain Peptides: Structure and Function" (K. W. McKerns and V. Pantić, eds.), pp. 397–426. Plenum, New York.

Flament-Durand, J., and Brion, J. P. (1985). Tanycytes: Morphology and functions. A review. *Int. Rev. Cytol.* **96**, 121–155.

Flerkó, B. (1954). Zur Hypothalamischgen Steuerung der gonadotrophen Function der Hypophyse. *Acta Morphol. Acad. Sci. Hung.* **4**, 457–492.

Flerkó, B. (1975). Perinatal androgen action and the differentiation of the hypothalamus. *In* "Growth and Development of the Brain" (M. A. B. Brasier, ed.), pp. 117–137. Raven Press, New York.

Flerkó, B., Merschentaler, I., and Sétáló, G. (1987). Short and ultrashort feedback control of gonatropin secretion. *In* "Endocrinology and Physiology of Reproduction," pp. 37–50. Plenum, New York.

Fontaine, M., and Olivereau, M. (1975). Some aspects of the organization and evolution of vertebrate pituitary. *Am. Zool.* **15**, Suppl. 1, 61–70.

Fontaine, Y. A., and Burzawa-Gerard, E. (1978). Biochemical and biological properties of fish gonadotropins and their subunits: Comparison with mammalian hormones. *In* "Structure and Function of Gonadotropins" (K. W. McKerns, ed.), pp. 361–380. Plenum, New York.

Frömter, E. (1993). Membrane transport phenomena: Pumps and coupled transporters involved in active and passive transport. *Biolo. Membr. Proc. Annu. Gen. Meet. Aust. Acad. Sci.,* Canberra, pp. 1–26.

Gage, P. (1993). Membrane Trasport Phenomena: Channels. *Biolo. Membr. Proc. Annu. Gen. Meet. Aust. Acad. Sci.,* Canberra, pp. 27–32.

Genbačev, O., and Pantić, V. (1972). LTH and STH in the pituitaries of the adult rats treated with oestrogen. *Jugosl. Physiol. Pharmacol. Acta (Belgrade)* **8**, (3), 309–316.

Genbačev, O., and Pantić, V. (1975). Pituitary cell activities in gonadectomized rats treated with estrogen. *Cell Tissue Res.* **157**, 273–282.

Genbačev, O., Pantić, V., and Ratković, M. (1977). Activities of luteotropic (LTH) and somatotropic (STH) cells in the pituitaries of rats treated with oestrogen on day 10 or 15 after birth. *In* "Problemi na sravnitelnata i ekperimental nata morfologija i embriologija," pp. 149–155. Akade. Nauki., Sofija, Bulgaria.

Giguere, V., and Labrie, F. (1982). Vasopressin potentiates cyclic AMP accumulation and ACTH release induced by corticotropin-releasing factor (CRF) in rat anterior cells in culture. *Endocrinology (Baltimore)* **111** (5), 1752–1754.

Gillies, O., and Loury, P. (1970). Corticotrophin releasing factor may be modulated by vasopressin. *Nature (London)* **278,** 463–464.

Goodman, R. H., Montiminy, M. R., Low, M. J., Taukada, T., Fink, S., Lechan, R. M., Wu, P., Jackson, Im, M. D., and Mandel, G. (1986). Biosynthesis of somatostatin, vasoactive intenstinal polypeptide and thyrotropin releasing hormone. *In* "Neuroendo-crine Molecular Biology" (G. Fink, A. J. Harmar, and K. W. McKerns, eds.), pp. 159–173. Plenum, New York.

Gourdji, D. (1980). Characterization of thyroliberin (TRH) binding sites and coupling with prolactin and growth hormone secretion in rat pituitary cell lines. *In* "Synthesis and Release of Adenohypophyseal Hormones" (M. Jutisz and K. W. McKerns, eds.), pp. 463–493. Plenum, New York.

Grimmelikhuijzen, G. I. P., Carstensen, K., Daimer, D., McFarlane, I. D., Moosler, A., Noithacken, H. P., Reinscheid, R. K., Rinehart, K. L., Schmutler, C., and Vollert, H. (1992). Neuropeptides in coelenterates: Structure, action and biosynthesis. *Conf. Eur. Comp. Endocrinol., 16th,* Padova, p. 20.

Guillemin, R., Ling, N., Lazarus, L., Burgus, R., Minick, S., Bloom, F., Nicoll, R., Siggins, G., and Segal, D. (1977). The endorphins, novel peptides of brain and hypophyseal origin, with opiate-like activity. Biochemical and biologic studies. *Ann. N.Y. Acad. Sci.* **297,** 131–156.

Guillemin, R., Brazeau, P., Bohler, P., Esch, F., Ling, N., and Wehrenberg, W. B. (1982). Growth hormone-releasing factor from a human pancreatic tumor that caused acromegaly. *Science* **218,** 585–587.

Halasz, I., Aird, F., and Redei, E. (1994). Effect of maternal adrenalectomy on hypothalamic CRF and glucocorticosteroid receptor mRNA and anterior pituitary POMC mRNA synthesis in male and female neonates. *Neuroendocrinology* **60,** Suppl. 1, 28.

Harmar, A. J., Pierotti, A. R., and Keen, P. (1986). Biosynthesis of the tachykins and somatostatin. *In* "Neuroendocrine Molecular Biology" (G. Fink, A. J. Harmar, and K. W. McKerns, eds.), pp. 147–158. Plenum, New York.

Haug, E. (1979). Progesteron suppression of estrogen stimulated prolactin secretion and estrogen receptor levels in rat pituitary cells. *Endocrinology (Baltimore)* **104**(2), 429–438.

Haug, E., Gautvik, K. M., Sand, O., Iversen, J. G., and Kriz, M. (1982). Interaction between thyrotropin-releasing hormone and prolactin-producing cells. *In* "Hormonally Active Brain Peptides: Structure and Function" (L. W. McKerns and V. Pantić, eds.), pp. 537–565. Plenum, New York.

Herbert, E., Philips, M., Hinman, M., Roberts, J. L., Budarf, M., and Pequette, T. I. (1980). Processing of the common precursor to ACTH and endorphin in mouse pituitary tumor cells and monolayer culture from mouse anterior pituitary. *In* "Synthesis and Release of Adenohypophyseal Hormones" (M. Jutisz and K. W. McKerns, eds.), pp. 237–261. Plenum, New York.

Herlant, M. (1964). The cells of the adenohypophysis and their functional significance. *Int. Rev. Cytol.* **17,** 299–382.

Herlant, M. (1967). Histophysiology of human anterior pituitary. *Methods Achiev. Exp. Pathol.* **3,** 250–382.

Hofman, M. A., and Swaab, D. F. (1991). Sexual dimorphism and the human brain: Myth and reality. *Exp. Clin. Endocrinol.* **98**(2), 161–170.

Hökfelt, T., Fuxe, I., Johansson, O., Jeffcoate, S., and White, A. (1975). Distribution of thyrotropin-releasing hormone (TRH) in the central nervous system as revealed with immunocytochemistry. *Eur. J. Pharmacol.* **34,** 389–392.

Hökfelt, T., Elfvin, L. G., Elde, R., Schultzberg, M., Goldstein, M., and Luft, R. (1977a). Occurrence of somatostatin-like immunoreactivity in some peripheral sympathetic noradrenergic neurons. *Proc. Natl. Acad. Sci. U.S.A.* **74,** 3587–3591.

Hökfelt, T., Elde, R., Johansson, O., Terenius, L., and Stein, L. (1977b). The distribution of enkephalin immunoreactive cell bodies in the rat central nervous system. *Neurosci. Lett.* **5,** 25–31.

Hökfelt, T., Elde, R., Fuxe, O., Ljungdahl, R., Goldstein, M., Luft, R., Efendic, S., Nilsson, G., Terenius, L., Ganten, D., Jeffcoate, S. L., Rehfeld, J., Said, S., Perez de la Mora, M., Possani, L., Tapia, R., Teran, L., and Palacios, R. (1978). Aminergic and peptidergic pathways in the nervous system, with special reference to the hypothalamus. *In* "The Hypothalamus" (S. Reichin, R. J. Baldessarini, and J. B. Martin, eds.), pp. 69–134. Raven Press, New York.

Hökfelt, T., Johansson, O., Ljungdahl, A., Lundberg, M. J., and Schultyberg, M. (1980). Peptidergic neurons. *Nature (London)* **284,** 515–521.

Holmquist, B. I., Carlberg, M., and Ekström, P. (1992). The galaninergic system in the brain of the salmon. *Conf. Eur. Comp. Endocrinol. 16th,* Padova, pp. 22.

Hough, C. J., Hargrave, P. A., and Chaiken, I. M. (1980). On the biosynthetic origin of neurophysin-neurohypophyseal peptide hormone complex. *In* "Biosynthesis, Modification and Processing of Cellular and Viral Polyproteins" (G. Koch and D. Richter, eds.), pp. 29–42. Academic Press, New York.

Hristić, M., and Pantić, V. (1973). ACTH cells of rats treated with adrenal steroids or ACTH. *Arch. Biol. Nauka* **22**(1–4), 87–92.

Hristić, M., and Pantić, V. (1978). Hypothalamic nuclei of rats treated with corticosteroids. *In* "Neurosecretion and Neuroendocrine Activity." Evolution, Structure and Function," pp. 205–208. Springer-Verlag, Berlin.

Hristić, M., Pantić, V., and Nickerson, P. (1978). Influence of maternal adrenalectomy on pituitary ACTH cells of fetal and neonatal rats. *Acta Vet. (Belgrade)* **28,**(6), 231–241.

Hristić, M., Pantić, V., and Kalafatić, D. (1987). ACTH cells and adrenal cortex in adult rats treated with a single dose of hydrocortisone during the neonatal or juvenile period of development. *Acta Vet. (Belgrade)* **37**(2–3), 93–100.

Hymer, W. C., Page, R., Kelsey, C., Augustine, E. C., Wiffinger, W., and Ciolkosz, M. (1980). Separated somatotrophs, their use in vitro and in vivo. *In* "Synthesis and Release of Adenohypophyseal Hormone" (M. Jutisz and K. W. McKerns, eds.), p. 125–166. Plenum, New York.

Inoué, S. (1989). Ultrastructure of basement membranes. *Int. Rev. Cytol.* **117,** 57–98.

Ivanišević-Milovanović, O., Stevanović-Lončar, H., Karakašević, A., and Pantić, V. (1990). Plasma adrenocorticotropic hormone, serum estradiol and progesterone concentrations and catecholamine content in ovarian tissues of female rats exposed to either continuous light or darkness. *Acta Vet. (Belgrade)* **40**(5–6), 243–252.

Ivanišević-Milovanović, O., Pantić, V., Demajo, M., and Lončar-Stevanović, H. (1993). Catecholamines in hypothalamus, ovaries and uteri of rats with precocious puberty. *J. Endocrinol. Invest.* **16,** 769–773.

Janzer, R. C., and Raff, M. C. (1987). Astrocytes induce blood-brain-barrier properties in endothelial cells. *Nature (London)* **325,** 253–257.

Jingami, H., Matsukura, S., Numa, S., and Imura, H. (1985). Effects of adrenalectomy and dexamethasone administration on the level of prepro-corticotropin-releasing factor messenger ribonucleic acid (mRNA) in the hypothalamus and adrenocorticotropin $\beta$-lipotropin precursor mRNA in the pituitary in rats. *Endocrinology (Baltimore)* **117,** 1314–1320.

Joosse, J. (1990). Neuropeptides: Unity and diversity. *Conf. Eur. Comp. Endocrinol. 15th,* Leuven, p. 88.

Kalra, S. P. (1976). Tissue levels of luteinizing hormone releasing hormone in the preoptic area and hypothalamus and serum concentration of gonadotropins following anterior hypothalamic deafferentation and estrogen treatment of female rat. *Endocrinology (Baltimore)* **99,** 101.

Kalra, S. P. (1982). Mode of opioid and catecholamine LH secretion. *In* "Hormonally Active Brain Peptides: Structure and Function" (K. W. McKerns and V. Pantić, eds.), pp. 141–155. Plenum, New York.

Kalra, S. P., and Kalra, P. S. (1984). Hypothalamic microenvironment controlling LH RH secretion. *In* "Regulation of Target Cell Responsiveness" (K. W. McKerns, A. Aakvaag, and V. Hansson, eds.), Vol. 2, pp. 127–155. Plenum, New York.

Kalra, S. P., Kalra, P. S., and Simpkins, J. W. (1981). Regulation of LH RH secretion by gonadal steroids and catecholamines. *In* "Reproductive Processes and Contraception" (K. W. McKerns, ed.), pp. 27–45. Plenum, New York.

Kawakami, M., Yoshioka, E., Konmda, N., Arita, J., and Visessuvan, S. (1978a). Data on the sites of the stimulatory feedback action of gonadal steroids indispensable for luteinizing hormone release in rat. *Endocrinology (Baltimore)* **102**(3), 791–798.

Kawakami, M., Arita, J., Yoshioka, E., Visessuvan, S., and Akema, T. (1978b). Data on the sites of the stimulatory feedback action of gonadal steroids indispensable for follicle stimulating hormone release in the rat. *Endocrinology (Baltimore)* **103,** 752–770.

Kelley, K. W., and Dantzer, R. (1990). The endocrine and immune system as partner in host defense. *Conf. Eur. Comp. Endocrinol. 15th,* Leuven, p. 3.

Kelly, M. J., Kuhnt, U., and Wuttke, W. (1980). Hyperpolarization of hypothalamic parvocellular neurons by 17-β-estradiol and their identification through intracellular staining with Procion Yellow. *Exp. Brain Res.* **40,** 440–447.

King, J. C., and Anthony, E. L. P. (1984). LH RH neurons and their projections in humans and other mammal species comparison. *Peptides (N.Y.)* **5,** Suppl., 195–207.

King, J. C., Tobet, S. A., Snavely, F. L., and Arimura, A. A. (1982). LH RH immunopositive cells and their projections to the median eminence and organum vasculosum of the lamina terminalis. *J. Comp. Neurol.* **209,** 287–300.

Kirschner, M. W. (1978). Microtubule assembly and nucleation. *Int. Rev. Cytol.* **54,** 1–71.

Knigge, K. M., Joseph, S. A., Sladek, J. K., Notter, M. F., Morris, M., Sundberg, D. K., Holzwarth, M. A., Hoffman, G. F., and O'Brien, I. (1976). Uptake and transport activity of the median eminence of the hypothalamus. *Int. Rev. Cytol.* **45,** 383–408.

Koch, B., and Lutz-Bucher, B. (1986). Characterization, regulation and functional activity of specific vasopressin receptors in the anterior pituitary gland. *In* "Neuroendocrine Molecular Biology" (G. Fink, A. J. Harmar, and K. W. McKerns, eds.), pp. 249–260. Plenum, New York.

Koch, Y., Elkabes, S., and Fridkin, M. (1984). Degradation of luteinizing hormone-releasing hormone by rat pituitary plasma membrane associated enzymes. *In* "Hormonal Control of the Hypothalamo-Pituitary-Gonadal Axis" (K. W. McKerns and Z. Naor, eds.), pp. 115–126. Plenum, New York.

Koch, Y., Elkabes, S., and Fridkin, M. (1986). Degradation of luteinizing hormone-releasing hormone (LHRH) by pituitary plasma membrane and by pituitary cells in culture. *In* "Neuroendocrine Molecular Biology" (G. Fink, A. J. Harmar, and K. W. McKerns, eds.), pp. 309–323. Plenum, New York.

Kovács, M., Mezö, I., Teplán, I., Hollósi, M., Kajtár, J., and Flerkó, B. (1993). New GABA-containing analogues of human growth hormone releasing hormone (1–30)-amide. II. Detailed *in vivo* biological examinations. *J. Endocrinol. Invest.* **16**(10), 799–805.

Krulich, L. (1979). Central neurotransmitters and the secretion of prolactin, GH, LH and TSH. *Annu. Rev. Physiol.* **41,** 603–615.

Krulich, L., Guijada, M., Wheaton, J., Illner, P., and McCann, S. M. (1977). Localization of hypophysiotropic neurohormones by assay of sections from various brain areas. *Fed. Proc. Fed. Am. Soc. Exp. Biol.* **36,** 1953–1959.

Kuchel, P. (1993). Membrane transport proteins: Molecular mechanisms. *Biol. Membr. Proc. Annu. Gen. Meet. Aust. Acad. Sci.* Canberra, pp. 33–49.

Kuo, C. C., and Bean, B. P. (1993). G-protein modulation of ion permeation through N-type calcium channels. *Nature (London)* **365,** 258–262.

Labrie, F., Borgeat, P., Godbout, M., Barden, N., Brauileu, M., Lagacé, L., Massicotte, J., and Vailleux, R. (1980). Mechanisms of action of hypothalamic hormones and interaction with sex steroids in the anterior pituitary gland. *In* "Synthesis and Release of Adenohypophyseal Hormones" (M. Jutisz and K. W. McKerns, eds.), pp. 415–439. Plenum, New York.

Land, H., Schulz, G., Schmale, H., and Richter, D. (1982). Nucleotide sequence of cloned cDNA coding the bovine arginine vasopressin-neurophysin II precursor. *Nature (London)* **295,** 299–303.

Laudon, M., Grossman, D. A., and Ben Jonathan, N. (1990). Prolactin releasing factor: Cellular origin in the intermediate lobe of pituitary. *Endocrinology (Baltimore)* **126,** 3185.

Lechan, R. M., Wu, P., Jackson, D., Wolf, H., Cooperman, S., Mandel, G., and Goodman, R. H. (1986). Thyrotropin-releasing hormone precursors: Characterization in rat brain. *Science* **231,** 159–161.

Lederis, K. P., Okawara, J., Richter, D., and Morley, S. D. (1990). Evolutionary aspects of corticotropin releasing hormone. *In* "Progress in Comparative Endocrinology" (A. Epple, C. G. Scanes, and M. H. Stretson, eds.), pp. 462–470. Wiley-Liss, New York.

Leenders, H. J., de Konig, H. P., Ponten, S. P., Jenks, B. G., and Roubos, E. W. (1992). Differential effects of coexisting regulatory factors dopamine, GABA and NPY on secretion of α-MSH from melanotrope cells of amphibian *Xenopus laevis. Conf. Eur. Comp. Endocrinol., 16th,* Padova, p. 137.

Lengvari, I., Liposits, Z., Vigh, S., Schally, A. V., and Flerkó, B. (1985). The origin and ultrastructural characteristics of corticotropin-releasing factor (CRF)-immunoreactive nerve fibers in the posterior pituitary of the rat. *Cell Tissue Res.* **240,** 467–471.

Le Roth, D., Liotta, A. S., Roth, J., Shiloach, J., Lewis, M. E., Pert, C. B., and Krieger, D. T. (1982). Corticotropin and endorphin-like materials are native to unicellular organisms. *Proc. Natl. Acad. Sci. U.S.A.* **79,** 2086.

Levi, G., Duband, J. L., and Thiery, J. P. (1990). Mode of cell migration in the vertebrate embryo. *Int. Rev. Cytol.* **123,** 201–252.

Li, C. H. (1972). Hormones of the adenohypophysis. *Proc. Am. Physiol. Soc.* **116**(5), 365–382.

Li, C. H., Dixon, J. S., Lo, T. B., Schmidt, K. D., and Pankov, Y. A. (1970). Study on pituitary lactogenic hormone. XXX. The primary structure of the sheep hormone. *Arch. Biochem. Biophys.* **141,** 705–737.

Lightman, S. I. (1994). Gene regulation in the PVN. *Neuroendocrinology* **60,** Suppl. 1, S11.1

Lincoln, D. W., and Russell, J. A. (1986). Oxytocin and vasopressin secretion: New perspectives. *In* "Neuroendocrine Molecular Biology" (G. Fink, A. J. Harmar, and K. W. McKerns, eds.), pp. 185–209. Plenum, New York.

Lintner, K., Toma, F., Piriou, F., Fromageot, P., and Formandjian, S. (1982). Studies on the conformation of neuropeptides. *In* "Hormonally Active Brain Peptides: Structure and Function" (K. W. McKerns and V. Pantić, eds.), pp. 45–69. Plenum, New York.

Liposits, Z., Lengvari, I., Vigh, S., Schally, A. V., and Flerkó, B. (1983). Immunohistological detection of degenerating CRF-immunoreactive nerve fibers in the median eminence after lesion of paraventricular nucleus of the rat. A light and electron microscopical study. *Peptides (N.Y.)* **4**, 941.

Liposits, Z., Uhr, R. M., Harrison, R. W., Gibbs, F. P., Paull, W. K., and Bohn, M. C. (1987). Ultrastructural localization of glucocorticoid receptor (GR) in hypothalamic paraventricular neurons synthesizing corticotropin releasing factor (CRF). *Histochemistry* **87**, 407–412.

Lowry, P. J., and Scott, A. P. (1975). The evolution of vertebrate corticotropin and melanocyte stimulating hormone. *Gen. Comp. Endocrinol.* **26**, 1–16.

Mains, R. E, Eipper, B. A., and Ling, N. (1977). Common precursor to corticotropin and endorphins. *Proc. Natl. Acad. Sci. U.S.A.* **74**, 3014–3018.

Mann, P. L. (1988). Membrane oligosacharides: Structure and function during differentiation. *Int. Rev. Cytol.* **112**, 67–96.

Manolov, S., and Ovtscharoff, W. (1982). Structure and cytochemistry of the chemical synapsis. *Int. Rev. Cytol.* **77**, 243–284.

Martens, G. J. M. (1992). Molecular biology of G-protein-coupled receptors. *Conf. Eur. Comp. Endocrinol. 16th,* Padova, p. 26.

Martin, J. B., Durand, D., Gurd, W., Failla, G., Audet, J., and Braseau, P. (1978). Neuropharmacological regulation of episodic growth hormones and prolactin secretion in the rat. *Endocrinology (Baltimore)* **102**(1), 106.

Martinović, P. N., and Pavić, B. (1960). Functional pituitary transplants in rats. *Nature (London)* **185**, 155–156.

Martinović, P. N., Živković, N., and Pavić, B. (1966). The response of the ovary of the rat heterotropically placed anterior pituitary transplants. *Gen. Comp. Endocrinol.* **7**, 215–223.

Martinović, P. N., Ivanišević, O., and Martinović, J. V. (1968a). Induction of hyperluteinization and precocious opening of the vagina in rats with transverse cut in the hypothalamus made shortly after birth. *Nature (London)* **217**, 866–867.

Martinović, P. N., Ivanišević, O. K., and Martinović, J. V. (1968b). The effect on the onset of puberty of whole body irradiation of infant female rats with and without hypothalamic lesions. *Experientia* **21**, 839–840.

Martinović, P. N., Pantić, V., Žgurić, M., and Ivanišević, O. (1969). Etude de l'hypothalamus et de l'hypophyse de rates avec hyperluteinisation provoque par lesion hypothalamique transversale. *Reun. Assoc. Anat.* **54**, 34–35.

Martinović, P. N., Ivanišević, O., Žgurić, M., and Pantić, V. (1970). Hyperluteinisation in rat with transverse cut and bilateral punctures in the hypothalamus made shortly after birth. *Acta Anat.* **75**, 141.

Martinović, P. N., Pantić, V., Žgurić, M., Martinović, J., and Ivanišević, O. (1977). Precocious opening of the vagina, accompanied in some cases by ovulation, induced in rats by mechanically produced lesions in the region of the pituitary stalk. *Arch. Biol. Nauka* **27**(3–4), 123–132.

Martinović, P. N., Milovanović, O., Hristić, M., Mušicki, B., and Pantić, V. (1982). Ultra structure of prolactin and gonadotropic cells in rats with lesions in caudal hypothalamus. *Biol. Cell* **45**, 18–23.

Maurer, R. A., Stone, R. T., and Gorski, J. (1978). The biosynthesis of prolactin. *In* "Structure and Function of the Gonadotropins" (K. W. McKerns, ed.), pp. 213–234. Plenum, New York.

McCann, S. M., Snyder, G. D., Ojeda, S. R., Lumpkin, M. D., Ottlecs, A., and Samson, W. K. (1984). Role of peptides in the control of gonadotropin secretion. *In* "Hormonal Control of the Hypothalamo-Pituitary-Gonadal Axis" (K. W. McKerns and Z. Naor, eds.), pp. 3–25. Plenum, New York.

McCann, S. M., Samson, W. K., Agulla, M. C., Bedran, D., Castro, J., Ono, N., Lumpkin,

M. D., and Khorram, O. (1986). The role of brain peptides in the control of anterior pituitary hormone secretion. In "Neuroendocrine Molecular Biology" (G. Fink, A. J. Harmar, and K. W. McKerns, eds.), pp. 101–111. Plenum, New York.

Meininger, V., and Binet, A. (1989). Characteristics of microtubules at the different stages of neuronal differentiation and maturation. Int. Rev. Cytol. 114, 21–79.

Mena, F., Enjalbert, A., Garbonell, L., Priam, M., and Kordon, C. (1976). Effect of suckling on plasma prolactin and hypothalamic monoamine levels in the rat. Endocrinology (Baltimore) 99, 445–451.

Merchenthaler, I., Göres, T., Sétáló, G., Petrusz, P., and Flerkó, B. (1984). Gonadotropin-releasing hormone (GnRH) neurons and pathways in the rat brain. Cell Tissue Res. 237, 15–29.

Merchenthaler, I., Cuiller, M. D., Negro-Vilar, A., Petrusy, P., and Flerkó, B. (1988). The pre-LH RH system of the rat brain. Effect of changes in the endocrine background. Brain Res. 20, 713–720.

Merchenthaler, I., Sétáló, G., Petrusz, P., Negro-Vilar, A., and Flerkó, B. (1989a). Identification of hypophysiotropic luteinizing hormone-releasing hormone (LH RH) neurons by combined retrograde labeling and immunocytochemistry. Exp. Clin. Endocrinol. 94(1/2), 133–140.

Merchenthaler, I., Sétáló, G., Csontos, C., Petrusz, P., Flerkó, B., and Negro-Vilar, A. (1989b). Combined retrograde tracing and immunocytochemical identification of luteinizing hormone-releasing hormone and somatostatin-containing neurons projecting to the median eminence of the rat. Endocrinology (Baltimore) 125,(6), 2812–2821.

Mescher, A. L., and Munaim, S. I. (1988). Transferrin and the growth-promoting effect of nerves. Int. Rev. Cytol. 110, 1–26.

Mess, B., and Józsa, R. (1989). Localization of different releasing hormone immunoreactive (LHRH) neurons and their projections in central nervous system. Endocrinol. Exp. 23, 305–320.

Meusy-Dessolle, N. (1975). Variation quantitative de la testosterone plasmatique chez le porc male, de la naissance à l'age adult. C. R. Hebd. Seances Acad. Sci. 281, 1875–1878.

Mezö, I., Kovaács, M., Szöke, B., Szabó, E. Y., Horváth, J., Makara, G. B., Rappay, Gy., Tamás, J., and Teplán, I. (1993). New GABA-containing analogues of human growth hormone-releasing hormone (1–30)-amide. I. Synthesis and in vitro biological activity. J. Endocrinol. Invest. 16(10,) 793–798.

Miholjčić, B., Radić, Lj., and Pantić, V. (1986). Thymus sensitivity to hydrocortisone. Acta Vet. (Belgrade) 36(5–6), 277–287.

Mikami, S. (1986). Immunocytochemistry of the avian hypothalamus and adenohypophysis. Int. Rev. Cytol. 103, 189–248.

Millar, R. P., and King, J. A. (1987). Structural and functional evolution of gonadotropin-releasing hormone. Int. Rev. Cytol. 106, 149–182.

Miller, R. A. (1991). Aging and immune function. Int. Rev. Cytol. 124, 187–215.

Milovanović, O., Pantić, V., and Hristić, M. (1988). The posterior hypothalamus and the hyperluteinization of the ovary. Bull.—Acad. Serbe Sci. Arts, Cl. Sci. Nat. Math. 30, 47–57.

Mitsuma, T., Hirooka, Y., Yuasa, K., and Nogimori. (1990). Effect of thyroid hormone, TRH and TSH on pro-TRH concentration in various organs of rats. Endocrinol. Exp. 24(4), 395–402.

Moss, R. L. (1979). Actions of hypothalamic-hypophysiotropic hormones on the brain. Annu. Rev. Physiol. 41, 617–631.

Murai, I., and Ben Jonathan, N. (1990). Acute stimulation of prolactin release by estradiol: Mediation by the posterior pituitary. Endocrinology (Baltimore) 126, 3179–3184.

Nakai, Y., Plant, T. M., Hess, D. L., Keogh, E. J., and Knobil, E. (1978). On the sites of negative and positive feedback actions of estradiol in the control of gonadotropin secretion in the rhesus monkey. *Endocrinology (Baltimore)* **101,** 1008.

Nakane, P. K. (1970). Classification of anterior pituitary cell types with immunoenzyme histochemistry. *J. Histochem. Cytochem.* **18,** 9–20.

Nakanishi, S., Inone, A., Kita, T., Nakamura, M., Chang, A. C. Y., Cohen, S. N., and Numa, S. (1979). Nucleotide sequence of cloned cDNA for bovine corticotropin-$\beta$-lipotropin precursor. *Nature (London)* **278,** 423–427.

Naor, Z., and Childs, G. V. (1986). Binding and activation of gonadotropin-releasing hormone. Receptors in pituitary and gonadal cells. *Int. Rev. Cytol.* **103,** 147–187.

Naumenko, E. V. (1984). Role of brain noradrenaline in the effects of pre- and early postnatal stress on the adrenocortical function in adults. *In* "Neuroendocrine Correlates of Stress" (K. W. McKerns and V. Pantić, eds.), pp. 63–80. Plenum, New York.

Nemeskeri, A., and Halasz, B. (1989). Cultured fetal rat pituitary cells kept in synthetic medium are able to initiate synthesis of trophic hormones. *Cell Tissue Res.* **255,** 645–650.

Nemeskeri, A., Sétáló, G., Kacsoh, B., and Halasz, B. (1990). Fetal pituitary graft is capable of initiating hormone synthesis in median eminence removed from adult rat. *Endocrinol. Exp.* **24**(3), 283–292.

Newgreen, D. F., and Erickson, C. A. (1986). The migration of neural crest cells. *Int. Rev. Cytol.* **103,** 89–145.

Nickerson, P., Hristić, M., and Pantić, V. (1978). Influence on maternal adrenalectomy on the adrenal glands in neonatal rats. *Cell Tissue Res.* **189,** 277–286.

Nikolics, K., Kéri, G., Szöke, B., Horvaáthe, A., and Teplán, I. (1982). Biodegradation of luteinizing hormone-releasing hormone. *In* "Hormonally Active Brain Peptides: Structure and Function" (K. W. McKerns and V. Pantić, eds.), pp. 427–443. Plenum, New York.

Oliver, C., Eskay, R. L., Ben-Jonathan, N., and Porter, J. C. (1974). Distribution and concentration of TRH in the rat brain. *Endocrinology (Baltimore)* **96**(2), 540–546.

Osamura, R. Y., and Watanabe, K. (1985). Histogenesis of the cells of the anterior and intermediate lobes of human pituitary glands: Immunohistochemical studies. *Int. Rev. Cytol.* **95,** 103–129.

Osborne, N. N., and Neuhoff, V. (1980). Identified serotonin neurons. *Int. Rev. Cytol.* **67,** 259–290.

Palkovits, M. (1982). Recent data on neuropeptide mapping in the central nervous system. *In* "Hormonally Active Brain Peptides: Structure and Function" (K. W. McKerns and V. Pantić, eds.), pp. 279–306. Plenum, New York.

Palkovits, M.,, Graf, I., Herman, I., Boyvendég, J., Ács, Y., and Láng, T. (1978). Regional distribution of enkefalins, endorphins and ACTH in the central nervous system of rats determined by radioimmunoassay. *In* "Endorphins 78" (I. Gráf, M. Palkovits, and A. Z. Rónai, eds.), pp. 187–195. Akadémiai Kiadó, Budapest.

Palme, K. (1992). Molecular analysis of signaling elements. Relevance of eukaryotic signal transduction models. *Int. Rev. Cytol.* **132,** 223–283.

Pantić, V. (1974a). The cytophysiology of thyroid cells. *Int. Rev. Cytol.* **36,** 153–243.

Pantić, V. (1974b). Gonadal steroids and hypothalamo-pituitary-gonadal axis. *INSERM Symp.* **32,** 97–118.

Pantić, V. (1975). The specificity of pituitary cells and regulation of their activities. *Int. Rev. Cytol.* **40,** 153–195.

Pantić, V. (1980). Adenohypophyseal cell specificities and gonadal steriods. *In* "Synthesis and Release of Adenohypophyseal Hormones" (M. J. Jutisz and K. W. McKerns, eds.), pp. 336–362. Plenum, New York.

Pantić, V. (1981). Sensitivity of pituitary gonadotropic cells and gonads to hormones. *In* "Reproductive Processes and Contraception" (K. W. McKerns, ed.), pp. 47–89. Plenum, New York.

Pantić, V. (1982). Genesis and properties of pituitary ACTH, MSH, prolactin and growth hormone producing cells. *In* "Hormonally Active Brain Peptides: Structure and Function" (K. W. McKerns and V. Pantić, eds.), pp. 503–536. Plenum, New York.

Pantić, V. (1984a). Prolactin target cells in rats treated with gonadal steroids. *In* "Regulation of Target Cell Responsiveness" (K. W. McKerns, A. Aakvaag, and V. Hansson, eds.), Vol. 2, pp. 283–295. Plenum, New York.

Pantić, V. (1984b). Reaction of neuroendocrine cells to stress. *In* "Neuroendocrine Correlates of Stress" (K. W. McKerns and V. Pantić, eds.), pp. 289–323. Plenum, New York.

Pantić, V. (1988). Role of gonadal steroids in regulation of reproductive and productive properties of the animals. *Glas. Acad. Serbe Sci. Arts (Belgrade)* **52,** 39–89.

Pantić, V. (1990a). Gonadal steroids as modulators of genetically programmed development of neuroendocrine cells involved in regulation of growth and reproduction. *Bull.—Acad. Serbe Sci. Arts, Cl. Sci. Nat. Math.* **32,** 47–66.

Pantić, V. (1990b). Brain neurons, pituitary and their target cells involved in regulation of reproduction and gonadal steroids. Scientific Symposium in Honor of Claude A. Villee "Regulatory Processes in Biology," pp. 17–37. Saunders College Publishing C. L. Fond, D. B. Villee, Philadelphia.

Pantić, V. (1992). Brain neurons, pituitary and their target cells reaction to stress. *Mat. Srpska, Dep. Nat. Sci.* pp. 5–29.

Pantić, V., and Ekholm, R. (1963). Effect of thyrotropin on *in vivo* iodine binding in thyroid subcellular fractions. *Nature (London)* **198,** 903–905.

Pantić, V., and Genbačev, O. (1969). Ultrastructure of pituitary lactotropic cells of oestrogen treated male rats. *Z. Zellforsch. Mikrosk. Anat.* **95,** 280–289.

Pantić, V., and Genbačev, O. (1970). Ultrastructure of pituitary luteotropic (LTH) and somatotropic (STH) cells of rats neonatally treated with oestrogen. *Electron Microsc., Proc. Int. Congr., 7th,* Grenoble, *1970,* pp. 560–561.

Pantić, V., Genbačev, O. (1972). Pituitaries of rats neonatally treated with oestrogen. I. Luteotropic and somatotropic cells and hormones content. *Z. Zellforsch. Mikrosk. Anat.* **126,** 41–52.

Pantić, V., and Gledić, D. (1977a). Reaction of pituitary gonadotropic cells and testes to testosterone propionate (TP). *Bull.—Acad. Serbe Sci. Arts, Cl. Sci. Nat. Math.* **15,** 131–146.

Pantić, V., and Gledić, D. (1977b). Longterm effects of gonadal steroids on pig pituitary gonadotropic cells and testes. *Bull.—Acad. Serbe Sci. Arts, Cl. Sci. Nat. Math.* **16,** 91–109.

Pantić, V., and Gledić, D. (1978). Reaction of pituitary gonadotropic and germ cells of male rats and pigs neonatally treated with estrogen. *Gen. Comp. Endocrinol.* **34**(1), 54 (abstr.)

Pantić, V., and Hristić, M. (1975). ACTH cells of rats after head irradiation. *Int. J. Radiat. Biol. Relat. Stud. Phys., Chem. Med.* **28**(1), 53–60.

Pantić, V., and Kalušević, S. (1974). Thyroid of hypophysectomized rats after administration of thyrotropic hormone. *Acta Anat.* **90,** 569–580.

Pantić, V., and Lovren, M. (1977). Neurosecretory cells and gonadal activities in *Serranus scriba. Arh. Sci. Biol.* **27**(1–2), 9–13.

Pantić, V., and Lovren, M. (1978). The effect of female gonadal steroids on carp pituitary gonadotropic cells and oogenesis. *Folia Anat. Iugosl.* (Sarajevo) **7,** 25–34.

Pantić, V., and Sekulić, M. (1975). Neurosecretory cells in *Torpedo ocelata. Arh. Biol. Nauka* **26**(3–4), 109–114.

Pantić, V., and Sekulić, M. (1978a). The reaction on thyroid follicular and pituitary TSH cells of *Carassius carassius* to glycoprotein hormones. *Acta. Vet. (Belgrade)* **28**(1), 17–33.

Pantić, V., and Sekulić, M. (1978b). Neurosecretory and pituitary gonadotropic cells in *Torpedo ocelata* and *Acipenser ruthenus*. *Arh. Biol. Nauka* **30**(1–4), 21–27.

Pantić, V., and Sekulić, M. (1978c). Pituitary prolactin and somatotropic cells of teleostea treated with gonadal steroids or choriogonadotropin. *Acta. Vet. (Belgrade)* **28**, 71–80.

Pantić, V., and Šimić, M. (1974). Pars intermedia of deer pituitary. *Arh. Biol. Nauka* **26**(1–2), 15–18.

Pantić, V., and Šimić, M. (1977a). Sensitivity of the pituitary pars intermedia to castration or gonadal steroids. *Bull.—Acad. Serbe Sci. Arts, Cl. Sci. Nat. Math.* **11**, 67–80.

Pantić, V., and Šimić, M. (1977b). Effect of gonadal steroids on pituitary pars intermedia cells of some teleostea and rat. *Bull.—Acad. Serbe Sci. Arts, Cl. Sci. Nat. Math.* **16**, 23–40.

Pantić, V., and Šimić, M. (1978). Effect of gonadal steroids on pars intermedia cells of some teleostei, the rat and the pig. *Gen. Comp. Endocrinol.* **34**(1), 27 (abstr.).

Pantić, V., and Šimić, M. (1980). Development of pars intermedia and neurohypophysis in oestrogen treated rats. *Gen. Comp. Endocrinol.* **40**, 340.

Pantić, V., and Škaro, A. (1974). Pituitary cells of rooster and hens treated with a single dose of estrogen during embryogenesis or after hatching. *Cytologie* **9**(1), 72–83.

Pantić, V., Ožegović, B., and Milković, S. (1971a). Ultrastructure of transplantable pituitary tumor cells producing luteotropic and adrenocorticotropic hormones. *J. Microsc. (Paris)* **12**(2), 225–232.

Pantić, V., Genbačev, O., Milković, S., and Ožegović, B. (1971b). Pituitaries of rats bearing transplantable MtT mammotropic tumor. *J. Microsc. (Paris)* **12**(3), 405–415.

Pantić, V., Šijački, N., and Kolarić, S. (1978). The role of gonadal steroids in the regulation of behaviour and productive performance of pigs. *Acta Vet. (Belgrade)* **28**(1), 31–44.

Pantić, V., Gledić, D., and Martinović, J. V. (1980). Pituitary gonadotropic cells, serum concentration of gonadotropic and testicular hormones in adult rats neonatally treated with a single dose of estradiol. *Acta Vet. (Belgrade)* **30**(3–4), 101–114.

Pantić, V., Gledić, D., and Martinović, J. V. (1982). Testes, accessory sex glands and serum concentration of gonadotropins in old rats treated with gonadal steroids. *Bull.—Acad. Serbe Sci. Arts, Cl. Sci. Nat. Math.* **23**, 23–36.

Pavel, S., Goldstein, R., Ghinea, E., and Calb, M. (1977). Chromatographic evidence for vasotocin biosynthesis by cultured pineal ependymal cells from rat fetuses. *Endocrinology (Baltimore)* **100**, 205–208.

Pavić, D., Živković, M., Pantić, V., and Martinović, P. (1977). Glycoprotein synthesizing cells in the anterior pituitaries transplanted into hypophysectomized male rats. *Arh. Biol. Nauka* **27**(1–2), 1–8.

Pavlović, M., and Pantić, V. (1973). Nucleus preopticus and nucleus lateralis tuberis in Teleostea *Alburnus albidus* and *Alosa falax* in various stages of sexual cycle. *Arh. Biol. Nauka* **25**(1–2), 9–16.

Pavlović, M., and Pantić, V. (1975). The adenohypophysis in the Teleostea *Alburnus albidus* and *Alosa falax* in different phases of sexual cycle. *Acta Vet. (Belgrade)* **25**(4), 163–178.

Pelletier, G., and Leclerc, R. (1979). Immunohistochemical localisation of adrenocorticotropin in the rat brain. *Endocrinology (Baltimore)* **104**, 1426–1433.

Pelletier, G., Liao, N., Follea, N., and Govindan, M. V. (1988). Distribution of estrogen receptors in the rat pituitary as studied by in situ hybridization. *Mol. Cell. Endocrinol.* **56**, 29–33.

Perry, V. H., and Gordon, S. (1991). Macrophages and the nervous system. *Int. Rev. Cytol.* **125**, 203–237.

Pfaff, D. W., and Keiner, H. (1973). Atlas of estradiol concentrating cells in the central nervous system of the female rat. *J. Comp. Neurol.* **151,** 121–158.

Plattner, H. (1989). Regulation of membrane fusion during exocytosis. *Int. Rev. Cytol.* **119,** 197–286.

Plickinger, R. A. (1982). Evolutionary aspects of cell differentiation. *Int. Rev. Cytol.* **75,** 229–241.

Plotsky, P. M., Cuningham, F. T., and Widmaier, E. P. (1989). Catecholaminergic modulation of corticotropin-releasing factor and adrenocorticotropin secretion. *Endocr. Rev.* **10**(4), 437–456.

Polenov, A. L., Kornienko, G. G., and Belenky, M. A. (1986). The hypothalamo-hypophyseal system of the wild carp, *Cyprinus carpio* L. III. Changes in the anterior and posterior neurohypophysis during spawning. *Z. Mikrosk.-Anat.-Forsch.* **100**(6), 990–1106.

Puck, T. T., and Krystosek, A. (1992). Role of the cytoskeleton in genome regulation and cancer. *Int. Rev. Cytol.* **132,** 75–108.

Radić, Lj., and Pantić, V. (1987). Thymus of adult rats after the steroid treatment in neonatal period. *Gen. Comp. Endocrinol.* Abstr. 27.

Raeside, J. T., and Sigman, D. M. (1975). Testosterone levels in early fetal testes of domestic pigs. *Biol. Reprod.* **13,** 318–328.

Raynaud, J. P., Mercier-Bodard, C., and Baulieu, E. E. (1971). Rat estradiol binding plasma protein (EBP). *Steroids* **18,** 767–788.

Reisman, G., and Field, P. M. (1973). Sexual dimorphism in the neuropil of the preoptic area of the rat and its dependence on neonatal androgen. *Brain Res.* **54,** 1–29.

Rhodes, C. H., Morell, J. I., and Pfaff, D. W. (1981). Distribution of estrogen-concentrating, neurophysin-containing magnocellular neurons in rat hypothalamus as demonstrated by a technique combining steroid autoradiography and immunohystology in the same tissue. *Neuroendocrinology* **33,** 18–23.

Richter, D., Schmale, H., Ivel, R., and Schmidt, C. (1980). Hypothalamic mRNA directed synthesis of neuropeptides. Immunological identification of precursors to neurophysin II arginine vasopressin and to neurophysin I oxytocin. *In* "Biosynthesis, Modification and Processing of Cellular and Viral Polyprotein" (J. H. Koch and D. Richter, eds.), pp. 43–66. Academic Press, New York.

Ringstrom, S. J., and Schwartz, N. B. (1984). Examination of prolactin and pituitary-adrenal axis components as intervening variable in the adrenalectomy-induced inhibition of gonadotropin response to castration. *Endocrinology (Baltimore)* **111**(3), 880–887.

Rivier, J., Spiess, J., Thorner, M., and Vale, M. W. (1982). Characterization of growth hormone releasing factor from a pancreatic islet tumor. *Nature (London)* **300,** 276.

Roberts, J. L., Wilcox, J. N., and Blum, M. (1986). The regulation of proopiomelanocortin gene expression by estrogen in the rat hypothalamus. *In* "Neuroendocrine Molecular Biology" (G. Fink, A. J. Harmar, and K. W. McKerns, eds.), pp. 261–270. Plenum, New York.

Rodriguez, E. M., Oksche, A., Hein, S., and Yulis, C. R. (1992). Cell biology of the subcomissural organ. *Int. Rev. Cytol.* **135,** 39–121.

Rosselin, G., Rotsytejn, W., Laburthe, M., and Dubois, P. M. (1982). Is VIP a neuroregulator or a hormone? *In* "Hormonally Active Brain Peptides: Structure and Function" (K. W. McKerns and V. Pantić, eds.), pp. 367–395. Plenum, New York.

Roth, J., LeRoith, D., Shiloach, J., and Rabinowitz, C. (1984). Hormones and other messenger molecules: An approach to unity. *In* "Hormonal Control of the Hypothalamo-Pituitary-Gonadal Axis" (K. W. McKerns and Z. Naor, eds.), pp. 71–87. Plenum, New York.

Roubos, E. W., Jenks, B. G., and Martens, G. J. M. (1992). Cellular and molecular aspects of signal transduction by the pars intermedia of *Xenopus laevis*. *Conf. Eur. Comp. Endocrinol. 16th,* Padova, p. 31.

Sage, M., and Bern, H. A. (1971). Cytophysiology of the teleost pituitary. *Int. Rev. Cytol.* **31,** 339–376.

Sar, M., and Stumpf, W. E. (1980). Simultaneous localisation of oxytocin and neurophysin I or arginine vasopressin in hypothalamic neurons demonstrated by a combined technique of dry-mount autoradiography and immunohistochemistry. *Neurosci. Lett.* **17,** 179–184.

Sar, M., Culler, M. D., McGimsey, W. C., and Negro-Vilar, A. (1987). Immunocytochemical localization of the gonadotropin-releasing hormone-associated peptide of the LHRH precursor. *Neuroendocrinology* **45,** 172–175.

Schally, A. V., Leipold, B., and Richter, D. (1973). Hypothalamic regulatory hormone. At least nine substances from the hypothalamus control the secretion of pituitary hormones. *Science* **179,** 341.

Schally, A. V., Kastin, A. J., and Coy, D. H. (1976). LH-releasing hormone and its analogues: Recent basic and clinical investigations. *Int. Rev. Fertil.* **21,** 1–30.

Schally, A. V., Kastin, A. J., and Arimura, A. (1977). Hypothalamic hormones: The link between brain and body. *Am. Sci.* **65**(6), 712–715.

Scharrer, B. (1967). The neurosecretory neuron in neuronendocrine regulatory mechanisms. *Am. Zool.* **7,** 161–169.

Scharrer, B. (1968). Ultrastructural study of sites of release of neurosecretory material in blattarian insects. *Z. Zellforsch. Mikrosk. Anat.* **89,** 1–16.

Scharrer, B. (1974). The spectrum of neuroendocrine communication. *In* "Hypothalamus and Hormones," Int. Smp. Calgary, pp. 8–16. Karger, Basel.

Scharrer, E. (1952). The general significance of the neurosecretory cell. *Scientia* **87,** 176–182.

Scharrer, E. (1966). Principles of neuroendocrine integration. *Res. Publ. Assoc. Res. Nerv. Ment. Dis.* **43,** 1–35.

Schmale, H., and Richter, D. (1981). Tryptic release of authentic arginine vasopressin from a composite arginine avasopressin neurophysin II precursor. *Neuropeptides (Edinburgh)* **2,** 47–52.

Schmale, H., Leipold, B., and Richter, D. (1979). Cell free translation of bovine hypothalamic mRNA: Synthesis and processing of the preproneurophysins I and II. *FEBS Lett.* **108,** 311–316.

Schulz, R. W., van der Sanden, M. C. A., Bosma, P. T., van Dijk, W, Janssen, J., Zandbergen, M. A., Bogerd, J., Peute, J., and Goos, H. J. T. (1992). Teleost GnRHs and gonadotropins with special reference to the African catfish. *Conf. Eur. Comp. Endocrinol. 16th,* Padova, p. 33.

Schwanzel-Fukida, M. (1994). Ontogeny and migration of LHRH neurons in human embryos. *Neuroendocrinology* **60,** Suppl. 1, S16.4, p. 22.

Seeburg, P. H., and Adelman, J. P. (1984). Characterization of cDNA for precursor of human luteinising hormone releasing hormone. *Nature (London)* **311,** 666–668.

Sekulić, M., and Pantić, V. (1977). Neurosecretory cells in clasmobranchia and acipenserida and the sexual cycle. *Acta Anat. Index* **99**(1), 113.

Sétáló, G. (1994). Qualitative and quantitative studies of extracramal GnRH-immunoreactive neurons during ontogenetic development. *Neuroendocrinology* **60,** Suppl. 1, S16.2, p. 22.

Sétáló, G., Flérkó, B., Arimura, A., and Schally, A. V. (1978). Brain cells as producers of releasing and inhibiting hormones. *Int. Rev. Cytol.* Suppl. 7, 1–52.

Sherwood, N. M. (1986). Gonadotropin-releasing hormone: Differentiation of structure and function during evolution. *In* "Neuroendocrine Molecular Biology" (G. Fink, A. J. Harmar, and K. W. McKerns, eds.), pp. 67–84. Plenum, New York.

Shine, J. (1993). Membrane receptors: Extracellular regulation. *Biol. Membr., Proc. Annu. Gen. Meet. Aust. Acad. Sci.* Canberra, pp. 51–62.

Shino, M., Williams, G., and Rennels, E. G. (1972). Ultrastructural observation of pituitary release of prolactin in the rat by suckling stimulus. *Endocrinology (Baltimore)* **90**(1), 176–187.

Siperstein, E. R. (1963). Identification of the adrenocorticotropin production cells in the rat by autoradiography. *J. Cell Biol.* **17**, 521–546.

Siperstein, E. R., and Miler, K. J. (1973). Hypertrophy of the ACTH-producing cell following adrenalectomy. A quantitative electron microscopic study. *Endocrinology (Baltimore)* **93**, 1257–1268.

Škaro, A., and Pantić, V. (1976). Gonadotropic and luteotropic cells in chicken treated with estrogen after hatching. *Gen. Comp. Endocrinol.* **28**, 283–291.

Sofroniev, M. A., and Weindl, A. (1978a). Extrahypothalamic neurophysin-containing perikarya, fiber pathways and fiber clusters in the rat brain. *Endocrinology (Baltimore)* **102**, 334–337.

Sofroniev, M. A., and Weindl, A. (1978b). Projections from the paraventricular vasopressin and neurophysin containing neurons of the suprachiasmatic nucleus. *Am. J. Anat.* **153**, 391.

Spona, J. (1974). LH-RH actions modulated by sex steroids. *Endocrinol. Exp.* **8**, 19–29.

Stark, E., and Makara, G. B. (1982). Stress, corticoliberin (CRF) and glucocorticoids in the regulations of ACTH release. *In* "Hormonally Active Brain Peptides: Structure and Function" (L. W. McKerns and V. Pantić, eds.), pp. 157–179. Plenum, New York.

Stefano, G. B. (1990). Involvement of opioid neuropeptide in the immune system of invertebrates. *Conf. Eur. Comp. Endocrinol. 15th,* Leuven, p. 28.

Stewart, J. M. (1982). The design of peptide hormone analogs. *Trends Pharmacol. Sci.* **3**(7), 300–303.

Stewart, J. M. (1984). ACTH neurons, stress and behavior A synthesis. *In* "Neuroendocrine Correlates of Stress" (K. W. McKerns and V. Pantić, eds.), pp. 239–268. Plenum, New York.

Stillman, A. M., Recht, L. D., Rosario, S. L., and Seif, S. M. (1977). The effects of adrenalectomy and glucocorticoid replacement on vasopressin and vasopressin-neurophysin in the zona externa of the median eminence of the rat. *Endocrinology (Baltimore)* **101**, 42–49.

Stošić, N., and Pantić, V. (1966). Cyclic changes in deer pituitary. *Jugosl. Physiol. Pharmacol. Acta (Belgrade)*(2–3), 231–237.

Stošić, N., and Pantić, V. (1967). Investigations of red-deer hypophysis. *Congr. Biol. Gibier, 7th,* Jelen (Beograd), pp. 81–86.

Stošić, N., and Pantić, V. (1973). The hypothalamo-pituitary system of intact and thyroidectomized rats treated with thyroxine. *Jugosl. Physiol. Pharmacol. Acta (Belgrade)* **9**(3), 335–348.

Stošić, N., and Pantić, V. (1977). The hypothalamo-pituitary system and testes of rats treated with thyroxine 2–13 days after partus. *Acta Anat.* **99**, 127.

Stošić, N., Pantić, V., Pavlović-Hournac, M., and Radivojević, D. (1969). Systeme hypothalamo-hypophysaire des rats thyréoidectomise's porteurs de greffes intraoculaires de thyroide. *Z. Zellforsch. Mikrosk. Anat.* **102**, 554–569.

Stumpf, W. (1968). Estradiol concentrating neurons: Topography in the hypothalamus by dry mount autoradiography. *Science* **162**, 1001–1003.

Stumpf, W., and Sar, M. (1976). Steroid hormone target sides in the brain: The differentiated distribution of estrogen, progesteron, androgen and glucocorticoid. *J. Steroid Biochem.* **7**, 1163–1170.

Szego, C. M. (1978). Parallels in the modes of action of peptide and steroid hormones: Membrane effects and cellular entry. *In* "Structure and Function of Gonadotropins" (K. W. McKerns, ed.), pp. 431–472. Plenum, New York.

Tainhof, R., Laurent, F., Brandt, J. W., Michielsen, C. P., Wismans, P. G., and Roubos, E. W. (1992). Magnocellular-suprachiasmatic complex and background adaptation in *Xenopus laevis. Conf. Eur. Comp. Endocrinol. 16th,* Padova, p. 95.

Takai, Y., Kaibuchi, K., Kikuchi, A., and Kawata, M. (1992). Small GTP-binding proteins. *Int. Rev. Cytol.* **133,** 187–230.

Tarasawa, E., Naonan, J. J., Nass, T. E., and Loose, M. D. (1984). Posterior hypothalamic lesions advance the onset of puberty in the female rhesus monkey. *Endocrinology (Baltimore)* **115,** 2241–2250.

Taylor, WS. J., Chae, H. Z., Rhee, S. G., and Exton, J. H. (1991). Activation of the isosyme of phospholipase C by subunits of the Gq class of G proteins. *Nature (London)* **350,** 516–518.

Thureson-Klein, A. K., and Klein, R. L. (1990). Exocytosis from neuronal large dense coated vesicles. *Int. Rev. Cytol.* **121,** 67–126.

Tixier-Vidal, A., and De Vitry, F. (1979). Hypothalamic neurons in cell culture. *Int. Rev. Cytol.* **58,** 291–331.

Tonon, M. C., Leboulenger, F., Delarue, C., Jegou, S., Fresel, J., Leroux, P., and Vaudry, H. (1980). TRH as MSH-releasing factor in the frog. *In* "Synthesis and Release of Adenohypophyseal Hormones" (M. Jutisz and K. W. McKerns, eds.), pp. 731–751. Plenum, New York.

Török, A., Vigh, S., Sétáló, G., Gledić, D., Pantić, V., and Flerkó, B. (1982). The effect of perinatal oestrogen treatment on the trophohormone secreting cells of the anterior pituitary of the rat. *Arh. Biol. Nauka* **33**(2–3), 319–329.

Tougard, C. (1980). Immunocytochemical identification of LH and FSH-secreting cells at the light and electron microscope levels. *In* "Synthesis and Release of Adenohypophyseal Hormones" (M. Jutisz and K. W. McKerns, eds.), pp. 15–37. Plenum, New York.

Turpen, C., Morris, M., and Knigge, K. M. (1978). Serum levels of prolactin, LH and LH RH after ablasion of the medial basal hypothalamus. *Horm. Res.* **9,** 73–82.

Ugrumov, M. V. (1991). Developing hypothalamus in differentiation of neurosecretory neurons and pathways for neurohormone transport. *Int. Rev. Cytol.* **129,** 207–267.

Vale, W., Spiess, J., Rivier, C., and Rivier, J. (1981). Characterization of a 41 residue ovine hypothalamic peptide that stimulates secretion of corticotropin and β-endorphin. *Science* **213,** 1394.

Vallarino, M., Feuilloley, M., Vandesande, F., and Vaudry, H. (1990). Galantin-like immuno reactivity in the brain of the cartilaginous fish, *Scyliorhinus canicula. Conf. Eur. Comp. Endocrinol., 15th,* Leuven, p. 7.

Vanden Broeck, J., Dupon, C., Vandesande, F., and De Loof, A. (1992). Molecular cloning of evolutionary conserved insect G-protein coupled receptor cDNAs. *Conf. Eur. Comp. Endocrinol., 16th,* Padova, p. 199.

Vandesande, F., and Dierickx, K. (1975). Identification of the vasopressin producing and of the oxytocin producing neurons in the hypothalamic magnocellular neurosecretory system of the rat. *Cell Tissue Res.* **164,** 153.

Van Deurs, B. (1980). Structural aspects of brain barriers, with special reference to the permeability of the cerebral endothelium and choroidal epithelium. *Int. Rev. Cytol.* **65,** 117–191.

Van Oordt, P. G. W. J. (1968). The analysis and identification of the hormone-producing cells of the adenohypophysis. *In* "Perspectives in Endocrinology. Hormones in the Lives of Lower Vertebrates" (E. J. W. Jorgensen and B. Barker, eds.), pp. 405–467. Academic Press, New York.

Vaudry, H. (1992). The melanotrophs: Pituitary hormones and neuronal transmitters. *Conf. Eur. Comp. Endocrinol., 16th,* Padova, p. 43.

Vigh, S., Sétáló, G., Török, A., Pantić, V., Flerkó, B., and Gledić, D. (1978). Deficiency of FSH and LH cells in rats treated with oestradiol in the early postnatal life. *Bull.—Acad. Serbe Sci. Arts, Cl. Sci. Nat. Math.* **17,** 121–127.

Vilchey-Martinez, J. A., Arimura, A., Debeljuk, L., and Schally, A. V. (1974). Biphasic effect of estradiol benzoate on the pituitary responsiveness to LH RH. *Endocrinology (Baltimore)* **94,** 1300–1303.

Vincent, G., and Labrie, F. (1982). Vasopressin potentiates cyclic and accumulation and ACTH release induced by corticotropin-releasing factor (CRF) in rat anterior pituitary cells in culture. *Endocrinology (Baltimore)* **111**(5), 1752–1754.

Vincent, J. D., and Liedo, P. M. (1992). The prolactin cells: A model for neuroendocrine integration. *Conf. Eur. Comp. Endocrinol., 16th,* Padova, p. 44.

Wallis, M. (1988a). Mechanism of action of growth hormones. *In* "Hormones and Their Action. Part II" (B. A. Cooke, R. J. B. King, and H. V. Van der Molen, eds.), pp. 265–294. Elsevier, Amsterdam.

Wallis, M. (1988b). Mechanism of action of prolactin. *In* "Hormones and Their Action. Part II" (B. A. Cooke, R. J. B. King, and H. V. Van der Molen, eds.), pp. 295–315. Elsevier, Amsterdam.

Watanabe, J. G. (1987). Failure of luteinizing hormone-releasing hormone (LH RH) to affect the differentiation of LH cells in the rat hypophysial primordium in serum-free culture. *Cell Tissue Res.* **250,** 35–42.

Weiner, R. I., and Ganong, W. F. (1978). Role of brain monoamines and histamine in regulation of anterior pituitary secretion. *Pharmacol. Rev.* **68,** 905.

Wendelaar Bonga, S. E., and Pang, P. K. T. (1991). Control of calcium regulating hormones in the vertebrates: Parathyroid hormone, calcitonin, prolactin and stanniocalcin. *Int. Rev. Cytol.* **128,** 139–213.

Wiche, G., Oberkanins, C., and Himmler, A. (1991). Molecular structure and function of microtubule associated proteins. *Int. Rev. Cytol.* **124,** 217–273.

Wolfson, B., Manning, R. W., Davis, L. G., Arentzen, R., and Baldino, F. (1985). Co-localization of corticotropin releasing factor and vasopressin mRNA in neurons after adrenalectomy. *Nature (London)* **315,** 59–61.

Wu, P., Lechan, R. M., and Jackson, M. D. (1987). Identification and characterization of thyrotropin-releasing hormone precursor peptides in rat brain. *Endocrinology (Baltimore)* **121,** 108–115.

Yates, F. E., Russell, S. M., Dallman, M. F., Hedge, G. A., and McCann, S. M. (1971). Potentiation by vasopressin of corticotropin release induced by corticotropin-releasing factor. *Endocrinology (Baltimore)* **88,** 3–15.

Young, W. S., Mezey, E., and Siegel, R. E. (1986). Quantitative in situ hybridization histochemistry reveals increased levels of corticotropin-releasing factor mRNA after adrenalectomy in rats. *Neurosci. Lett.* **70,** 198–203.

Zanetta, J. P., Kuchler, S., Lehmann, S., Badache, A., Maschke, S., Marschal, P., Dufourcq, P., and Vincendon, G. (1992). Cerebellar lectins. *Int. Rev. Cytol.* **135,** 123–154.

Žgurić, M., Pantić, V., and Bogojević, B. (1968). Specificity of deer hypothalamus. *Arh. Biol. Nauka* **20**(1–2), 15–18.

# Angiogenesis: Models and Modulators

Gillian W. Cockerill, Jennifer R. Gamble, and Mathew A. Vadas
Hanson Center for Cancer Research, Institute of Medical and Veterinary
Research, Adelaide 5000, South Australia, Australia

Angiogenesis *in vivo* is distinguished by four stages: subsequent to the
transduction of signals to differentiate, stage 1 is defined as an altered proteolytic
balance of the cell allowing it to digest through the surrounding matrix. These
committed cells then proliferate (stage 2), and migrate (stage 3) to form aligned
cords of cells. The final stage is the development of vessel patency (stage 4),
generated by a coalescing of intracellular vacuoles. Subsequently, these structures
anastamose and the initial flow of blood through the new vessel completes the
process. We present and discuss how the available models most closely represent
phases of *in vivo* angiogenesis. The enhancement of angiogenesis by hyaluronic
acid fragments, transforming growth factor $\beta$, tumor necrosis factor $\alpha$, angiogenin,
okadaic acid, fibroblast growth factor, interleukin 8, vascular endothelial growth
factor, haptoglobin, and gangliosides, and the inhibition of the process by
hyaluronic acid, estrogen metabolites, genestein, heparin, cyclosporin A, placental
RNase inhibitor, steroids, collagen synthesis inhibitors, thrombospondin, fumagellin,
and protamine are also discussed.

KEY WORDS: Angiogenesis, Cell proliferation, Cell migration, Proteolytic balance,
Collagen synthesis inhibitors.

## I. Introduction

Endothelial cells are derived from pluripotent mesodermal precursors
during the process of vasculogenesis, which occurs in the extraembryonic

mesoderm of the yolk sac in both avian and mammalian embryos, and in selected organ systems (Risau and Lemmon, 1988; Pardanaud *et al.*, 1989). Angiogenesis is the development of the complex network of blood vessels that occurs following vasculogenesis, when endothelial cells proliferate and migrate throughout the embryo. The process of angiogenesis is important not only during embryological development, but during a variety of normal and pathological conditions in the adult, including ovulation, implantation, during mammary gland changes associated with lactation, bone formation, inflammation wound repair ( Jakob *et al.*, 1977; Gospodarowicz and Thakral, 1978; Nomura *et al.*, 1989; Brannström *et al.*, 1988; Knighton *et al.*, 1990), and tumor growth (Folkman, 1985; Furcht, 1986).

Light and electron microscopy studies, combined with *in situ* hybridization, of both the normal genesis of vessels during embryological development and during tumor angiogenesis have demonstrated a number of discrete events that occur during antiogenesis (Schoefl, 1963; Yamagami, 1970). Following a stimulus for neovascularization endothelial cells change their morphology and begin to degrade their surrounding basement membrane (Ausprunk and Folkman, 1977; Moscatelli *et al.*, 1980; Gross *et al.*, 1983). These "leading" cells must modulate the expression of their proteases to allow degradation of existing extracellular matrix (ECM) components, and the migrating cells following this front must be supported by the appropriate ECM to allow for their proliferation, migration, and differentiation into vascular tubes. This initial migration and proliferation is in a fibronectin-rich ECM, and during the later stages of angiogenesis, when cords of endothelial cells align, the cells express laminin, a matrix component associated with vascular maturation (Risau and Lemmon, 1988). Finally, the generation of vessel patency is achieved by the coalescing of intra- and intercellular vacuoles (Sabin, 1920; Lewis, 1925; Clark and Clark, 1937).

The dependence of tumor growth on angiogenesis is well documented (Folkman, 1990). This relationship has been demonstrated for many types of tumor, invasive breast cancer (Weidner *et al.*, 1991), non-small cell lung cancer (Macchiarini *et al.*, 1992), and prostate carcinoma (Weidner *et al.*, 1993). Studies using the pancreatic $\beta$ cells of animals transgenic for a hybrid oncogene (RIP1–Tag2) (Brinster *et al.*, 1993) would indicate that angiogenesis is an important step in carcinogenesis in this system (Folkman *et al.*, 1989a).

To investigate factors that influence angiogenesis and to gain a more fundamental understanding of the cellular processes involved in the generation of capillaries, it has been necessary to develop a number of models of angiogenesis.

## II. Models of Angiogenesis

### A. Chicken Chorioallantoic Membrane Assay

The chicken chorioallantoic membrane assay is a technique traditionally used by embryologists that involves analysis of the developmental potential of grafts transplanted onto the chorioallantoic membrane (CAM). Because the early chicken embryo lacks a complete immune system xenografts from mammalian species become established and grow. Vascularization of these grafts is rapid.

Sorgente and colleagues (1975) first described the inhibitory effects of cartilage grafts on vascular development using this model. Subsequently, Folkman and co-workers (1979) used the model to study tumor angiogenesis directly. Fertile eggs were incubated for 72 hr and prepared for grafting by removal of enough albumin to facilitate the placement of a graft without causing subsequent cramping and sticking to the shell membrane. A rectangular window was cut in the shell to place and access the graft or test substance on the CAM. Angiogenesis was scored 3–4 days after grafting. Angiogenesis was considered to have been induced if a spoke–wheel arrangement of vessels was generated, directed toward the graft. Substances were lyophilized onto coverslips, then applied to the CAM to examine the effects on angiogenesis (Folkman *et al.*, 1979).

Quantitation of angiogenesis using the CAM assay was initially done on a graded score of 0–4, by observation. Computer analysis was subsequently applied to score the total number of vessels and obtain a directional vector value (Voss *et al.*, 1984; Jakob and Voss, 1984). The use of labeled sulfate to follow the angiogenic process has also made quantitation more accurate (Spisni *et al.*, 1992). Apart from problems associated with quantitation, the most common problem is the result of false positives due to wounding or irritants generated during the initial setting up of the assay. Because an angiogenic response may be consequent to wound healing or inflammation (Mahaderan *et al.*, 1989), this problem is not surprising. The CAM assay is sensitive to modification by many factors, including gas content and pH. The most pronounced variation observed is of keratinization, which in turn has significant effects on the CAM response to stimulation (Ausprunk *et al.*, 1991). This method has been applied to a wide range of both inhibitors and inducers of angiogenesis, as discussed in subsequent sections of this article (Folkman and Klagsbrun, 1987).

A further development of this model has been the *in vitro* method of maintaining the chick embryo in culture (Auerbach *et al.*, 1974). Although this is an *in vitro* assay, it is closest to a whole animal assay because the entire embryo and its membranes remain intact. In this assay, the egg content is transferred to a petri dish, where development continues to take place. This model has the advantage that multiple grafts can be placed on one embryo, and the effects can be photographed over time. Quantitation is simplified by the fact that the *in vitro* CAM presents a two-dimensional monolayer, not subject to the distortion of the *in ovo* CAM assay. The advantage is that multiple grafts may then be placed on the one embryo, and they can more easily be photographed over time. A further modification of the *in vitro* CAM assay, in which the embryo is supported on Gladwrap stretched across the mouth of a beaker (Dunn *et al.*, 1981), has improved embryo survival. The advantages of increased viability are offset by the difficulty in photographing the results. This model is technically easier that the *in ovo* assay and is better suited to large-scale experiments. The addition of sterile silicon rings on the yolk sac membrane creates discrete observation windows and assists in quantitation (Takigawa *et al.*, 1990).

## B.  Corneal Neovascularization Model

As the cornea is normally avascular, induction of an angiogenic reaction is a true demonstration of neovascularization (Hendkin, 1978). The earliest studies of corneal neovascularization were in the rabbit (Gimbrone *et al.*, 1974), in which insertion of tumor cells or extracts placed within 2 mm of the cornea–scleral junction generated vascular sprouts within 36 hr. However, because of the absence of genetically similar strains, expense, and difficulty in handling, other species have been used for angiogenesis studies in the cornea, including guinea pigs, rats, and mice (Fournier *et al.*, 1981; Muthukkaruppan and Auerbach, 1979; Muthukkaruppan *et al.*, 1982). Although the use of mice overcame the strain variation problem their small size makes the introduction of slow-release polymer into the eye a procedure requiring microsurgical skill. Quantitation of corneal neovascularization is difficult owing to the variability arising from an inability to achieve uniform placement of the test substance. Consequently, reagents under test have been incorporated into ethylene–vinyl acetate pellets (Elvax) prior to implantation into the cornea (Gimbrone *et al.*, 1974; Risau, 1986). The implantation of tumor cells also requires the incorporation of those cells into an inert medium that allows for accurate placement (Ausprunk and Folkman, 1977). The expression of corneal-derived cytokines such as interleukin 8 (IL-8), which has been shown to be angiogenic, may also lead to some variability in assays of angiogenic factors (Strieter *et al.*, 1992).

Advances in image analysis (Proia *et al.*, 1988; Haynes *et al.*, 1989) have improved the capacity to quantitate using the corneal model. Often a computerized digitalyzer, for example the Optomax Image analysis system (Optomax, Hollis, NY) or similar, is used. This system consists of a high-sensitivity closed circuit television (CCTV) camera mounted on a Nikon Optiphot-2 microscope. The image is displayed on a color video monitor that is interfaced with a microprocessor. Histological slides stained with von Willebrand factor antibodies may be used to locate blood vessel formation. Sequential monitoring of neovascularization in individual animals makes it possible to evaluate progressive changes in the process (Folkval, 1991). Indeed, development of computer-assisted image analysis has made many models of angiogenesis more quantitative (Parke *et al.*, 1988).

## C. Pouch Assays

The hamster cheek pouch is considered to be an "immune privileged" site because allogeneic or xenogeneic grafts may grow without eliciting an immune response. The anterior eye chamber is another "immune privileged" site that has been used to study neovascularization of preneoplastic mammary tumor cells (Folkman *et al.*, 1989b). Quantitation of this model is by morphometric analysis of histologically prepared sections following angiogenesis. Tumor implants have also been used (Auerbach *et al.*, 1976) in this model, as have slow-release vectors to assess the effects of transforming growth factor (TGF-$\alpha$) (Schreiber *et al.*, 1986).

The dorsal air sac method was developed by Selye (1953), to monitor vascularization of tumor grafts. Dorsal air sacs are created by injecting 10–15 ml of air into the backs of rats, and the model modified by the insertion of a transparent window in the skin, through which the process may be monitored. Using this model, angiogenesis mediated by the injection of tumor cells (Sakamoto *et al.*, 1991) or endothelial cells (Schweigerer *et al.*, 1992) has been assayed in response to various reagents.

The method of subcutaneous implantation of polyvinyl acetate (PVA) sponge disks impregnated with angiogenic factors is in common use (Fajardo *et al.*, 1988). Flat sponges of PVA foam are cut into 11-mm disks and their flat sides are sealed with Millipore (Bedford, MA) filters. Prior to sealing, a core is cut where the test material is to be inserted. This core is sealed with a slow-release polymer, ethylene–vinyl acetate copolymer (Elvax) (Langer *et al.*, 1980), then reinserted into the sponge. The sponges are recovered 1–3 weeks after subcutaneous implantation. Xenon clearance has been shown to be a useful means of quantifying new blood vessel formation (Andrade *et al.*, 1987).

Several *in vivo,* or *in ovo,* angiogenesis assays rely on being able to deliver a discrete amount of effector substance or cells to a precise location. Currently reagents are imbedded in Elvax, and the rate of release of components is dependent on the thickness of the coating of Elvax, making it difficult to reproduce these inserts. Alginate, a glycuron extracted from brown seaweed algae, gels in the presence of calcium ions or other multivalent counterions by anisocooperatively forming junctions between contiguous blocks of $\alpha$1,4-L-glucuronan residues present in the polysaccharide. Growth of avian and mammalian chondrocytes in ionotrophically gelled alginate beads demonstrates the potential of using this model for an alternative delivery system in angiogenesis models (Guo *et al.*, 1989), or it may provide an alternative method for the slow release of effectors of angiogenesis (Downs *et al.*, 1982). Matrigel can also be injected subcutaneously in mice, and used as a vehicle to assess angiogenic activity of different compounds (Passanti *et al.*, 1992; Kibbey *et al.*, 1992). Although the subcutaneous injection of Matrigel alone is insufficient to induce focal angiogenesis when fibroblast growth factor (FGF)–heparin is mixed with the Matrigel, in-growth of vessels is observed within days. The Matrigel plug can be removed, and processed for vessel quantitation (Passanti *et al.*, 1992; Kibbey *et al.*, 1992).

## D.  Mesenteric Window Assay

The mesenteric window assay examines the effect of reagents on normally vascularized mammalian tissues. Angiogenesis in this model is mediated by autologous mast cells, and probably occurs frequently because mast cells are activated in tissue trauma, wound healing, inflammation, as well as in many clinical and experimental tumors (Enerback and Norrby, 1989). Although the mechanism of the mast cell-mediated angiogenic reaction is not completely understood it is known that preformed mast cell products such as heparin and histamine can be angiogenic (Norrby *et al.*, 1986, 1990; Garrison, 1990; Norrby and Sorbo, 1992; Sorbo and Norrby, 1992). Mast cell-mediated angiogenesis has also been reported using the CAM assay (Clinton *et al.*, 1988; Duncan *et al.*, 1992).

The mesenteric window assay is well suited to quantitative analysis. In addition to the number of vessels per unit length of tissue and the vascularized area, it permits quantitation of vascular density and total vascular quantity, as well as measurement of the branching pattern (Norrby *et al.*, 1990; Jakobsson and Norrby, 1991).

E.  Spontaneous Angiogenesis

When endothelial cells are maintained as a confluent monolayer for prolonged periods of time without replenishing the nutrients, capillary-like
vessels spontaneously form. This "spontaneous" tube formation takes
between 10 and 14 days after confluence (Folkman et al., 1979). During
spontaneous tube formation of human umbilical vein endothelial cells
(HUVECs) the majority of the cells are involved in chaotic cell death, as
nutrients become depleted. The capillary-like vessels generated from the
small percentage of the population that differentiates are anchored loosely
in the culture dish, with no formation of a monolayer (see Fig. 1). However,
endothelial cells of bovine origin spontaneously form capillary-like tubes
by a process of "sprouting" of a subpopulation of cells that form a reticular
network of vessels sitting on top of a monolayer of cells (Maciag et al.,
1982). The reason for this difference is not understood but may relate to
the fact that bovine endothelial cells have a lower growth factor requirement and may tolerate longer periods of time in growth factor-depleted
culture conditions, suggesting that the cell death seen in the human cell
cultures may be consequential and not important to angiogenesis in this
model. An important feature in both cases is that only a small proportion
of the initial cell population is involved in forming these tubelike structures. What distinguishes this small subpopulation, and at what point in
the proceedings these cells commit to differentiation, remains unknown.
In bovine aortic endothelial cell spontaneous tube formation, sprout formation precedes the generation of a capillary-like network on top of the
monolayer, and has been shown to involve the synthesis of type I collagen
(Cotta-Pereira et al., 1980). The possibility of modulating this early event
has come to light from studies in which the addition of 10–100 $\mu$g/ml of
type I collagen plus phorbol 12-myristate 13-acetate (PMA) to monolayers
of human neonatal foreskin capillary endothelial cells rapidly (3 hr after
addition) induced the initial sprouting patterns previously observed only in
bovine spontaneous tube formation ( Jackson and Jenkins, 1991). Further
reports on this intriguing model have not been forthcoming.
      We have observed that the rate of spontaneous tube formation is in-
creased when confluent cultures of endothelial cells are stressed by altered
pH or temperature (G. W. Cockerill, unpublished observation). This observation is consistent with the stress-related changes in heat shock protein
90 (hsp90), which have been shown to be mediated through alterations in
extracellular matrix (Ketis et al., 1993). Little is known about the effects
of modulation of hsp90 on angiogenesis. However, because hsp90 has
been shown to mediate the nuclear translocation of the estrogen receptor
it is interesting to propose this as a productive avenue of research.

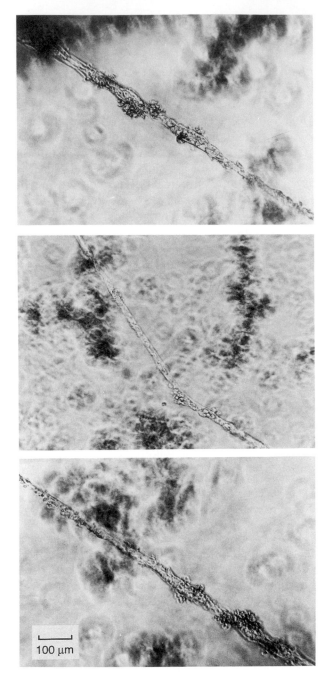

100 μm

FIG. 1  Spontaneous angiogenesis of HUVECs. Endothelial cell monolayers were allowed to exhaust the culture medium over a period of 10–14 days. Although most of the cells died a small subpopulation differentiated to form tubelike structures. These structures have lumena, as determined by serial sectioning, and were attached to the edge of the culture dish, and also at a single point in the center of the dish (middle).

## F. Three-Dimensional Gel Assays

Several three-dimensional gel assays have been developed, including collagen type I (Montesano and Orci, 1985, 1987; Montesano *et al.*, 1986), fibrin (Montesano *et al.*, 1987), fibronectin (Ingber and Folkman, 1989a), and Matrigel (Kubota *et al.*, 1988). Here we discuss the collagen and Matrigel model.

### 1. Type I Collagen Gels

The addition of PMA to endothelial cells seeded onto type I collagen induces invasive, capillary-like tubes (Montesano and Orci, 1985; Montesano *et al.*, 1887), as illustrated in Fig. 2, suggesting involvement of protein kinase C (PKC) in this process. Invasion is mediated by inducing expression and synthesis of type I collagenase, plasminogen activator (PA) activity, and stromelysin in endothelial cells in culture (Gross *et al.*, 1982; Moscatelli and Rifkin, 1988; Herron *et al.*, 1986). Kalebić and

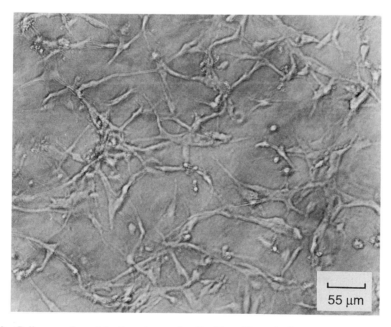

FIG. 2 Collagen gel model of angiogenesis. Capillary-like vessels were generated within 24 hr of seeding human umbilical vein endothelial cells onto thick type I collagen gels in the presence of PMA.

colleagues (1983) have also shown type IV collagenase is present in endothelial cells. Although these metalloproteinases are likely to be produced in their latent/inactive forms, it is possible that they are activated locally by plasmin produced by the action of coordinately expressed urokinase-type plasminogen activator (uPA).

Reduction of proteolytic activity may be achieved by several protease inhibitors. Plasminogen activator inhibitor 1 (PAI-1) is the major secreted inhibitor of bovine aortic endothelial cells (van Mourik et al., 1984), and has been shown to be expressed in virtually all endothelial cell types (Moscatelli and Rifkin, 1988). Tissue inhibitor of metalloproteinases (TIMP) has been shown to be synthesized by rabbit brain capillary endothelial cells, and is able to inhibit collagenase, stromelysin, and other proteases (Herron et al., 1986). Another level of control of angiogenesis may be afforded by the differential localization of expression of proteases and inhibitors, whereas the proteases are often cell surface associated, the PAI-1 is ECM associated, and TIMP may be secreted into the interstitial fluids.

Comparison between endothelial cell invasion and tumor cell invasion suggests that the proteolytic activity of endothelial cells may be cell associated (Moscatelli and Rifkin, 1988). This hypothesis is supported by the fact that uPA (Moscatelli, 1986), plasminogen, and plasmin bind to cultured endothelial cells (Bauer et al., 1992). Other components of the proteolytic cascade have also been localized to endothelial cell ECM. For example, PAI-1 is associated with the substratum and stabilized in its active form such that it can complex with tissue-type plasminogen activator (tPA) and inhibit its activity (Levin and Santell, 1987). Plasminogen activator inhibitor 1 deposited near endothelial cells may protect the capillary basement membrane and other matrix proteins from proteolysis by plasmin-generating enzymes. In addition, enzyme activation may be favored by the localized deposition of proenzymes and activators, which could result in an enhanced rate of enzyme activity and invasion. Protection from inactivation from secreted or ECM-associated inhibitors, and localization of enzyme activity to discrete regions allow proteolysis and thus angiogenesis to proceed in a specific direction (Moscatelli and Rifkin, 1988).

Tube formation in this model occurs within 24 hr and requires both transcription and translation (Montesano and Orci, 1985). Cells are seeded onto the top of a thick gel of collagen in the presence of PMA. The subsequent cell invasion of the matrix mimics the early events seen during angiogenesis in vivo. Capillary-like structures form throughout the gel, making quantitation difficult. Not all cells are stimulated to differentiate, and some remain as a monolayer on the surface of the gel.

## 2. Matrigel

The basement membrane is an important biological mediator of angiogenesis, and has been exploited in both *in vivo* and *in vitro* assays to assess the angiogenic activity of various factors (Madri *et al.*, 1983; Ingber and Folkman, 1989a,b; Form *et al.*, 1986). Matrigel is made by extracting the basement membrane matrix of Englebreth–Holm–Swarm (EHS) tumors taken from lathrytic mice (Kleinman *et al.*, 1982). At 4°C the extract is a viscous liquid that gels on warming to 37°C. The major components of this material are laminin, collagen IV, entactin/nidogen, heparan sulfate proteoglycan (Kleinman *et al.*, 1982), and growth factors (Taub *et al.*, 1990). The direct application of the material to angiogenesis was by Kubota and colleagues (1988). Figure 3 shows a typical response of human umbilical endothelial cells to this matrix. Within 1 hr the cells have rapidly migrated into a reticular network of aligned cells (Fig. 3a), after 2 hr the cells have started to flatten (Fig. 3b), and by 12–18 hr they have formed a network of capillary-like structures on the surface of the gel (Fig. 3c). These structures have a well-defined lumen that can be visualized by serial cross-section at the electron microscope level. Tube formation on Matrigel is a density-dependent phenomenon (Fig. 4). At too high a cell density a monolayer is formed, and at too low a cell density the cells do not contact each other, and in both instances tube formation is inhibited (Cockerill *et al.*, manuscript in preparation).

Alignment of the cells appears to be necessary for tube formation on Matrigel. However, many cell types are able to transiently form an aligned network on top of the Matrigel gels (Emonard *et al.*, 1987), but do not form structures with a lumen, indicating that alignment is necessary but not sufficient for tube formation. Figure 5 shows the time course of tube formation for HUVECs (Fig. 5, left) and a stromal fibroblast cell line (Fig. 5, right). Although the cells appear to align (Fig. 5A and B), only the HUVECs remain in the reticular pattern 2 hr after seeding Fig. 5C), whereas the stromal cells are clumping together in nodules (Fig. 5D). Whereas the HUVECs still display a network of capillary-like vessels after 24 hr (Fig. 5E), the stromal cells are in tight nodules (Fig. 5F). The inset shows the stromal cells 3 days after seeding, at which time the cells begin to migrate out of the nodules as solid cords of cells (Cockerill *et al.*, manuscript in preparation). Matrigel seems to support differentiation of many cell types. Mammary epithelial cells form nodes that produce casein (Seeley and Aggeler, 1991), and baby mouse kidney cells form nodes that eventually (after 6 days) form structures with lumena (Klein *et al.*, 1988). Sertoli cells form short, cordlike structures (Hadley *et al.*, 1990).

Alignment of endothelial cells on Matrigel does not require protein synthesis or gene expression (Bauer *et al.*, 1992). However, tube formation

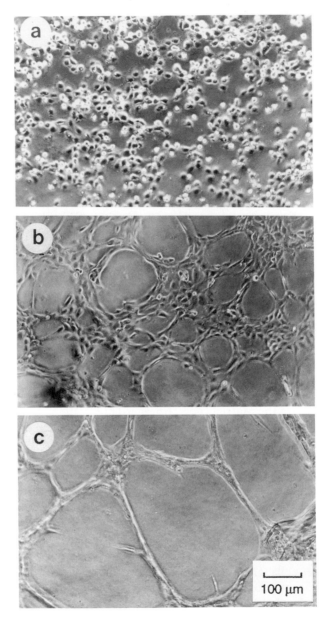

FIG. 3   Matrigel model of angiogenesis. HUVECs align within 60 min of being seeded onto thick Matrigel gels (a), and by 2–3 hr have flattened into a reticular network on the surface of the gel (b), and within 12–20 hr the cells have formed a mesh of capillary-like vessels (c).

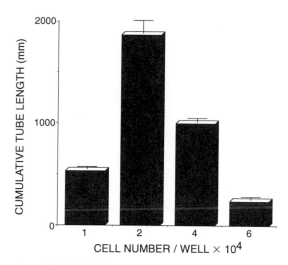

FIG. 4  Angiogenesis on Matrigel is density dependent. The extent of tube formation in HUVECs seeded onto Matrigel gels over a range of cell densities, was measured. The optimal cell density, using gels in microtiter trays, was $2 \times 10^4$ cells/well. At the highest density examined ($6 \times 10^4$ cells/well) we observed an approximately 6-fold reduction in angiogenesis. At the lowest cell density examined, the number of capillary-like vessels was reduced approximately 3.5-fold.

does require gene expression during the period of cell alignment, as the addition of transcriptional inhibitors during alignment abolishes tube formation whereas the addition of inhibitors after this event does not affect tube formation (Fig. 6). Gene induction by Matrigel has also been demonstrated in HUVECs (Sarma *et al.*, 1992), where contact with the matrix induces a primary response gene.

As in the type I collagen assay PKC mobilization is clearly required in the Matrigel model of angiogenesis (Bauer *et al.*, 1992; Kinsella *et al.*, 1992). Unlike the collagen gel model, the activation of PKC observed on Matrigel does not lead to invasion of the cells into the gel, suggesting that the balance of proteolytic enzymes is significantly different between the two models. The Matrigel model of angiogenesis more accurately represents conditions that are likely to occur as a late event in angiogenesis *in vivo*. Differentiation of endothelial cells on Matrigel may be blocked by pertussus toxins, indicating that the process requires G proteins (Bauer *et al.*, 1992). These cAMP-dependent G proteins have been shown in other systems to be important is cell–cell interactions and development (Devrotes, 1989). It remains to be seen if the G protein-coupled receptor cloned by differential screening of endothelial cell libraries with and without PMA activation (Hla and Maciag, 1990) has a role in angiogenesis.

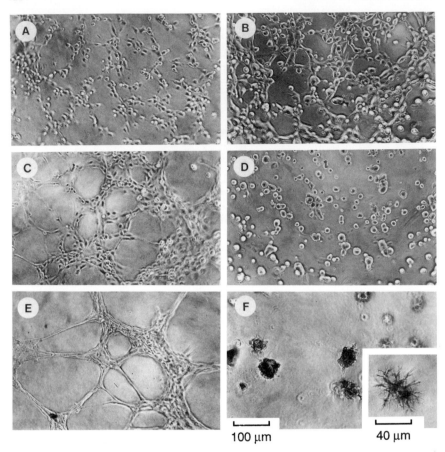

FIG. 5   Tube formation on Matrigel is cell specific. *Left:* The time course of angiogenesis seen in normal HUVECs. *Right:* The time course of events when stromal fibroblasts are seeded onto Matrigel under the same conditions. After 1 hr both cell types appear to have aligned (A and B). At 2 hr the HUVECs are flattening (C), whereas the stromal fibroblasts are clumping together (D). At 20 hr postseeding, the HUVECs have formed a reticulum of vessel-like structures (E) whereas the fibroblasts have aggregated into dense balls (F). Interestingly, at 3 days the stromal fibroblasts were beginning to migrate out of the cell clusters (inset). Serial sectioning of these structures showed only cords of cells, with no apparent lumen.

Similar to other models of angiogenesis, the Matrigel-induced angiogenesis may be inhibited using analogs of proline to inhibit collagen synthesis (Grant *et al.,* 1991).

Unlike the collagen gel model of angiogenesis the majority of cells seeded onto Matrigel gels will differentiate and enter into angiogenesis. The basement membrane extract of the EHS sarcoma was the starting

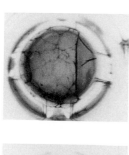 carrier
control

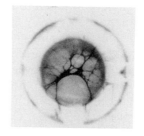 cell
control

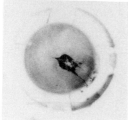 2.5µM DRB
0-18 hrs

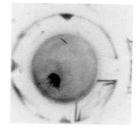 2.5µM DRB
0-1 hr

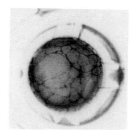 2.5µM DRB
4-18 hrs

FIG. 6  Angiogenesis on Matrigel requires gene expression.  HUVECs were plated onto
Matrigel and allowed to form capillary networks. By using the reversible transcriptional
inhibitor (DRB) we were able to show that inhibition of transcription during the first hour,
although not affecting alignment of the cells, completely abolished the development of
capillary-like structures. The addition of the inhibitor after alignment and flattening had no
effect on the ability of the cells to form capillary-like structures.

material for the purification of the calcium-binding basement membrane
protein 40 (BM-40)/SPARC (Dziadek *et al.*, 1986). Although the steady
state levels of SPARC mRNA have been shown to increase during spontaneous tube formation, levels do not change during angiogenesis on Matrigel. Because SPARC has been shown to be able to arrest cells in cycle
(Funk and Sage, 1991) it is interesting to suggest that this may be a
mechanism for establishing synchrony in this model, explaining why almost all the cells on Matrigel differentiate.

Matrigel is subject to batch-to-batch variation. It is likely that small variations in components greatly affect cell adhesion, motility, and proliferation, and all contrive to alter the differentiation of endothelial cells on Matrigel. We have observed differences in adhesion and also in levels and types of proteoglycan between batches. An altered combination of matrix components may lead to differences in the malleability of the final gel. Studies indicate that this is a factor that could potentially alter cell response (Vernon *et al.*, 1992). One of the advantages of its use is that the cell response is more homogeneous and, because the capillary-like structures generated form on the surface of the gel, the model is relatively easy to quantitate (Grant *et al.*, 1989).

To date, the factors tested in *in vitro* and *in vivo* Matrigel assays show activities similar to those observed in the CAM model (Grant *et al.*, 1989; Sakamoto *et al.*, 1991; Passanti *et al.*, 1992; Kibbey *et al.*, 1992).

## 3. Future Directions

Laminin antibodies that block binding of endothelial cells to laminin or Matrigel demonstrated a requirement for cell adhesion prior to growth and differentiation (Grant *et al.*, 1989). From these studies it was also suggested that the rapidity of angiogenesis on Matrigel versus collagen gels was a result of the need to synthesize a basement membrane on collagen gel that was already present on Matrigel (Grant *et al.*, 1989). Application of YIGSR-NH$_2$ laminin peptide to a monolayer of endothelial cells resulted in 30% of the population developing a ringlike structure, suggested by the authors as paralleling lumen formation (Grant *et al.*, 1989). Could this be a significant way of distinguishing a tube-competent endothelial subpopulation? If this suggestion is supported then this phenomenon would be appropriate to apply to subtractive hybridization, or the more recently developed differential display technology to clone those genes that mediate these events.

A number of workers have previously suggested a role for integrins in the *in vitro* angiogenic behavior of endothelial cells on extracellular matrix (Grant *et al.*, 1989; Basson *et al.*, 1990). The addition of monoclonal antibodies against $\alpha 6$ and $\beta 1$ to endothelial cells seeded onto Matrigel completely blocked angiogenesis on this matrix (Bauer *et al.*, 1992). Because laminin is the major component of this gel and the $\alpha 6 \beta 1$ integrin is shown to be the major receptor for laminin (Sonnenberg, 1988), it is likely that antibodies prohibit cell attachment to such a degree as to prevent the formation of capillary-like structures. We have shown that this model of angiogenesis is density dependent (Fig. 3), therefore it is not clear if inhibition of cell adhesion to Matrigel per se blocks angiogenesis, or whether by reducing the cell number finally attached to the gel (Bauer *et* al., 1992), they have inhibited angiogenesis by reducing the cell density. In similar studies, polyclonal antibodies to the entire vitronectin ($\alpha$v$\beta$3) and fibronectin ($\alpha$5$\beta$1) receptors totally inhibited cell adhesion to Matrigel and hence capillary formation in Matrigel (Davis et al., 1993). The use of $\alpha$5-integrin monoclonal antibodies in the same studies either had no effect or enhanced tube formation (Bauer et al., 1992; Davis et al., 1993). A clue to the mechanism of this effect comes from a study by Gamble et al. (1993), in which selectively restricting the adhesive repertoire of endothelial cells for the specific matrix to which they are exposed, the authors were able to show an enhancement of angiogenesis. On collagen gels in the presence of PMA, $\alpha$2$\beta$1 antibodies that block the collagen receptor enhanced tube formation, whereas no effect was seen on fibrin gels (Gamble et al., 1993). Conversely, antibodies directed against the major fibrin receptor $\alpha$v$\beta$3 enhanced tube formation on fibrin gels, but had no effect on collagen. Thus, restricting the usage of fibronectin receptors ($\alpha$5$\beta$1) on Matrigel by the use of anti-$\alpha$5 antibodies may explain the enhancement observed in the Bauer et al. studies (1992). The potential use of anti-integrin antibodies as therapeutic agents has recently been demonstrated in the CAM assay using anti-$\alpha_v\beta_3$ antibodies (Brooks et al., 1994).

Matrigel induces a motile phenotype in endothelial cells seeded onto the gel. Within 5 min of contacting the gel the endothelial cell is covered with microspikes that may be visualized by fluorescently labeled phalloidin stains (Fig. 7). Whereas on plastic the endothelial cytoskeleton rapidly polymerizes and forms a complex network of filaments observable after 20–30 min, after several hours on Matrigel the only cytoskeletal architecture one can detect with phalloidin is at cell junctions and in the microspikes on the cell surface. Not until late in tube formation on Matrigel can actin filaments, running parallel to the tube, be observed (Fig. 8). We postulate that the establishment of a motile phenotype is likely to be a prerequisite of endothelial cell differentiation. This proposal is supported by studies that show that the angiogenin-binding protein is a 42-kDa cell surface actin-related molecule (Hu et al., 1991; Moroianu et al., 1993). In addition, several other more potent angiogenic factors have actin-binding capacity and the angiogenin has been shown to be able to induce actin polymerization at suboptimal concentrations for spontaneous polymerization (Hu et al., 1993). The fact that angiogenic factors such as basic FGF (bFGF) and tumor necrosis factor $\alpha$ (TNF-$\alpha$) can stimulate angiogenesis by receptor-mediated mechanisms, but can also bind actin, suggests that actin binding may provide a more general mechanism for mediating angiogenesis. Alternatively, it may provide a mechanism for mediating an early event in the process, subsequent to which receptor-mediated mechanisms may play the major role. These observations may distinguish an early event in angiogenesis, which demands further research.

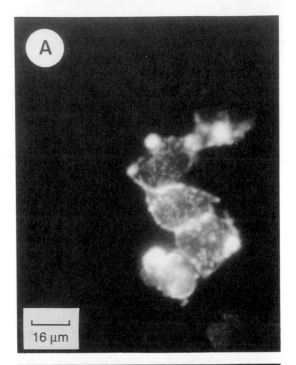

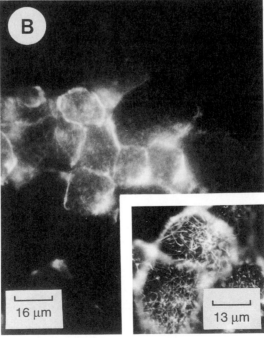

## G. Coculture Models of Angiogenesis

Angiogenesis *in vivo* rarely occurs within an environment free of other cell types. The influence of other cell microenvironments on microvessel formation and the expression of tissue-specific endothelial properties are being increasing realized (Auerbach *et al.*, 1987; Butcher *et al.*, 1980). In a more recent model, angiogenesis is induced by culturing brain capillary endothelial cells on collagen gels in a chamber above confluent tumor cells, such that both cell types are bathed in the same medium (Okamura *et al.*, 1992; Abe *et al.*, 1993). Other groups have been successful in the use of a variety of cell types such as esophageal cancer cells (Okamura *et al.*, 1992) and keratinocytes (Ono *et al.*, 1992) to induce angiogenesis.

Demonstration of the abilities of astroglial cells to induce angiogenesis in brain capillary endothelial cells (Lattera *et al.*, 1990) has provided a model for the study of neural microvessel development, and blood–brain barrier formation. In this model, 24 hr after the seeding of C6 astroglial cells in a culture, endothelial cells are seeded at twice the density. Angiogenesis in this model requires both gene expression and protein synthesis (Lattera and Goldstein, 1991), and was induced within 24 hr of coculture. Furthermore, the induction of angiogenesis required direct cell–cell contact, as no enhancement was observed when the two cultures were bathed in the same growth medium in a Boyden chamber, where they were not in direct contact. Quantitation of this model is facilitated by being able to differentially stain the cell types involved in the culture model. In addition, computer-assisted analysis of fluorescently stained photographs enables the assay to be relatively accurate.

## H. Aortic Ring Model

The radial growth of microvessels is easily monitored in rings of aortas imbedded in three-dimensional (thick) gels, using standard phase microscopy. The end point can be histologically processed and sections cut for morphometric analysis (Nicosia and Ottinetti, 1990). Aortic rings of rat aorta embedded into collagen or fibrin gels in the absence of exogenous

---

FIG. 7  Actin staining of early events in angiogenesis on Matrigel. At optimal cell density on Matrigel, HUVECs aligned within 1 hr (see Fig. 3a). The cells were polarized end-to-end. Actin staining with Bodipy–phalloidin showed polymerized actin cytoskeleton at cell junctions and in microspikes (A). At high cell density, the cells attached to the matrigel as a monolayer and did not form extensive capillary-like structures. Actin staining of densely seeded cells after 1 hr (B) showed no end-to-end polarity and polymerized actin cytoskeleton surrounded most of the cells. The inset in (B) shows the staining of microspikes on the cells, characteristic of the motile phenotype.

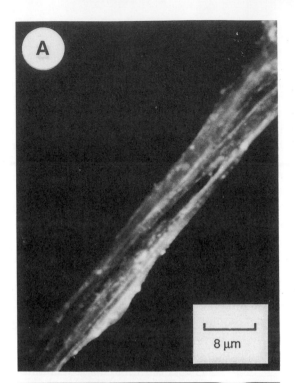

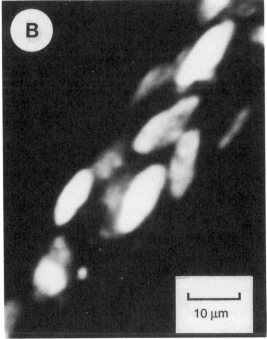

growth factors can also generate a complex array of microvessels. This demonstrates the potential usefulness of this more defined model of angiogenesis (Nicosia and Ottinetti, 1990).

## I. Human Amnion

The basement membrane of the human amnion may be used as a growth substrate for microvascular endothelium (Madri *et al.*, 1983; Furie *et al.*, 1984). Endothelial cells differentiate on this substrate but do not invade (Madri *et al.*, 1983; Furie *et al.*, 1984), as observed in the Matrigel model.

## J. Summary

Whereas the collagen gel assay and the aortic ring model are best suited to investigating the effects of reagents on invasion of interstitial collagens, the human amnion and Matrigel model are best suited to examining the effects of mediators on later events in angiogenesis. All of these models address subtly different aspects of angiogenesis. To examine thoroughly the likely effects of a reagent, multiple models should be used. The *in vivo* model most appropriate for a particular reagent needs to be employed.

## III. Enhancers and Inhibitors

## A. Enhancers and Angiogenesis

Classically, angiogenic factors have been defined as those that act directly on the endothelial cells to stimulate motility and mitosis, and as those that act indirectly to induce host cells to release growth factors that then target endothelial cells (Folkman and Klagsbrun, 1987). Development of reagents that enhance angiogenesis would have direct relevance in the management of severe wounds, and would facilitate many situations. Although some factors have been purified from highly vascular tumors, more

---

FIG. 8   Fluorescent staining of capillary-like structures on Matrigel. The parallel arrays of actin filaments can be observed in the capillary-like structures formed by HUVECs on Matrigel. (a) A capillary stained with Bodipy–phalloidin. (b) The nuclei of a tube stained with propidium iodide, demonstrating the multicellular nature of these structures.

recently a factor has been isolated from the blood of patients with systemic vasculitis (Cid *et al.*, 1993).

## 1. Transforming Growth Factor β

Transforming growth factor β is angiogenic *in vivo* (Roberts *et al.*, 1986). However, whereas TGF-β induced tube formation when microvascular endothelial cells were placed in thick collagen gels (Madri *et al.*, 1988), it inhibited proliferation (Frater-Schroder *et al.*, 1986) and migration (Heinmark *et al.*, 1986) in other cell culture systems. Furthermore, in cultured bovine capillary endothelial cells TGF-β decreased the amount of cell-associated and secreted PAI-1, decreasing cell invasion into collagen matrices and through amniotic membranes (Muller *et al.*, 1987; Mignatti *et al.*, 1989). Because TGF-β is a potent chemoattractant for macrophages (Wahl *et al.*, 1987), it is possible that the TGF-β-induced neovascularization is a consequence of angiogenic components produced from attracted macrophages.

## 2. Tumor Necrosis Factor α

Antibodies to TNF-α have been shown to neutralize the angiogenic activity of thioglycolate-treated macrophages in the chick chorioallantoic membrane assay, and also in the type I collagen gel assay (Liebovich *et al.*, 1987), and in the rabbit corneal model (Frater-Schroder *et al.*, 1987). With both TNF-α and TGF-β angiogenesis is associated with an inflammatory process, unlike bFGF-induced angiogenesis (Esch *et al.*, 1985).

## 3. Angiogenin

Angiogenin is a 14-kDa protein initially identified in HT 29 adenocarcinoma cells (Fett *et al.*, 1985). It was later found to be in adult liver, and at lower concentrations in many normal tissues as well as in serum (Weiner *et al.*, 1987; Shapiro *et al.*, 1987). Angiogenin shows 35% homology with pancreatic RNase (Shapiro *et al.*, 1987). However, its RNase activity is limited compared to pancreatic RNase, as it can break down tRNA, but only partially cleaves 18S and 28S ribosomal RNAs. Site-directed mutagenesis to determine the significant residues for RNase activity shows an increase in RNase activity with a concomitant loss in angiogenic activity (Shapiro *et al.*, 1989; Harper *et al.*, 1989). Further studies suggest that the RNase activity site in angiogenin is essential, but not sufficient, for its angiogenic activity, and that a second site on the molecule is also required (Hallahan *et al.*, 1991). Most recently, it was shown that the specific endocytosis of angiogenin is followed by nuclear translocation

(Moroianu and Riordan, 1994). Nuclear translocation has also been demonstrated to occur with the FGFs and endothelial cell growth factor (EGF) (Bouche *et al.*, 1987; Baldine *et al.*, 1990; Sano *et al.*, 1990), suggesting that this may be a common pathway in the mechanism of angiogenesis. Angiogenin can bind specifically to the endothelial cell and this binding is specifically inhibited by RNase inhibitor (Badet *et al.*, 1989). In addition, placental RNasin binds to angiogenin hundreds of times more efficiently than other RNases (Fox and Riordan, 1990). An angiogenin-binding protein with properties consistent with being an angiogenin receptor component has been identified in a transformed endothelial cell line, GM7373 (Hu *et al.*, 1993). It is a 42-kDa cell surface protein that is released by exposure of cells to heparin, heparin sulfate, or angiogenin. This protein has been shown to be a cell surface actin (Hu *et al.*, 1993). Angiogenin was able to induce the polymerization of actin at suboptimal concentrations required for spontaneous polymerization. This ability may be central to its mechanism of action, because such events could result in shape change and detachment, and precipitate subsequent events such as migration and proliferation, which lead to capillary formation. In support of this idea, reorganization of extracellular actin has been observed during the growth and formation of the corneal endothelium (Klagsbrun and D'Amore, 1991).

## 4. Fibroblast Growth Factor

The heparin-binding fibroblast growth factor (FGF) family, acidic FGF (aFGF/FGF-1) and basic FGF (bFGF/FGF2), are among the growth factors that act directly on vascular cells to induce endothelial cell growth and angiogenesis (Burgess and Maciag, 1989; Ausprunk and Folkman, 1977; Abraham *et al.*, 1986; Gospodarowicz *et al.*, 1984; Thomas *et al.*, 1985; Yanagisawa-Miwa *et al.*, 1992). Whereas aFGF is found primarily in normal tissues (Risau *et al.*, 1988) and in vascular SMC (Winkles *et al.*, 1988), bFGF has a wide distribution (Lobb *et al.*, 1986; Schweigerer *et al.*, 1987). Because FGFs lack a signal sequence for secretion, their normal mode of release is not fully understood (Burgess and Maciag, 1989). Basic FGF is, however, associated with the ECM components, and its most important stored form is thought to be complexed to heparin sulfate proteoglycan (Folkman *et al.*, 1988). It has been hypothesized that poor perfusion in tumors may result in ischemia, acidosis, and tissue damage, which in turn may release FGF from the cells and ECM stores, and subsequently stimulate angiogenesis (D'Amore and Thompson, 1987).

Fibroblast growth factor may be detected after endothelial injury (Gadjusek and Carbon, 1989; McNeil *et al.*, 1989), and are present in the subendothelial matrix (Vlodavsky *et al.*, 1987). However, several other genes are observed to induce intimal hyperplasia, such as platelet-derived growth factor (PDGF) or TGF-$\beta$ (Nabel *et al.*, 1993b). Studies using direct gene transfer of a secreted form of aFGF (Nabel *et al.*, 1993a) showed the induction of intimal thickenings 21 days after gene transfer. Neovascularization of this intimal thickening was observed, suggesting that the FGFs could potentially cause neovascularization in similar preatherosclerotic lesions. Acidic FGF stimulates hyperplasia and neovascularization of the hyperplastic intima, suggesting that smooth muscle hyperplasia alone is insufficient for the formation of new capillaries.

The locomotion of cells of endothelial origin is suppressed by TGF-$\beta$ but is enhanced by bFGF (Sato and Rifkin, 1988; Madri *et al.*, 1988). The locomotion of many normal and transformed cells of epithelial and mesenchymal origin can also be induced by members of the FGF family, insulin-like growth factor (IGF), PDGF, TGF-$\alpha$, TNF-$\alpha$, colony-stimulating factors (CSFs), interleukin 8 (IL-8), and interferons, in addition to complement and some matrix proteins (reviewed in Stoker and Gherardi, 1991).

Mignatti and colleagues (1989) have shown that bovine capillary endothelial cell migration through human amnion basement membrane was inhibited with antibodies to bFGF. This FGF-induced migration could also be inhibited by inhibitors of both plasmin and metalloproteinases as well as antibodies to tPA and type I and IV collagenase, demonstrating that both tPA, plasmin, and specific metalloproteinases are involved in the bFGF-induced invasion associated with angiogenesis.

## 5. Vascular Endothelial Growth Factor/Vascular Permeability Factor

The vascular endothelial growth factor (VEGF) family of proteins, also referred to as vascular permeability factor (VPF), exists as dimeric glycoproteins of $M_r$ 34K–46K that affect capillary permeability, and stimulate endothelial cell growth *in vitro*, and angiogenesis *in vivo* (Keck *et al.*, 1989; Connolly *et al.*, 1989; Ferrara *et al.*, 1992). Vascular endothelial growth factor has been characterized in several tumors of different species (Ferrara *et al.*, 1992), and is structurally related to PDGF with 18% identity between VEGF and the PDGF B chain. Vascular endothelial growth factor, like PDGF, can bind heparin and can be eluted off at low salt (Ferrara *et al.*, 1992; Senger *et al.*, 1990). Monoclonal antibodies to VEGF inhibited the growth of tumors in nude mice (Kim *et al.*, 1993). Four splice variants of VEGF may exist in four different homodimeric molecular species (Leung *et al.*, 1989; Houck *et al.*, 1991; Tisher *et al.*, 1991). A variety of transformed cell lines express the VEGF mRNA and secrete VEGF (Senger *et al.*, 1986; Rosenthal *et al.*, 1990). *In situ* hybridization studies demonstrate high levels of VEGF mRNA in highly vascularized glioblastoma multiforme and capillary hemangioblastoma (Berse *et al.*, 1992;

Shweiki *et al.*, 1992; Plate *et al.*, 1992; Berkman *et al.*, 1993). Monoclonal antibodies capable of blocking VEGF-induced angiogenesis *in vivo* and *in vitro* were used to assess the effect of VEGF on tumor growth (Kim *et al.*, 1992).

## 6. Interleukin 8

Interleukin 8 is a cytokine involved in leukocyte–vascular endothelial cell interactions such as the invasion of neutrophils through a vessel wall model via $\beta$2-integrin attachment (Huber *et al.*, 1991), which more recently has been shown to have angiogenic properties (Koch *et al.*, 1992). It has also been implicated in angiogenic disease states such as psoriasis and rheumatoid arthritis (Brennan *et al.*, 1990, Seitz *et al.*, 1991; DeMarco *et al.*, 1991; and Schroeder and Christophers, 1989). Indeed, the rheumatoid synovium is a major source of IL-8. Similar to other angiogenic factors, IL 8 was shown to bind heparin and to have potent angiogenic activity when implanted into the rat cornea (Koch *et al.*, 1992). It also induced proliferation and chemotaxis of human endothelial cells (Koch *et al.*, 1992). However, whereas 2–40 ng of IL-8 induced corneal vascularization, 400 ng did not induce significant angiogenesis in this model, suggesting that in areas where high concentrations of IL-8 are produced (in areas of acute severe inflammation) neovascularization may not occur (Robbins *et al.*, 1984). Differing dose-dependent actions of IL-8 have also been demonstrated for neutrophil chemotaxis. Hence, high levels of IL-8 induce neutrophil chemotaxis (Yoshimura *et al.*, 1987; Larsen *et al.*, 1989) but low levels result in selective lymphocyte chemotaxis (Yoshimura *et al.*, 1987). In support of the idea that IL-8 may have a direct role as an inducer of neovascularization, studies have shown that recombinant IL-8 (rIL-8) can induce endothelial chemotaxis and proliferation (Koch *et al.*, 1992). These findings raise the possibility that TNF-$\alpha$- or IL-1B-induced angiogenesis in the cornea may be mediated by induction of endogenous IL-8. Also, other factors may be produced in the cornea that mediate angiogenesis. For example, bFGF may be released from corneal extracellular matrix by the action of heparitinase (Folkman *et al.*, 1988), which may be important in mediating corneal repair.

## 7. Phosphatase Inhibitors

Vanadate potentiates the effect of growth factor-induced angiogenesis (Spisni *et al.*, 1992). Okadaic acid also induces angiogenesis in the chick chorioallantoic membrane model, with a minimum effective dose of 5 fmol/egg, and half-maximal dose being 90 fmol/egg (Oikawa *et al.*, 1992). Okadaic acid exerts an angiogenic activity an order of magnitude stronger than PMA but the time course of induction is slower than for PMA, suggesting a differing mechanism of action (Suganumu *et al.*, 1988). Okadaic acid inhibits type 1 and 2A protein phosphatases, resulting in an increase in phosphoproteins within the cell (Haystead *et al.*, 1989; Sassa *et al.*, 1989; Yatsunami *et al.*, 1991). Unlike PMA, okadaic acid stimulates the production of prostaglandin $E_2$ in rat peritoneal macrophages (Ohuchi *et al.*, 1989) and potentiates the ability of TGF-$\beta$1 to upregulate uPA expression (Falcone *et al.*, 1993). The delayed upregulation of c-*fos*, transin, and urokinase by okadaic acid has also been demonstrated in mouse keratinocyes (Holladay *et al.*, 1992). It is likely that some proteases such as urokinase and collagenase are involved in angiogenic induction by okadaic acid because the expression of these two protease activities was induced by either okadaic acid or TPA (Holladay *et al.*, 1992; Kim *et al.*, 1990; Levy *et al.*, 1991; Whitham *et al.*, 1986).

## 8. Haptoglobin

Sera from patients with systemic vasculitis had the capacity to stimulate angiogenesis *in vitro* (Cid *et al.*, 1993), using the Matrigel model of angiogenesis. Haptoglobin was identified as one of the components of these sera able to mediate the angiogenic effect. Furthermore, antibodies to this protein partially inhibited the angiogenic activity of these sera. The angiogenic activity of haptoglobin was confirmed in two *in vivo* models; implanted disk and subcutaneous injection with Matrigel. This suggests that the increased levels of haptoglobin in chronic inflammatory conditions may play a role in tissue repair, and it may offset the effects of ischemia in systemic vasculitis by promoting the development of collateral vessels. Histopathological studies of affected tissues from systemic vasculitis patients often demonstrate new reparative vessels (Olsson *et al.*, 1990). It is not yet clear whether the enhanced angiogenic effect of haptoglobin from vasculitis sera is due to quantitative differences or to the presence of different haptoglobin with higher angiogenic activity.

## 9. Hyaluronic Acid Fragments

Fragments of hyaluronic acid between 4 and 25 disaccharides in length have been shown to be angiogenic in the corneal model (West *et al.*, 1985). Similarly sized hyaluronic acid fragments are also known to influence the binding to, and effect the interactions between, fibronectin and collagen (Yamada, 1981), and have been shown to cause aggregation of proteoglycans (Hascall and Heinegard, 1974) and self-association of the molecule to a considerable degree (Morris *et al.*, 1980). These events were similarly shown to be inhibited by the same sized hyaluronic acid fragments that inhibit angiogenesis (Eriksson *et al.*, 1983; Morris *et al.*, 1980). These findings support the idea that angiogenesis may be regulated at the level of extracellular matrix, and that factors that influence its composition and integrity may influence the differentiation process at work in angiogenesis.

## 10. Synergism between Gangliosides and Fibroblast Growth Factor

At suboptimal doses of angiogenic factors, the addition of gangliosides promoted angiogenesis (Ziches *et al.*, 1989). Molecules with a high sialic acid content, such as GT1b (bisialoganglioside), are more efficient at influencing the biological response of capillary endothelial cells than are molecules with lower sialic acids, such as GM1 (monosialoganglioside) (Alessandri *et al.*, 1986; Ziches *et al.*, 1989). Further studies show that gangliosides can synergize with bFGF and promote endothelial growth, motility, and survival (De Cristan *et al.*, 1990).

## B. Inhibitors of Angiogenesis

As our knowledge of angiogenesis increases, so do the approaches used to inhibit this process. The existing inhibitors of angiogenesis target a variety of functions such as cell proliferation, migration, matrix-metabolizing mechanisms, matrix production, and cell–cell recognition.

## 1. Fumagellin

The observation that a fungal contaminant in an endothelial cell culture produced an agent that perturbed the growth of the cells around the contaminant led Folkman *et al.* in conjunction with the Tekada Chemical Company to the isolation of fumagellin. Fumagellin inhibited angiogenesis in the CAM assay, but the levels required for prevention of solid tumor growth were cytotoxic. Synthesis of a more potent analog, *o*-(chloroacetylcarbomoyl) : fumagellol (AGM-1470), provided a safe and effective alternative with few side effects *in vivo*. However, although AGM-1470 inhibited endothelial proliferation *in vitro* it did not inhibit tumor cell growth. More recently another potent analog, TNP-470, did have the capacity to inhibit tumor growth *in vivo* (Ingber *et al.*, 1990; Kusaka *et al.*, 1994). The actions of TNP-470 appeared to be mediated through its ability to inhibit endothelial cell growth (Kusaka *et al.*, 1994). Current studies are directed toward generating a nontoxic analog effective in tumor regression.

## 2. Inhibitors of Collagen Synthesis

Several reagents that modify the synthesis of collagen have been used to inhibit angiogenesis. The proline analog L-azetidine-2-carboxylic acid (LACA) prevents the triple-helical formation of collagen, and has been shown to induce regression of growing capillaries in the CAM model (Ingber and Folkman, 1988) and inhibit branching or tortuosity in the rat mesenteric window model (Norrby, 1993). Another proline analog, *cis*-hydroxyproline, has been shown to block the synthesis and deposition of collagen in basement membrane and reduce the growth of rat mammary tumors (Lewko *et al.*, 1981; Wicha *et al.*, 1981). However, the ability to inhibit angiogenesis does not always correlate with the ability to act as an antitumor agent. Others have shown that LACA is ineffective as an antitumor agent (Klohs *et al.*, 1985). LACA cannot be hydroxylated and the newly synthesized polypeptides of procollagen do not fold into stable triple-helical conformations. When a critical number of prolyl residues have been substituted by the analog the thermal stability of the molecule is decreased, leading to alterations in the extracellular matrix, which is less able to support the normal proliferative capacity of the cell (Jimenez and Rosenbloom, 1974; Uitto and Prockop, 1974; Tay *et al.*, 1983). Titano-cene dichloride, a reagent that inhibits the biosynthesis of collagens, has also been shown to be an active antitumor agent by suppressing angiogenesis (Bastaki *et al.*, 1994).

### 3. Protamine

Protamine, a 4.3-kDa arginine-rich protein, acts at a cellular site not associated with the FGF receptor, because although it inhibits the mitogenic effect of FGF it potentiates the mitogenic effect of epidermal growth factor (EGF) (Neufeld and Gospodarowicz, 1987; Majewski *et al.*, 1984). In addition, it has been shown to inhibit cross-linking between angiogenin and actin (Taylor and Folkman, 1982), an event directly related to the mechanism of action of this potent angiogenic factor. Protamine is not used for the control of neovascularization because of its unacceptable cytotoxicity (Folkman, 1985).

### 4. Cyclosporin A

The immunosupressant drug cyclosporin A (CsA), administered as a long-term treatment during renal transplantation has been shown to have angiostatic properties in the rat mesenteric window model (Norrby, 1992). The mechanism of its angiostatic activity is not understood. However, a clue as to the mechanism of its actions may be derived from the demonstration that the CsA-sensitive transcription factor NFAT (nuclear factor of activated T cells) is present in endothelial cells (Cockerill *et al.*, 1994). The angiostatic effects of CsA may, therefore, be mediated through effects of the drug on genes regulated by the CsA-sensitive transcription factor NFAT.

## 5. Cartilage-Derived Factors

Extracts of cartilage, one of the few avascular tissues in the body, can inhibit angiogenesis (Eisenstein *et al.*, 1975; Brem and Folkman, 1975; Langer *et al.*, 1976, 1980). A protein, with sequence homology in the $NH_2$-terminal region to collagenase inhibitor was purified from bovine scapular cartilage (Moses *et al.*, 1990). This protein inhibited proliferation and migration *in vitro* and angiogenesis *in vivo* in the CAM assay (Moses *et al.*, 1990). Because the dissolution of interstitial collagens is an important step in angiogenesis (Langer *et al.*, 1980; Rifkin *et al.*, 1982), the presence of collagenase inhibitors in cartilage explains its resistance to invasion and vascularization.

## 6. Heparin, Steroids, and Heparin–Steroid Conjugates

The control of angiogenesis with synthetic heparin substitutes was first demonstrated by Folkman and co-workers (1988). The angiostatic activity of heparin and nonanticoagulant heparin fragments was shown to be enhanced by administration of steroids (Crum *et al.*, 1985; Folkman and Ingber, 1987; Ingber *et al.*, 1986). Their mechanism of action was thought to be via induction of plasminogen activator inhibitor (PAI-1), thus affecting the breakdown of basement membrane (Blei *et al.*, 1993). The efficacy of these drugs was increased again by conjugating the two moieties. The covalent linking of a nonanticoagulating derivative of heparin (heparin adipic hydrazide) to antiangiogenic steroid (cortisol) via a labile bond generated a drug able to concentrate cortisol inside the vascular endothelium. The heparin moiety was able to target to the sulfated polyanion receptor on the cell surface, followed by endocytosis and release of cortisol inside the cell. The antiproliferative effect of these conjugates was far greater than that of cortisol and heparin administered in their unconjugated form (Thorpe *et al.*, 1993). The drugs were also shown to reduce vascularization of subcutaneous sponge implants and retard the growth of subcutaneous Lewis lung carcinoma by 65% (Thorpe *et al.*, 1993).

## 7. Platelet Factor 4

The platelet $\alpha$-granule protein PF4 was shown to inhibit angiogenesis (Taylor and Folkman, 1982), as was recombinant human PF4 (Maione *et al.*, 1990), and the CAM assay. Furthermore, PF4 completely suppressed the growth factor-dependent proliferation of human umbilical vein endothelial cells in culture (Maione *et al.*, 1990). Analysis of small peptides of the molecule suggests that the angiostatic activity was associated with the heparin-binding domain of the molecule, and addition of heparin in experimental implants abrogated the effects of PF4. Platelet factor 4 has also been shown to have collagenase inhibitor activity (Hiti-Harper *et al.*, 1978).

## 8. Linomide

When given systemically to mice, linomide reduces primary and secondary tumor growth and metastasis of murine B16 melanoma cells (Kallard, 1986; Harning and Szalay, 1988; Passanti *et al.*, 1992; Vukanović *et al.*, 1993). The low toxicity of linomide, and its androgen-independent ability to inhibit tumor angiogenesis and hence suppress tumor growth, make it a putative clinically useful drug. Currently, its long-term effects are under investigation.

## 9. Placental RNase Inhibitors

Although RNase inhibitors are currently not feasible clinical reagents, as a result of their rapid clearance, they have significant antiangiogenic activities *in vitro*. It may be possible to conjugate these reagents with a protective protein to render them clinically useful. Placental RNasin binds to angiogenin hundreds of times more efficiently than it does to other ribonucleases (Fox and Riordan, 1990), suggesting a possible mechanism of action of this class of reagent. Studies using the corneal model and the subcutaneous implantation model demonstrate reduction of FGF- and orthovanadate-enhanced angiogenesis (Shapiro and Vallee, 1987; Polakowski *et al.*, 1993). RNasin prevented tumor growth of C755 mammary tumor cells. Furthermore, its antitumorigenic activity correlated with its effect on tumor-induced angiogenesis, suggesting that the ability of RNasin to inhibit tumor growth was due to its ability to inhibit angiogenesis (Polakowski *et al.*, 1993).

## 10. Hyaluronic Acid

Although hyaluronic acid (HA) fragments can be angiogenic, high molecular weight hyaluronic acid inhibits the vascularization of chick embryo limb bud (Feinberg and Beebe, 1983), and conversely the differentiation and vascular ingrowth are associated with an increase in tissue hyaluronidase activity (Toole, 1976; Belsky and Toole, 1983). Also, it has been shown that hyaluronic acid can reduce the rate of development of granulation tissue and newly formed capillaries around subcutaneous implants (Balazs and Darykiewicz, 1975). Studies by West and co-workers (1985) showed that the removal of HA may not only represent the removal of an inhibitor of angiogenesis, but that the degradative products of HA may be angiogenic.

## 11. Inhibitors of Oligosaccharide Processing

In a study of oligosaccharide processing, inhibitors of capillary formation in the fibronectin-induced model of capillary formation showed that the synthesis of hybrid-type oligosaccharides is required for capillary formation *in vitro*. During this process there is an increase in the synthesis of monosialated and fucosialated glycans on asparagine-linked oligosaccharides (Nguyen *et al.*, 1992). This observation may explain the mechanism whereby angiogenesis has been inhibited by antibodies directed against sialyl-Lewis-X and sialyl-Lewis-A (Lowe *et al.*, 1990; Phillips *et al.*, 1990; Walz *et al.*, 1990).

## 12. Tumor Suppressor Genes

Modulation of angiogenesis has been a possible function propounded for tumor suppressor genes (Bouck *et al.*, 1986; Sager, 1986). Demonstration that the expression of a 148-kDa protein in the culture medium of BHK 21/c113 (baby hamster kidney) cells was related to an active tumor suppressor gene, and that this protein inhibited angiogenesis in the corneal assay, supports this idea (Rastincjad *et al.*, 1989). The function of this gene is clearly not specific to hamster, as it can be complemented by chromosome 1 from normal human fibroblasts (Stoler and Bouck, 1985). The identity of the inhibitor is as yet unknown. Antibodies to the protein show no cross-reactivity to known antigens of this size. Weak cross-reactivity to collagen type IV was observed. However, the BHK inhibitor did not show the expected sensitivity to collagenase.

## 13. Thrombospondin

Good and co-workers were the first to identify thrombospondin (TSP) as being an inhibitor of angiogenesis (1990) when the amino acid sequence of an antiangiogenic tumor suppressor gene (Rastinejad *et al.*, 1989) was found to be similar to thrombospondin. Further studies showed that purified human TSP, isolated from platelets, was able to block neovascularization in the rat corneal model, and inhibits chemotaxis of capillary endothelial cells toward angiogenic factors (Good *et al.*, 1990). Its role as an angiogenic inhibitor was further supported by the elegant studies of O'Shea and Dixit (1988), who showed the presence of TSP to be adjacent to mature quiescent vessels, but absent from actively growing sprouts. This relationship was subsequently demonstrated *in vitro* (Iruela-Arispe *et al.*, 1991). In addition, the role of TSP as an angiogenic inhibitor is further supported by the inability of endothelial cells in fast-growing hemangiomas to make TSP (Sage and Bornstein, 1982), and the ability of antibodies to TSP to increase angiogenesis *in vitro* (Iruela-Arispe *et al.*, 1991). Thrombospondin mRNA has been shown to be downregulated in endothelial cells forming tubes in culture (Canfield *et al.*, 1986).

The mechanism of action of TSP is unclear, but is postulated to be related to modulation of adhesion interaction and growth because TSP can mediate cell–cell interactions, and may also play a role in cell–substrate interactions. For endothelial cells TSP can be deadhesive. When endothelial cells are spread on other substrates the focal contacts can be broken by exposure to soluble TSP (Murphy-Ullrich and Hook, 1989; Murphy-Ullrich *et al.*, 1991). Thrombuspondin has also been shown to inhibit endothelial cell growth (Bagavandoss and Wilks, 1990; Tarabolett *et al.*, 1990; Murphy-Ullrich *et al.*, 1992). Further studies showed that both the $NH_2$-terminally truncated TSP, and a series of peptides from the procollagen-like region of the molecule, also blocked angiogenesis (Tolsma *et al.*, 1993). In more recent studies, TSP-containing fibrin and collagen matrices were able to promote angiogenesis in rat aortic explants on Matrigel (Nicosia and Tuszynski, 1994). These investigators showed that TSP directly stimulated the growth of aortic culture-derived myofibroblasts, which in turn promoted microvessel formation when cocultured with the aortic explants. This result is inconsistent with the interpretation of the *in vivo* studies, which show matrix-bound TSP in mature vessels, and report its absence in actively growing sprouts (O'Shea and Dixit, 1988).

## 14. Estrogen Metabolites

The endogenous estrogen metabolite 2-methoxyestradiol inhibits angiogenesis and suppresses tumor growth (Fotsis *et al.*, 1994). This derivative is shown to inhibit cell proliferation and migration, and angiogenesis *in vitro*. It has also been shown to inhibit neovascularization and tumor growth in mice. This is the first steroid derivative to be active without heparin or sulfated cyclodextrins, indicating a different mechanism of action. 2-Methoxyestradiol has negligible interaction with the estrogen receptor (MacClusky *et al.*, 1983). Although its mechanism of action is not fully understood, it has been shown to induce urokinase-type plasminogen activator, suggesting that modulation of endothelial cell proteolysis may be responsible, in part, for the inhibitory action of this compound (Fotsis *et al.*, 1993; 1994).

## 15. Genistein

Genistein was isolated from the urine of vegetarians, and was shown to inhibit angiogenesis and cell proliferation (Fotsis *et al.*, 1993; Schweigerer *et al.*, 1992). This reagent was also shown to inhibit the production of plasminogen activator and plasminogen activator inhibitor in vascular endothelial cells, suggesting a role in matrix metabolism. Genistein precursors are present in soy products, and its role as an angiogenic inhibitor correlates with the epidemiological data showing cultures consuming high soy diets (traditional Japanese) having a lower incidence of vascular tumors (Setchell and Adlercreutz, 1988; Adlercreutz *et al.*, 1991; Muir *et al.*, 1987).

## 16. Synergism between Polysaccharides and Estrogen

Sulfated polysaccharide–peptidoglycan complex, isolated from *Athrobacter*, inhibited embryonic and tumor-induced angiogenesis and the growth of solid tumors (Inoue *et al.*, 1988). More recently, this reagent has been shown to synergize with Tamofexin and α-estrogen, and to reduce angiogenesis to a greater extent (Tanaku *et al.*, 1991).

## 17. Angiostatin

The observation that some tumor masses were able to suppress tumor growth has recently led to the isolation of a 38-kDa inhibitor of angiogenesis, named angiostatin (O'Reilly *et al.*, 1994). This molecule is able to specifically inhibit endothelial cell proliferation, inhibit neovascularization, and the growth of metastases. Angiostatin shares considerable homology to an internal fragment of plasminogen, which corresponds to the first four Kringle regions of the molecule (Lerch *et al.*, 1990). The mechanism of its action is not yet known. It is interesting to note that angiostatin shares structural homology to hepatocyte growth factor (HGF), a glycoprotein suggested to act as a paracrine mediator of angiogenesis (Grant *et al.*, 1993). This raises the intriguing possibility that angiostatin could compete with HGF for its receptor, c-met (Tsarfaty *et al.*, 1992).

## IV. Concluding Remarks

The aim of further research must surely be to devise a more satisfactory regime of treatment to enhance angiogenesis where it would be beneficial, such as in wound healing, and to abrogate the process in solid tumors, where clearly their progress is dependent on the maintenance of a competent vascular supply.

Therapeutic modulation of angiogenesis is shown to be more effective through regimes that combine effective agents. To allow a more relevant evaluation of reagents with potential angiogenic responses it will be important to develop more sophisticated *in vitro* models that more closely parallel the *in vivo* situation. Searching for a single gene that determines this complex process is perhaps a simplistic and naive approach. It seems more likely that greater advances are to be made in understanding the factors that influence those common molecules that we know are altered during angiogenesis. Understanding the factors that alter the extracellular milieu and alter gene expression during early events in angiogenesis will greatly assist the development of clinical regimes that modulate angiogenesis.

## Acknowledgments

The authors wish to thank Dr. Peter Cockerill for many helpful discussions and for critically reviewing this article.

## References

Abe, T., Okamura, K., Ono, M., Kohno, K., Miri, T., Hori, S., and Kuwano, M. (1993). Induction of vascular endothelial tubular morphogenesis by human glioma cells. A model system for tumour angiogenesis. *J. Clin. Invest.* **92,** 54–61.

Abraham, J. A., Mergia, A., Whang, G. L., Tumoto, A., Friedman, J., Hjerrild, K. A., Gospodarowicz, D., and Fiddes, J. (1986). Nucleoside sequence of a bovine clone encoding the angiogenic protein, basic fibroblast growth factor. *Science* **233,** 545–548.

Adlercreutz, H., Hongo, H., Higashi, A., Fotsis, Y., Hamalainen, E., and Okada, H. (1991). Urinary excretion of lignans and isoflavoid phytoestrogens in Japanese men and women consuming a traditional Japanese diet. *Am. J. Clin. Nutr.* **54,** 1093–1100.

Alessandri, G., Raja, K. S., and Guillino, P. M. (1986). Gangliosides promote the angiogenic response. *Invasion Metastasis* **6,** 145–165.

Andrade, S. P., Fan, T.-P. D., and Lewis, G. P. (1987). Quantitative *in-vivo* studies on angiogenesis in a rat sponge model. *Br. J. Exp. Pathol.* **68,** 755–766.

Auerbach, R., Kubai, L., Knighton, D., and Folkman, J. (1974). A simple procedure for the long term cultivation of chick embryo. *Dev. Biol.* **41,** 391–394.

Auerbach, R., Kubai, L., and Sidky, Y. (1976). Angiogenesis induction by tumors, embryonic tissues, and lymphocytes. *Cancer Res.* **36,** 3435–3440.

Auerbach, R., Lu, W. G., Pardon, E., Gumkowski, F., Kaminiska, G., and Kaminski, M. (1987). Specificity of adhesion between murine tumour cells and capillary endothelium; an *in vitro* correlate of preferential metastasis *in vivo*. *Cancer Res.* **47,** 1492–1496.

Ausprunk, D. H., and Folkman, J. (1977). Migration and proliferation of endothelial cells preformed and newly formed blood vessels during tumour angiogenesis. *Microvasc. Res.* **14,** 53–65.

Ausprunk, D. H., Dethlefseu, S. M., and Higgins, E. R. (1991). Distribution of fibronectin, laminin and type IV collagen during development of blood vessels in chick chorioallantoic membrane. *In* "The Development of the Vascular System" (R. W. Feinberg, G. K. Sherrer, and R. Auerbach, eds.), pp. 93–108. Karger, Basel.

Badet, J., Soncin, F., Guitton, J. D., Lamare, O., Cartright, T., and Baritault, D. (1989). Specific binding of angiogenin to calf pulmonary artery endothelial cells. *Proc. Natl. Acad. Sci. U.S.A.* **86,** 8427–8431.

Bagavandoss, P., and Wilks, J. W. (1990). Specific inhibition of endothelial cell proliferation by thrombospondin. *Biochem. Biophys. Res. Commun.* **170,** 867–872.

Balazs, E. A., and Darykiewicz, Z. (1975). *In* "Biology of Fibroblast" (E. Kulonae and J. Pikkaraineaen, eds.), pp. 237–252. Academic Press, New York.

Baldine, V., Roman, A.-M., Bosc-Bierne, I., Amalric, F., and Bouche, G. (1990). Translocation of bFGF to the nucleus is G1 phase cell cycle specific in bovine aortic endothelial cells. *EMBO J.* **9**, 1511–1517.

Basson, C. T., Knowles, W. J., Bell, L., Abelson, S. M., Castronova, V., Liotta, L. A., and Madri, J. A. (1990). Spatiotemporal segregation of endothelial cell inegrins and non-integrin extracellular matrix-binding proteins during adhesion events. *J. Cell. Biol.* **110**, 789–801.

Bastaki, M., Missirlis, E., Klouras, N., Karakiulakis, G., and Maragoudakis, E. (1994). Suppression of angiogenesis by the antitumor agent titanocene dichloride. *Eur. J. Pharmacol.* **251**, 263–269.

Bauer, J., Margolis, M., Schreiner, C., Edgell, C.-J., Azizkhen, J., Lazarowski, E., and Juliano, R. L. (1992). *In vitro* model of angiogenesis using a human endothelial derived permanent cell line: Contribution of induced gene expression, G proteins and integrins. *J. Cell. Physiol.* **153**, 437–449.

Belsky, E., and Toole, B. P. (1983). Hyaluronate and hyaluronidase in the developing chick embryo kidney. *Cell Differ.* **12**, 61–66.

Berkman, R. A., Merrill, M. J., Reinhold, W. C., Monacci, W. T., Saxena, A., Robertson, J. T., Ali, I. U., and Oldfield, E. H. (1993). Expression of vascular permeability factor/vascular endothelial growth factor gene in central nervous system neoplasms. *J. Clin. Invest.* **91**, 153–159.

Berse, B., Brown, L. F., Van Der Water, L., Dvorak, H. F., and Senger, D. R. (1992). Vascular permeability factor (vascular endothelial growth factor) gene is expressed differentially in normal tissues, macrophages and tumours. *Mol. Biol. Cell.* **3**, 211–220.

Blei, F., Wilson, E. L., Mignatti, P., and Rifkin, D. B. (1993). Mechanism of action of angiogenic steroids: Suppression of plasminogen activator activity via stimulation of plasminogen activator inhibitor synthesis. *J. Cell. Physiol.* **155**, 568–578.

Bouche, G., Gas, N., Prats, H., Baldin, V., Tauger, J.-P., Teissie, J., and Amalric, F. (1987). Basic fibroblast growth factor enters the nucleolus and stimulates the transcription of ribosomal genes in BAE cells undergoing G0-G1 transition. *Proc. Natl. Acad. Sci. U.S.A.* **84**, 6770–6774.

Bouck, N. P., Stoler, A., and Polverini, P. J. (1986). Coordinate control of anchorage independence, actin cytoskeleton, and angiogenesis by human chromosome 1 in hamster-human hybrids. *Cancer Res.* **46**, 5101–5105.

Brannström, M., Woessner, J. F., Jr., Koos, R. D., Sear, C. H., and LeMaire, W. J. (1988). *Endocrinology (Baltimore)* **122**, 1715–1721.

Brem, H., and Folkman, J. (1975). Inhibition of tumor angiogenesis mediated by cartilage. *J. Exp. Med.* **141**, 427–439.

Brennan, F. M., Zachariae, C. O., Chantry, D., Larsen, C. G., Turner, M., Maini, R. N., Matsushima, K., and Feldmann, M. (1990). Detection of interleukin 8 biological activity in synovial fluids from patients with rheumatoid arthritis and production of interleukin 8 mRNA by isolated synovial cells. *Eur. J. Immunol.* **20**, 2141–2144.

Brinster, R. L., Chen, H. Y., Messing, A., van Dyke, T., Levin, A. J., and Palmiter, R. D. (1993). Transgenic mice harboring SV40 T-antigen genes develop characteristic brain tumors. *Cell (Cambridge, Mass.)* **37**, 367–379.

Brooks, P. C., Clark, R. A. F., Cheresh, D. A. (1994). Requirement of vascular integrin $\alpha v \beta 3$ for angiogenesis. *Science* **264**, 569–571.

Burgess, W. H., and Maciag, T. (1989). A heparin-binding (fibroblast) growth factor family of proteins. *Annu. Rev. Biochem.* **58**, 575–606.

Butcher, E., Scollary, R. G., and Weissman, I. L. (1980). Organ specificity of lymphocyte migration: Mediation by highly selective lymphocyte interaction with organ-specific determinants on high endothelial venules. *Eur. J. Immunol.* **10**, 556–561.

Canfield, A. E., Schor, A. M., Schor, S. L., and Grant, M. E. (1986). The biosynthesis of extracellular-matrix components by bovine retinal endothelial cells displaying distinctive morphological phenotypes. *Biochem. J.* **235,** 375–383.

Cid, M., Grant, D. S., Hoffman, G. S., Auerbach, R., Fauci, A. S., and Kleinman, H. K. (1993). Identification of haptoglobin as an angiogenic factor in sera from patients with systemic vasculitis. *J. Clin. Invest.* **91,** 977–985.

Clark, E. R., and Clark, E. L. (1937). Observations on living mammalian lymphatic capillaries, their relationship to the blood vessels. *Am. J. Anat.* **60,** 253–298.

Clinton, M., Loud, W. F., Williamson, F. B., Duncan, J. I., and Thompson, W. D. (1988). Effect of the mast cell activator compound 48/80 and heparin on angiogenesis in the chick chorioallantoic membrane. *Int. J. Microcirc.: Clin Exp.* **7,** 315–326.

Cockerill, G. W. (1995). In preparation.

Connolly, D. T., Heuvelman, D. M., Nelson, R., Olander, J. V., Epplesey, B. L., Din, J. J., Siegel, W. R., Leimgruber, R. M., and Feder, J. (1989). Tumour vascular permeability factor stimulates endothelial cell growth and angiogenesis. *J. Clin. Invest.* **84,** 1470–1478.

Cotta-Pereira, G., Sage, H., Bornstein, P., Ross, R.., and Schwartz, S. (1980). Studies of morphologically atypical ("sprouting") cultures of bovine aortic endothelial cells. Growth characteristics and connective tissue protein synthesis. *J. Cell. Physiol.* **102,** 183–191.

Crum, R., Szabo, S., and Folkman, J. (1985). A new class of steroids inhibits angiogenesis in the presence of heparin or a heparin fragment. *Science* **230,** 1375–1378.

D'Amore, P. A., and Thompson, R. W. (1987). Mechanisms of angiogenesis. *Annu. Rev. Physiol.* **49,** 453–464.

Davis, C. M., Danehower, S. C., Laurenza, A., and Malony, J. L. (1993). Identification of a role of the vitronectin receptor and PKC in the induction of endothelial cell vascular formation. *J. Cell. Biochem.* **51,** 206–218.

De Cristan, G., Morbidelli, L., Alessandri, G., Ziche, M., Cappa, A. P. M., and Guillino, P. M. (1990). Synergism between gangliosides and basic fibroblast growth factor in favoring survival, growth, and motility of capillary endothelium. *J. Cell. Physiol.* **144,** 505–510.

DeMarco, D., Kunkel, S. L., Strieter, R. M., Basha, M., and Zivier, R. B. (1991). Interleukin-1 induced gene expression of neutrophil activating protein (interleukin 8) and monocyte chemotactic peptide in human synovial cells. *Biochem. Biophys. Res. Commun.* **174,** 411–416.

Devrotes, P. (1989). *Dictyostelium discoideum:* A model system for cell-cell interaction in development. *Science* **245,** 1054–1058.

Downs, E. C., Robertson, N. E., Ris, T. L., and Plunkett, M. L. (1982). Calcium alginate beads as a slow release system for delivering angiogenic molecules *in vivo* and *in vitro*. *J. Cell. Physiol.* **152,** 422–429.

Duncan, J. I., Brown, F. I., McKinnon, A., Long, W. F., Williamson, F. B., and Thompson, W. P. (1992). Pattern of angiogenic response to mast cell granule constituents. *Int. J. Microcirc.: Clin. Exp.* **11,** 21–33.

Dunn, B. E., Fitzharn, S. T. P., and Barnett, B. D. (1981). Effects of varying chamber construction and embryo pre-incubation age on survival and growth of chick embyros in shell-less culture. *Anat. Rec.* **199,** 33–43.

Dziadek, M., Paulsson, M., Aumalley, M., and Timpl, R. (1986). Purification and tissue distribution of a small protein (BM-40) extracted from a basement membrane tumor. *Eur. J. Biochem.* **161,** 455–464.

Eisenstein, R., Kuettner, K. E., Neopolitan, C., Soble, L. W., and Sorgente, N. (1975). The resistance of certain tissues to invasion. III. Cartilage extracts inhibit the growth of fibroblasts and endothelial cells in culture. *Am. J. Pathol.* **81,** 337–348.

Emonard, H., Calle, A., Grimaud, J.-A., Peyrol, S., Castronova, V., and Noel, A. (1987).

Interactions between fibroblasts and a reconstitutes membrane matrix. *Invest. Dermatol.* **89**, 156–163.

Enerback, L., and Norrby, K. (1989). The mast cell. *Curr. Top. Pathol.* **79**, 169–204.

Eriksson, S., Fraser, J. R., Laurent, T. C., Pertoft, H., and Smedrod, B. (1983). Endothelial cells are a site of uptake and degradation of hyaluronic acid in the liver. *Exp. Cell Res.* **144**, 223–228.

Esch, F., Baird, A., Ling, N., Ueno, N., Hill, F., Denoroy, L., Klepper, R., Gospodarowicz, D., Bohlen, P., and Guillemin, R. (1985). Primary structure of bovine pituitary basic fibroblast growth factor (FGF) and comparison with the amino-terminal sequence of bovine brain acidic FGF. *Proc. Natl. Acad. Sci. U.S.A.* **82**, 6507–6511.

Fajardo, L. F., Kowalski, J., Khwan, H. H., Prionas, S. P., and Allison, A. C. (1988). The disc angiogenesis system. *Lab. Invest.* **58**, 718–724.

Falcone, D. J., McCaffrey, T. A., Halmovitz-Friedman, A., and Garcia, M. (1993). Transforming growth factor $\beta$1 stimulated macrophage urokinase expression and release of matrix bound basic fibroblast growth factor. *J. Cell. Physiol.* **155**, 595–605.

Feinberg, R. N., and Beebe, D. C. (1983). Hyaluronate in vasculogenesis. *Science* **220**, 1177–1179.

Ferrara, N., Houck, K., Jakema, L., and Leung, D. W. (1992). Molecular and biological properties of the vascular endothelial growth factor family of proteins. *Endocrinology (Baltimore)* **13**, 18–32.

Fett, J. W., Strydom, D. L., Lobb, R. R., Alderman, E. M., Bethume, J. L., Riordan, J. F., and Vallee, B. L. (1985). Isolation and characterization of angiogenin, and angiongenic protein from human carcinoma cells. *Biochemistry* **24**, 5480–5486.

Folkman, J. (1985). Angiogenesis and its inhibitors. *In* "Important Advances in Oncology" (V. T. DeVita, Jr., S. Hellman, and S. A. Rosenberg, eds.), pp. 42–62. Lippincott, Philadelphia.

Folkman, J. (1990). What is the evidence that tumors are angiogenesis dependant? *J. Natl. Cancer Inst.* **82**, 4–6.

Folkman, J., and Ingber, D. E. (1987). Angiostatic steroids. Method of discovery and mechanisms of action. *Ann. Surg.* **206**, 374–383.

Folkman, J., and Klagsbrun, M. (1987). Anigogenic factors. *Science* **235**, 442–447.

Folkman, J., Haudenschild, B. C., and Zetter, B. R. (1979). Long-term culture of capillary endothelial cells. *Proc. Natl. Acad. Sci. U.S.A.* **76**, 5217–5221.

Folkman, J., Klagsbrun, M., Sasse, J., Wadzinski, M., Ingber, D., and Vlodavsky, I. (1988). A heparin-binding angiogenic factor—basic fibroblast growth factor—is stored within basement membrane. *Am. J. Pathol.* **130**, 393–400.

Folkman, J., Watson, K., Ingber, D., and Hanahan, D. (1989a). Induction of angiogenesis during the transition from hyperplasia to neoplasia. *Nature (London)* **339**, 58–61.

Folkman, J., Weisz, P. B., Joullie, M. M., Li, W. W., and Ewing, W. R. (1989b). Control of angiogenesis with synthetic heparin substitutes. *Science* **243**, 1490–1493.

Folkval, K. H. (1991). A method to quantify neovascularization in the mouse cornea. *Ophthalmic Res.* **23**, 935–939.

Form, D. M., Pratt, B. M., and Madri, J. A. (1986). Endothelial cell proliferation during angiogenesis. *In vitro* modulation by basement membrane components. *Lab. Invest.* **55**, 521–530.

Fotsis, T., Pepper, M. S., Aldercrutz, H., Fleischmann, G., Hase, T., Montesano, R., and Schweigerer, L. (1993). Genistein, a dietary-derived inhibitor of *in vitro* angiogenesis. *Proc. Natl. Acad. Sci. U.S.A.* **90**, 2690–2694.

Fotsis, T., Zhang, Y., Pepper, M. S., Aldercreutz, H., Montesano, R., Nauroth, P. P., and Schweigerer, L. (1994). The endogenous oestrogen metabolite 2-methoxyoestradiol inhibits angiogenesis and suppresses tumour growth. *Nature (London)* **368**, 237–239.

Fournier, G. A., Lutty, G. A., Watt, S., Fensellar, A., and Patz, A. (1981). A corneal micropocket assay for angiogenesis in the rat eye. *Invest. Ophthamol. Visual Sci.* **21,** 351–354.

Fox, E. A., and Riordan, J. F. (1990). Molecular biology of angiogenesis. *In* "Molecular Biology of the Cardiovascular System" (S. Chien, ed.), pp. 139–154. Lea & Febiger, Philadelphia.

Frater-Schroder, M., Muller, G., Birchmeier, M., and Bohlen, P. (1986). Transforming growth factor-beta inhibits endothelial cell proliferation. *Biochem. Biophys. Res. Commun.* **137,** 295–302.

Frater-Schroder, M., Risau, W., Hallmann, R., Gautsch, P., and Bohlen, P. (1987). Tumour necrosis factor type $\alpha$, a potent inhibitor of endothelial cell growth *in vitro* is angiogenic *in vivo. Proc. Natl. Acad. Sci. U.S.A.* **84,** 5277–5281.

Funk, S. E., and Sage, H. (1991). The $Ca^{2+}$-binding glycoprotein SPARC modulates cell cycle progression in bovine aortic endothelial cells. *Proc. Natl. Acad. Sci. U.S.A.* **88,** 2648–2652.

Furcht, L. T. (1986). Critical factors controlling angiogenesis: Cell products, cell matrix, and growth factors. *Lab. Invest.* **55,** 505–509.

Furie, M. B., Cramer, E. B., Naprstek, B. L., and Silverstein, S. C. (1984). Cultured capillary endothelial cell monolayers that restrict the transendothelial passage of macro-molecules and electrical current. *J. Cell Biol.* **98,** 1022–1041.

Gadjusek, C. M., and Carbon, S. (1989). Injury-induced release of basic fibroblast growth factor from bovine aortic endothelium. *J. Cell. Physiol.* **139,** 570–579.

Gamble, J. R., Mathias, L. J., Meyer, G., Kaur, P., Russ, G., Faull, R., Berndt, M. C., and Vadas, M. A. (1993). Regulation of *in vitro* capillary tube formation by anti integrin antibody. *J. Cell Biol.* **121,** 931–943.

Garrison, J. C. (1990). Histamine, bradykinin, 5-hydroxytryptamine, and their antagonists. *In* "The Pharmacological Basis of Therapeutics" (A. G. Gilman, T. W. Rall, A. S. Nies, and P. Taylor, eds.), 8th ed., p. 578. Pergamon, New York.

Gimbrone, M., Cotran, R. S., Leapman, S. B., and Folkman, J. (1974). Tumor growth and neovascularisation: An experimental model using the rabbit corneal. *J. Natl. Cancer Inst.* *(U.S.)* **52,** 413–427.

Good, D. J., Polverini, R. J., Rastinejad, F., Le Beau, M. N., Leuras, R. S., Frazer, W. A., and Bouck, N. P. (1990). A tumor suppressor dependent inhibitor of angiogenesis is immunologically and functionally indistinguishable from a fragment of thrombospondin. *Proc. Natl. Acad. Sci. U.S.A.* **87,** 6624–6628.

Gospodarowicz, D., and Thakral, K. K. (1978). Production of a corpus luteum angiogenic factor responsible for proliferation of capillaries and neovascularization of the corpus luteum. *Proc. Natl. Acad. Sci. U.S.A.* **75,** 847–851.

Gospodarowicz, D., Chen, J., Lui, G. M., Baird, A., and Bohlen, P. (1984). Isolation by heparin-sepharose affinity chromatography of brain fibroblast growth factor: Identity with pituitary fibroblast growth factor. *Proc. Natl. Acad. Sci. U.S.A.* **81,** 6963–6967.

Grant, D. S., Tashiro, K. I., Segui-Real, B., Yamada, Y., Martin, G. R., and Kleinman, H. K. (1989). Two different laminin domains mediate the differentiation of human endothelial cells into capillary-like structures *in vitro. Cell (Cambridge, Mass.)* **58,** 933–943.

Grant, D. S., Lelkes, P. I., Fukuda, K., and Kleinman, H. K. (1991). Intracellular mechanism involved in basement membrane induced blood vessel differentiation. *In Vitro Cell Dev. Biol.* **27A,** 327–336.

Grant, D. S., Kleinman, H. K., Goldberg, I. D., Bhargara, M. M., Nickoloff, B. J., Kinsella, J. L., Polverini, P., and Rosen, E. M. (1993). Scatter factor induces blood vessel formation in-vivo. *Proc. Natl. Acad. Sci. U.S.A.* **90,** 1937–1941.

Gross, J. L., Moscatelli, D., Jaffe, E. A., and Rifkin, D. B. (1982). Plasminogen activator and collagenase production by cultured capillary endothelial cells. *J. Cell Biol.* **95**, 974–981.

Gross, J. L., Moscatelli, D., and Rifkin, D. B. (1983). Increased capillary endothelial cell protease activity in response to angiogenic stimuli *in vitro*. *Proc. Natl. Acad. Sci. U.S.A.* **80**, 2623–2627.

Guo, J., Jourdian, G. W., and MacCullum, D. K. (1989). Culture and growth characteristics of chondrocytes encapsulated in alginate beads. *Connect. Tissue Res.* **19**, 277–297.

Hadley, M. A., Weeks, B. S., and Kleinman, H. K. (1990). Laminin promotes formation of cord-like structures by Stertoli cells *in vitro*. *Dev. Biol.* **140**, 318–327.

Hallahan, T. W., Shapiro, R., and Vallee, B. L. (1990). Dual site mode for the organogenic activity of angiogenin. *Proc. Natl. Acad. Sci. U.S.A.* **88**, 2222–2226.

Harning, R., and Szalay, J. (1988). A treatment for metastasis of murine ocular melanoma. *Invest. Ophthalmol. Visual Sci.* **29**, 1505–1510.

Harper, J. N., Fox, E. A., Shapiro, R., and Vallee, B. L. (1989). Mutagenesis of residues flanking Lys-40 enhances the enzymatic activity and reduces the angiogenic potency of angiogenin. *Biochemistry* **29**, 7297–7302.

Hascall, V. C., and Heinegard, D. (1974). Aggregation of cartilage proteoglycans. II. Oligosaccharide competitors of the proteoglycan-hyaluronic acid interaction. *J. Biol. Chem.* **249**, 4242–4249.

Haynes, W. L., Proia, A. D., and Klintworth, G. K. (1989). Effect of inhibitors of angiogenesis on corneal neovascularization in the rat. *Invest. Ophthalmol. Sci.* **30**, 1588–1593.

Haystead, T. A. J., Sim, A. T. R., Carling, D., Honnor, R. C., Tsukitani, Y., Cohen, P., and Hardie, D. G. (1989). Effects of the tumour promoter okadaic acid on intracellular protein phosphorylation and metabolism. *Nature (London)* **337**, 78–81.

Heinmark, R. L., Twardzik, D. R., and Schwartz, S. T. (1986). Inhibition of endothelial regeneration by type-beta like transforming growth factor from platelets. *Science* **233**, 1078–1080.

Hendkin, P. (1978). Ocular neovascularization. The Krill memorial lecture. *Am. J. Ophthalmol.* **85**, 287–301.

Herron, G. S., Werb, Z., Dwyer, K., and Banda, M. J. (1986). Secretion of metalloproteinases by stimulated capillary endothelial cells. *J. Biol. Chem.* **261**, 2810–2814.

Hiti-Harper, J., Wohl, H., and Harper, E. (1978). Platelet factor 4: An inhibitor of collagenase. *Science* **199**, 991–992.

Hla, T., and Maciag, T. (1990). An abundant transcript induced by differentiating human endothelial cells encodes a polypeptide with structural similarities to G-protein-coupled receptors. *J. Biol. Chem.* **265**, 9308–9313.

Holladay, K., Fujiki, H., and Bowden, T. (1992). Okadaic acid induced expression of both early and secondary response genes in mouse kerotinocytes. *Mol. Carcinog.* **5**, 16–24.

Houck, K. A., Ferrara, N., Winer, J., Cachianes, L. G., Li, B., and Leung, D. W. (1991). The vascular endothelial growth factor. Identification of a fourth molecular species and characterisation of alternative splicing of RNA. *Mol. Endocrinol* **5**, 1806–1814.

Hu, G.-F., Change, S.-I., Riordan, J. F., and Vallee, B. L. (1991). An angiogenin-binding protein from endothelial cells. *Proc. Natl. Acad. Sci. U.S.A.* **88**, 2227–2231.

Hu, G.-F., Strydom, D. J., Fett, J. W., Riordan, J. F., and Vallee, B. L. (1993). Actin is the binding protein for angiogenin. *Proc. Natl. Acad. Sci. U.S.A.* **90**, 1217–1221.

Huber, A. R., Kunkel, S. L., Todd, R. F., II, and Weiss, S. J. (1991). Regulation of transendothelial neutrophil migration by endogenous interleukin 8. *Science* **254**, 99–102.

Ingber, D., and Folkman, J. (1988). Inhibition of angiogenesis through modulation of collagen metabolism. *Lab. Invest.* **59**, 44–51.

Ingber, D. E., and Folkman, J. (1989a). How does extracellular matrix control capillary morphogenesis? *Cell (Cambridge, Mass.)* **58**, 803–805.

152 GILLIAN W. COCKERILL *ET AL.*

Ingber, D. E., and Folkman, J. (1989b). Mechanicochemical switching between growth and differentiation during fibroblast growth factor-stimulated angiogenesis *in vitro:* Role of extracellular matrix. *J. Cell Biol.* **109**, 317–330.

Ingber, D. E., Madri, J., and Folkman, J. (1986). A possible mechanism of inhibition of angiogenesis by angiostatic steroids: Induction of capillary basement membrane dissolution. *Endocrinology (Baltimore)* **119**, 1768–1775.

Ingber, D. E., Prusty, D., Fragioni, J. V., Cragoe, E. J., Jr., Lechene, C., and Schwartz, M. A. (1990). Control of intracellular pH and growth by fibronectin in capillary endothelial cells. *J. Cell Biol.* **110**, 1803–1811.

Inoue, K., Korenga, H., Tanaka, N. G., Sakamoto, N., and Kadoya, S. (1988). The sulfated polysaccharide-peptidoglycan complex potently inhibits embryonic angiogenesis and tumour growth in the presence of cortisone acetate. *Carbohydr. Res.* **181**, 135–142.

Iruela-Arispe, M. L., Borstein, P., and Sage, H. (1991). Thrombospondin exerts an anti-angiogenic effect on cord formation by endothelial cells *in vitro. Proc. Natl. Acad. Sci. U.S.A.* **88**, 5026–5030.

Koch, A. E., Polverini, P. J., Kunkel, S. L., Harlow, L. A., DiPietro, L. A., Elner, V. M., Elner, S. G., and Stricter, R. M. (1992). Interleukin-8 as a macrophage derived mediator of angiogenesis. *Science* **258**, 1798–1801.

Jackson, C. J., and Jenkins, K. L. (1991). Type I collagen fibrils promote rapid vascular tube formation upon contact with the apical side of cultured endothelium. *Exp. Cell Res.* **192**, 319–323.

Jakob, W., and Voss, K. (1984). Utilization of image analysis for the quantitation of vascular response in the chick chorioallantoic membrane. *Exp. Pathol.* **26**, 93–99.

Jakob, W., Jentzch, K. D., Mauersberger, B., and Oehme, P. (1977). Demonstration of angiogenesis activity in the corpus luteum of cattle. *Exp. Pathol.* **13**, 231–236.

Jakobsson, A., and Norrby, K. (1991). Kinetics of a mammalian angiogenic response. *Cell Proliferation* **24**, 59.

Jimenez, S. A., and Rosenbloom, J. (1974). Stability of collagens containing analogus of proline or lysine. *Arch. Biochem. Biophys.* **163**, 459–465.

Kalebić, T., Garbisa, S., Glaser, B., and Liotta, L. A. (1983). Basement membrane collagen; degradation by migrating endothelial cells. *Science* **221**, 281–283.

Kallard, T. (1986). Effect of immunomodulators LS-2616 on growth and metastasis of the murine B16-F10 melanoma. *Cancer Res.* **46**, 3018–3022.

Keck, P. J., Hauser, S. D., Krivi, G., Sanzo, K., Warren, T., Feder, J., and Connolly, D. T. (1989). Vascular permeability factor, an endothelial cell mitogen related to PDGF. *Science* **246**, 1309–1311.

Ketis, N. V., Lawler, J., and Bendena, W. G. (1993). Extracellular matrix components affect the pattern of protein synthesis of endothelial cells responding to hyperthermia. *In Vitro Cell Dev. Biol.* **29A**, 768–772.

Kibbey, M. C., Grant, D. S., Auerbach, R., and Kleinman, H. K. (1992). The SIKVAV site of laminin promotes angiogenesis and tumour growth in an *in vivo* matrigel model. *J. Natl. Cancer Inst.* **84**, 1633–1638.

Kim, K.-J., Li, B., Houck, K., Winer, J., and Ferrera, N. (1992). The vascular endothelial growth factor proteins: Identification of biological relevant regions by neutralizing monoclonal antibodies. *Growth Factors* **7**, 53–64.

Kim, K.-J., Bing, L. I., Winer, J., Armanin, M., Gillett, N., Phillips, H. S., and Ferrara, N. (1993). Inhibition of vascular endothelial growth factor-induced angiogenesis suppresses growth *in vitro. Nature (London)*, **362**, 841–844.

Kim, S. J., Lafyatis, R., Kim, K. Y., Angel, P., Fujiki, H., Karin, M., Sporn, M. B., and Roberts, A. B. (1990). Regulation of collagenase gene expression by okadaic acid, an inhibitor of protein phosphorylation. *Cell Regul.* **1**, 269–278.

Kinsella, J. L., Grant, D. S., Weeks, B. S., and Kleinman, H. K. (1992). Protein kinase C regulates endothelial cell tube formation on basement membrane Matrigel. *Exp. Cell Res.* **199**, 56–62.

Klagsbrun, M., and D'Amore, P. A. (1991). Regulators of angiogenesis. *Annu. Rev. Physiol.* **53**, 217–239.

Klein, G., Langegger, M., Timpl, R., and Ekblom, P. (1988). Role of laminin A-chain in the development of epithelial cell polarity. *Cell (Cambridge, Mass.)* **55**, 331–341.

Kleinman, H. K., McGarvey, M. L., Liotta, L. A., Robey, P. G., Tryggvason, K., and Martin, G. R. (1982). Isolation and characterisation of type IV procollagen, laminin, and heparin sulfate proteoglycan from the EHS sarcoma. *Biochemistry* **21**, 6188–6193.

Klohs, W. D., Steinkampf, R. W., Wicha, M. S., Merlins, A. E., Tunas, J. B., and Leopold, W. R. (1985). Collagen production inhibitors evaluated as antitumour-agents. *JNCI, J. Natl. Cancer Inst.* **75**, 353–359

Knighton, D. R., Phillips, G. D., and Fiegel, V. D. (1990). Wound healing angiogenesis: Indirect stimulation by basic fibroblast growth factor. *J. Trauma* **30** (Suppl. 12), S134–S144.

Koch, A. E., Ploverini, P. J., Kunkel, S. L., Harlow, L. A., DiPietro, L. A., Elner, V. M., Elner, S. G., and Strieter, R. M. (1992). Interleukin-8 as a macrophage-derived mediator of angiogenesis. *Science* **258**, 1798–1800.

Kubota, Y., Kleinman, H. K., Martin, G. R., and Lawley, T. J. (1988). Role of laminin and basement membrane in the morphological differentiation of human endothelial cells in capillary-like structure. *J. Cell Biol.* **107**, 1589–1598.

Kusaka, M., Sudo, K., Matsutani, E., Kozai, Y., Fijita, T., Ingber, D., and Folkman, J. (1994). Cytostatic inhibition of endothelial cell growth by the angiogenesis inhibitor TNP-470 (AGM-1470). *Br. J. Cancer* **69**, 212–216.

Langer, R., Brem, H., Falterman, K., Klein, M., and Folkman, J. (1976). Isolation of a cartilage factor that inhibits tumor neovascularization. *Science* **193**, 70–72.

Langer, R., Conn, H., Vacanti, J., Haudenschild, J. C., and Folkman, J. (1980). Control of tumour growth in animals by infusion of an angiogenic inhibitor. *Proc. Natl. Acad. Sci. U.S.A.* **77**, 4331–4335.

Larsen, C. G., Anderson, A. O., Apella, O., Oppenheim, J. J., and Matsushima, K. (1989). The neutrophil activating protein (NAP-1) is also chemotactic for T lymphocytes. *Science* **243**, 1464–1466.

Lattera, J., and Goldstein, G. W. (1991). Astroglial-induced *in vitro* angiogenesis: Requirements for RNA and protein synthesis. *J. Neurochem.* **57**, 1231–1239.

Lattera, J., Guérin, C., and Goldstein, G. W. (1990). Astrocytes induce neural microvascular endothelial cells to form capillary-like structures *in vitro*. *J. Cell. Physiol.* **144**, 204–215.

Lerch, D. G., Rickli, E. E., Legier, W., and Gillessen, D. (1980). Localization of individual lysine binding regions in human plasminogen and investigation of their complex-forming properties. *Eur. J. Biochem.* **107**, 7–13.

Leung, D. W., Cachiane, G., Kuang, W.-J., Goeddel, D. V., and Ferrara, N. (1989). Vascular endothelial growth factor is a secreted angiogenic mitogen. *Science* **246**, 1306–1309.

Levin, E. G., and Santell, L. (1987). Association of plasminogen activator inhibitor (PAI-1) with the growth substratum and membrane of human endothelial cells. *J. Cell Biol.* **105**, 2543–2549.

Levy, J. P., Fujiki, H., Angel, P., and Bowden, G. T. (1991). An inactive mutant TPA responsive element acts as an active okadaic acid response element in mouse keratinocytes. *Proc. Am. Assoc. Cancer Res.* **32**, 291.

Lewis, W. H. (1925). The outgrowth of endothelium and capillaries in tissue culture. *Bull. Johns Hopkins Hosp.* **48**, 242–253.

Lewko, W. M., Liotta, L. A., Wicha, M. S., Vanderhaar, B. K., and Kidwell, W. R. (1981). Sensitivity of *N*-nitrosomethylurea-induced rat mammary tumour to *cis*-hydroxyproline, an inhibitor of collagen production. *Cancer Res.* **41,** 2855–2862.

Liebovich, S. J., Polverini, P. J., Shephard, H. M., Wiseman, D. M., Shively, V., and Nuseir, N. (1987). Antibodies to TNFa neutralise the activity in conditioned medium of thioglycollate-induced peritroneal macrophages. *Nature (London)* **329,** 630–632.

Lobb, R., Sasse, J., Sullivan, R., Shing, Y., D'Amore, P., Jacobs, J., and Klagsbrun, M. (1986). Purification and characterisation of heparin-binding endothelial cell growth factors. *J. Biol. Chem.* **261,** 924–1928.

Lowe, J. B., Stoolman, L. M., Nair, R. P., Larsen, R. D., Berherd, T. C., and Marks, R. M. (1990). ELAM-1-dependant cell adhesion to vascular endothelium determined by a transfected human fucosyltransferase cDNA. *Cell (Cambridge, Mass.)* **63,** 475–484.

Macchiarini, P., Fontanini, G., Harden, M. J., Squartini, F., and Aggelli, C. A. (1992). Relationship of neovascularisation to metastasis of non-small cell lung cancer. *Lancet* **340,** 145–146.

MacClusky, W. J., Barnea, E. C., Clark, C. R., and Naftolin, F. (1983). *In* "Catechol Estrogens" (G. R. Merrian and M. B. Lipsett, eds.), pp. 151–165.

Maciag, T., Kadish, J., Wilkins, L., Stemerman, M. B., and Weinstein, R. (1982). Organizational behaviour of human vein endothelial cells. *J. Cell Biol.* **94,** 511–520.

Madri, J. A., Williams, S. K., Wyatt, T., and Mezzio, C. (1983). Capillary endothelial cell cultures: Phenotype modulation by matrix components. *J. Cell Biol.* **97,** 153–166.

Madri, J. A., Pratt, B. M., and Tucker, A. M. (1988). Phenotypic modulation of endothelial cells by transforming growth factor-$\beta$ depends upon the composition and organization of the extracellular matrix. *J. Cell Biol.* **106,** 1375–1384.

Mahaderan, V., Hart, R., and Lewis, G. P. (1989). Factors influencing blood supply in wound granuloma quantitated by a new *in-vivo* technique. *Cancer Res.* **49,** 415–419.

Maione, T. E., Gray, G. S., Petro, J., Hunt, A. J., Donner, A. L., Bauer, S. I., Carson, H. F., and Sharper, K. J. (1990). Inhibition of angiogenesis by recombinant human platelet factor-4 and related peptides. *Science* **247,** 77–79.

Majewski, S., Kamisinski, M. J., Szurto, A., Kuminska, G., and Melejczk, J. (1984). Inhibition of tumor induced angiogenesis by systemically administered protamine sulfates. *Int. J. Cancer* **33,** 831–833.

McNeil, P. L., Muthukrishnan, L., Warder, E., and D'Amore, P. A. (1989). Growth factors are released by mechanically wounded endothelial cells. *J. Cell Biol.* **109,** 811–822.

Mignatti, P., Tsuboi, R., Robbins, E., and Rifkin, D. B. (1989). In vitro angiogenesis on the human amniotic membrane: Requirement for basic fibroblast growth factor-induces proteinases. *J. Cell Biol.* **108,** 671–682.

Montesano, R., and Orci, L. (1985). Tumor-promoting phorbol ester induce angiogenesis *in vitro. Cell (Cambridge, Mass.)* **42,** 469–477.

Montesano, R., and Orci, L. (1987). Phorbol ester induces angiogenesis *in vitro* from large vessel endothelial cell. *J. Cell Physiol.* **139,** 284–291.

Montesano, R., Vasalli, J. D., Baird, A., Guilleman, R., and Orci, L. (1986). Basic fibroblast growth factor induces angiogenesis *in vitro. Proc. Natl. Acad. Sci. U.S.A.* **83,** 7297–7301.

Montesano, R., Pepper, M. S., Vassalli, J. D., and Orci, L. (1987). Phorbol ester induces cultured endothelial cells to invade a fibrin matrix in the presence of fibrinolytic inhibitors. *J. Cell. Physiol.* **132,** 460–466.

Moroianu, J., and Riordan, J. F. (1994). Nuclear translocation of angiogenin in proliferating endothelial cells is essential to its angiogenic activity. *Proc. Natl. Acad. Sci. U.S.A.* **91,** 1677–1681.

Morris, E. R., Rees, D. A., and Welsh, E. J. (1980). Conformational and dynamic interactions in hyaluronate solutions. *J. Mol. Biol.* **138,** 383–400.

Moscatelli, D. (1986). Urokinase-type and tissue-type plasminogen activators have different distributions in cultured bovine capillary endothelial cells. *J. Cell. Biochem.* **30,** 19–29.

Moscatelli, D., Jaffe, E., and Rifkin, D. B. (1980). Tetradecanoyl phorbol acetate stimulates latent collagenase production by cultured human endothelial cells. *Cell (Cambridge, Mass.)* **20,** 343–351.

Moscatelli, D., and Rifkin, D. B. (1988). Membrane and matrix localisation of proteinase: A common theme in tumour cell invasion and angiogenesis. *Biochim. Biophys. Acta* **948,** 67–85.

Moses, M. A., Sudhalter, J., and Langer, R. (1990). Identification of an inhibitor of neovascularisation from cartilage. *Science* **248,** 1408–1410.

Muir, C., Waterhouse, J., Powell, M. T., and Whelan, S. (1987). "Cancer Incidence in Five Continents," Vol. 5. Int Agency Res Cancer, Lyon, France.

Muller, G., Behrems, J., Nussbaumer, U., Bohlen, P., and Birchmeier, W. (1987). Inhibitory action of transforming growth factor $\beta$ on endothelial cells. *Proc. Natl. Acad. Sci. U.S.A.* **84,** 5600–5604.

Murphy-Ullrich, J. E., and Hook, M. (1989). Thrombospondin modulates local adhesion in endothelial cells. *J. Cell Biol.* **109,** 1309–1319.

Murphy-Ullrich, J. E., Lightener, V. A., Aukhil, I., Yan, Y. Z., Erikson, H. P., and Hook, M. (1991). Focal adhesion integrity is down regulated by the alternatively spliced domain of human tenascin. *J. Cell Biol.* **115,** 1127–1136.

Murphy-Ullrich, J. E., Schultz-Cheny, S., and Hook, M. (1992). TGFb complexes with thrombospondin. *Mol. Biol. Cell.* **3,** 181–188.

Muthukkaruppan, V. R., and Auerbach, R. (1979). Angiogenesis in the mouse cornea. *Science* **205,** 1416–1418.

Muthukkaruppan, V. R., Kubui, I., and Auerbach, R. (1982). Tumour-induced neovascularization in the mouse eye. *JNCI, J. Natl. Cancer Inst.* **69,** 699–708.

Nabel, E. G., Yang, Z.-Y., Plautz, G., Forough, R., Zhan, X., Haudenschild, C. C., Maciag, T., and Nabel, G. J. (1993a). Recombinant fibroblast growth factor-1 promotes intimal hyperplasia and angiogenesis in arteries *in vivo. Nature (London)* **362,** 844–846.

Nabel, E. G., Yang, Z. Y., Lipsay, S., San, H., Haudenschild, C. C., and Nabel, G. J. (1993b). Recombinant platelet-derived growth factor B gene expression in porcine arteries induce intimal hyperplasia *in vivo. J. Clin. Invest.* **91,** 1822–1829.

Neufeld, G., and Gospodarowicz, D. (1987). Protamine sulfate inhibits mitogenic activities of the extracellular matrix and fibroblast growth factor, but potentiates that of epidermal growth factor. *J. Cell. Physiol.* **132,** 287–294.

Nguyen, M., Folkman, J., and Bischoff, J. (1992). 1-Deoxymannojirimycin inhibits capillary tube formation *in vitro. J. Biol. Chem.* **267,** 26157–26165.

Nicosia, R. F., and Ottinetti, A. (1990). Growth of microvessels in serum-free matrix culture of rat aorta. *Lab. Invest.* **63,** 115–122.

Nicosia, R. F., and Tuszynski, G. P. (1994). Matrix bound thrombospondin promotes angiogenesis *in vitro. J. Cell Biol.* **124,** 184–193.

Nomura, S., Hogan, B. L. M., Wills, A. J., Heath, J. K., and Edwards, D. R. (1989). Developmental expression of tissue inhibitor of metalloproteinase (TIMP) RNA. *Development (Cambridge, UK)* **105,** 575–583.

Norrby, K. (1992). Cyclosporin is angiostatic. *Experientia* **48,** 1135–1138.

Norrby, K. (1993). L-Proline, LACA, inhibits mast cell induced angiogenesis. *Int. J. Microcirc.: Clin. Exp.* **12,** 119–129.

Norrby, K., and Sorbo, J. (1992). Heparin enhances angiogenesis by a systemic mode of action. *Int. J. Exp. Pathol.* **73,** 147–155.

Norrby, K., Jacobsson, A., and Sorbo, J. (1986). Mast-cell mediated angiogenesis: A novel experimental model using the rat mesentery. *Virchows Arch. B* **52,** 195–206.

Norrby, K., Jakobsson, A., and Sorbo, J. (1990). Quantitative angiogenesis in spreads of intact rat mesenteric windows. *Microvasc. Res.* **39**, 341–348.

Ohuchi, K., Tamura, T., Oshashi, M., Watanabe, M., Hirasawa, N., Tsunifuji, S., and Fujiki, H. (1989). Okadaic acid and dinophysistoxin 1, non-TPAptype tumor promoters stimulate E2 production in rat peritoneal macrophages. *Biochim. Biophys. Acta* **1013**, 86–91.

Oikawa, T., Suganuma, M., Ashino-Fuse, H., and Shimamura, M. (1992). Okadaic acid is a potent angiogenesis inducer. *Jpn. J. Cancer Res.* **83**, 6–9.

Okamura, K., Morimoto, A., Hamanaka, R., Ono, M., Kohno, K., Uchida, Y., and Kuwano, M. (1992). A model system for tumour angiogenesis: Involvement of transforming growth factor-α in tube formation of human microvascular endothelial cells induced by oesophageal cancer cells. *Biochem. Biophys. Res. Commun.* **186**, 1471–1479.

Olsson, A., Elling, P., and Elling, H. (1990). Serological and immunohistochemical determination of von Willebrand factor antigen in serum and biopsy specimens from patients with arteritis temporalis and polymyalgia rheumatica. *Clin. Exp. Rheumatol.* **8**, 55–58.

Ono, M., Okamura, K., Nakayama, Y., Tomita, M., Sato, Y., Nomatsu, Y., and Kuwano, M. (1992). Induction of human microvascular endothelial tubular morphogenesis by human keratinocytes: Involvement of transforming growth factor-α. *Biochem. Biophys. Res. Commun.* **189**, 601–609.

O'Reilly, M. S., Holmgren, L., Shing, Y., Chen, C., Resenthal, R. A., Moses, M., Lane, W. S., Cao, Y., Sage, E. H., and Foldman, J. (1994). Angiostatin: A novel angiogenesis inhibitor that mediates the suppression of metastases by a Lewis Lung Carcinoma. *Cell* **79**, 315–328.

O'Shea, K. S., and Dixit, V. (1988). Unique distribution of extracellular matrix component thrombospondin in the developing mouse embryo. *J. Cell Biol.* **107**, 2737–2748.

Pardanaud, L., Yassine, F., and Dieterlen-Lievre, F. (1989). Relationship between vasculogenesis, angiogenesis, and haemopoiesis during avian ontogony. *Development (Cambridge, UK)* **105**, 473–485.

Parke, A., Bhattacherjee, P., Palmer, R. M., and Lazzarus, N. R. (1988). Characterisation and quantification of copper sulfate-induced vascularization of the rabbit cornea. *Am. J. Pathol.* **130**, 173–180.

Passanti, A., Taylor, R. M., Pili, R., Guo, Y., Long, P. V., Haney, J. A., Pauly, R. R., Grant, D. S., and Martin, G. R. (1992). A simple quantitative method for assessing angiogenesis and antiangiogenic agents using reconstituted basement membrane heparin, and FGF. *Lab. Invest.* **67**, 519–529.

Pepper, M. S., Spray, D. C., Chanson, M., Montesano, R., Orci, L., and Meda, P. (1989). Junctional communication is induced in migrating capillary endothelial cells. *J. Cell Biol.* **109**, 3027–3038.

Phillips, M. L., Nudelman, E., Gaeta, F. C. A., Perez, M., Singhal, A. K., Hakomori, S.-I., and Paulson, J. C. (1990). ELAM-1 mediates cell adhesion by recognition of a carbohydrate ligand, sialyl-Lex. *Science* **250**, 1130–1132.

Plate, K. H., Breir, G., Weich, H. A., and Risau, W. (1992). Vascular endothelial growth factor is a potential tumour angiogenesis factor in human gliomas *in vivo*. *Nature (London)* **359**, 845–847.

Polakowski, I. J., Lewis, M. K., Mulhukkaruppan, J. R., Erdman, B., Kubai, L., and Auerbach, R. (1993). A ribonuclease inhibitor expresses anti-angiogenic properties and leads to reduced tumor growth in mice. *Am. J. Pathol.* **143**, 507–517.

Proia, A. D., Chandler, D. B., Haynes, W. L., Smith, G. F., Suvarnamani, C., Erkel, F. H., and Klintworth, G. K. (1988). Quantitation of corneal neovascularisation using computerised image analysis. *Lab. Invest.* **58**, 473–479.

Rastinejad, F., Polverini, P. J., and Bouck, N. P. (1989). Regulation of the activity of a new inhibitor of angiogenesis by a cancer suppressor gene. *Cell* (*Cambridge, Mass.*) **56**, 345–355.

Rifkin, D. B., Gross, S. L., Moscatelli, D., and Jaffe, E. (1982). *In* "Pathobiology of the Endothelial Cell" (H. L. Nossel and H. J. Vogel, eds.), pp. 191–197. Academic Press, New York.

Risau, W. (1986). Developing brain produces an angiogenesis factor. *Proc. Natl. Acad. Sci. U.S.A.* **83**, 3855–3859.

Risau, W., and Lemmon, V. (1988). Changes in the vascular extracellular matrix during embryonic vasculogenesis and angiogenesis. *Dev. Biol.* **125**, 441–450.

Risau, W., Gautschi-Sova, P., and Bohlen, P. (1988). Endothelial cell growth factors in embryonic and adult chick brain are related to human acidic fibroblast growth factor. *EMBO J.* **7**, 959–962.

Robbins, S. V., Cotran, R. S., and Kumar, V. (1984). Inflammation and repair. *In* "Pathological Basis of Disease" (W. B. Sanders, ed.), pp. 40–84.

Roberts, A. B., Sporn, M. B., Assoian, R. K., Smith, O. M., Roche, L. A., Falanga, V., Lehrl, J. H., and Fauci, A. S. (1986). Transforming growth factor type beta: Rapid induction of fibrosis and angiogenesis *in vivo* and stimulation of collagen formation *in vitro*. *Proc. Natl. Acad. Sci. U.S.A.* **83**, 4167–4171.

Rosenthal, R., Megyesi, J. F., Henzel, W. J., Ferrara, N., and Folkman, J. (1990). Conditioned medium from mouse sarcoma 180 cells contains vascular endothelial growth factor. *Growth Factors* **4**, 53–59.

Sabin, F. R. (1920). Studies on the origin of blood vessels and of red corpuscles as seen in living blastoderm of the chick during the second day of incubation. *Carnegie Contrib. Embryol.* **9**, 213–259.

Sage, H., and Bornstein, P. (1982). Endothelial cells from umbilical vein and a hemangioendothelioma secrete basement membrane largely to the exclusion of interstitial procollagen. *Arteriosclerosis* (*Dallas*) **2**, 27–36.

Sager, R. (1986). Genetic suppression of tumour formation: A new frontier in cancer research. *Cancer Res.* **46**, 1573–1580.

Sakamoto, N., Iwahana, M., Tanaka, N. G., and Osada, Y. (1991). Inhibition of angiogenesis and tumour growth by a synthetic laminin peptide CDPGYIGSR-NH2. *Cancer Res.* **51**, 903–906.

Sano, H., Forough, R., Maier, J. A., Case, J. P., Jackson, A., Engleka, K., Maciag, T., and Wilder, R. L. (1990). Detection of high levels of heparin binding growth factor-1 (acidic fibroblast growth factor) in inflammatory arthritic joints. *J. Cell Biol.* **110**, 1417–1426.

Sarma, V., Wolf, F. W., Marks, R. M., Shows, T. B., and Dixit, V. M. (1992). Cloning of a novel tumor necrosis factor-α-inducible primary response gene that is differentially expressed in development and capillary tube-like formation *in vitro*. *J. Immunol.* **148**, 3302–3311.

Sassa, T., Richter, W. W., Uda, N., Suganume, N., Suguri, H., Yoshizawa, S., Hirota, M., and Fujiki, H. (1989). Apparent activation of protein kinases by okadaic acid class tumor promoters. *Biochem. Biophys. Res. Commun.* **159**, 939–944.

Sato, Y., and Rifkin, D. B. (1988). Autocrine activities of basic fibroblast growth factor: Regulation of endothelial cell movement, plasminogen activator synthesis and DNA synthesis. *J. Cell Biol.* **107**, 1199–1205.

Schoefl, G. I. (1963). Studies on inflammation. III. Growing capillaries: Their structure and permeability. *Virchows Arch. A: Pathol. Anat.* **337**, 97–141.

Schreiber, A. B., Winkler, M. E., and Derynk, R. (1986). Transforming growth factor alpha: A more potent angiogenic factor than epidermal growth factor. *Science* **232**, 1250–1253.

Schroeder, J. M., and Christophers, E. (1989). Secretion of novel and homologous neutrophil-activating peptides by LPS-stimulated human endothelial cells. *J. Immunol.* **142**, 244–251.

Schweigerer, L., Neufeld, G., Friedman, J., Abraham, J. A., Fiddes, J. C., and Gospodarowicz, D. (1987). Capillary endothelial cells express basic fibroblast growth factor, a mitogen
that promotes their own growth. *Nature (London)* **325,** 257–259.

Schweigerer, L., Christeleit, K., Fleischmann, G., Adlerceutz, H., Wahala, K., Hase, T.,
Schwab, M., Ludwig, R., and Fotsis, T. (1992). Identification in human urine of a natural
growth inhibitor of cells derived from solid paediatric tumors. *Eur. J. Clin. Invest.* **22,**
260–264.

Shweiki, D., Itin, A., Soffer, D., and Keshnet, E. (1992). Vascular endothelial growth factor
induced by hypoxia may mediate hypoxia-initiated angiogenesis. *Nature (London)* **359,**
843–845.

Seeley, K. A., and Aggeler, J. (1991). Modification of milk protein synthesis through alteration of the cytoskeleton in mouse mammary epithelial cells cultured on a reconstituted
basement membrane. *J. Cell. Physiol.* **146,** 117–130.

Seitz, M., Dewald, B., Gerber, N., and Baggolini, M. (1991). Enhanced production of
neutrophil-activating peptide-1/interleukin-8 in rheumatoid arthritis. *J. Clin. Invest.* **87,**
463–469.

Selye, H. (1953). On the mechanism through which hydrocortisone effects the resistence
of tissue to injury. *JAMA, J. Am. Med. Assoc.* **152,** 1207–1213.

Senger, D., Perruzzi, C. A., Feder, J., and Dvorak, H. F. (1986). A highly conserved
vascular permeability factor secreted by a variety of human and rodent tumor cell lines.
*Cancer Res.* **46,** 5629–5632.

Senger, D. R., Connolly, D. T., and Van De Water, L., Feder, J., and Orona, U. (1990).
Purification and $NH_2$-terminal amino acid sequence of guinea pig tumor-vascular permeability factor. *Cancer Res.* **50,** 1774–1778.

Setchell, K. D. R., and Adlercreutz, H. (1988). *In* "Role of Gut Flora in Cytotoxicity and
Cancer" (I. R. Rowlands, ed.), pp. 315–345. Academic Press, London.

Shapiro, R., and Vallee, B. L. (1987). Human placental RNase inhibitor abolishes both
angiogenic and ribonucleolytic activities of angiogenin. *Proc. Natl. Acad. Sci. U.S.A.* **84,**
2238–2241.

Shapiro, R., Strydom, D. J., Olson, K. A., and Vallee, B. L. (1987). Isolation of angiogenin
from normal human plasma. *Biochemistry* **26,** 5141–5146.

Shapiro, R., Fox, E. A., and Riordan, J. F. (1989). Role of lysines in human angiogenin:
Chemical modification and site directed mutagenesis. *Biochemistry* **28,** 7401–7408.

Sonnenberg, A. (1988). The laminin receptor on platelets is the integrin VLA-6. *Nature
(London)* **336,** 487–489.

Sorbo, J., and Norrby, K. (1992). Mast-cell histamine expands the microvasculature spatially.
*Agents Actions, Spec. Conf. Issue,* pp. C387–C389.

Sorgente, N., Kuetter, K. E., and Soble, L. W. (1975). The resistance of certain tissues to
invasion. II. Evidence for extractable factors in cartilage which inhibits invasion by
vascularized mesenchyme. *Lab. Invest.* **32,** 217–222.

Spisni, E., Mamica, F., and Tomasi, V. (1992). Involvement of prostanoids in the regulation
of angiogenesis by polypeptide growth factors. *Prostaglandins, Leukotrienes Essent. Fatty
Acids* **47,** 11–115.

Stoker, M., and Gherardi, E. (1991). Regulation of cell movement in the mitogenic cytokines.
*Biochim. Biophys. Acta* **1022,** 81–102.

Stoler, A., and Bouck, N. P. (1985). Identification of a single chromosome in the normal
human genome essential for suppression of hamster cell transformation. *Proc. Natl. Acad.
Sci. U.S.A.* **82,** 570–574.

Strieter, R. M., Kunkel, S. L., Elner, J. M., Martonyi, C. L., Kock, A. E., Polverini,
P. J., and Elner, S. G. (1992). IL8 a corneal factor that enhances angiogenesis. *Am. J.
Pathol.* **14,** 1279–1284.

Suganumu, M., Fujiki, H., Surguri, H., Yoshizawa, S., Hirota, M., Makayasu, M., Ojika, M., Wakamatsu, K., Yamada, K., and Sugimura, T. (1988). Okadaic acid: An additional non-phorbol-12-tetradecanoate-13-acetate-type tumor promoter. *Proc. Natl. Acad. Sci. U.S.A.* **85**, 1768–1771.

Takigawa, M., Enomoto, M., Nishida, Y., Pan, H.-O., Kinoshita, A., and Suzuki, F. (1990). Tumour angiogenesis and polyamines: $\alpha$-difluoromethylornithine, an irreversible inhibitor of ornithine decarboxylase, inhibits B16 melanoma-induced angiogenesis *in ovo* and the proliferation of vascular endothelial cells *in vitro. Cancer Res.* **50**, 4131–4138.

Tanaku, N. G., Sakamoto, N., Korenaga, H., Inoue, K., Ogamura, H., and Osada, Y. (1991). The combination of bacterial polysaccharides and tamofexin inhibits angiogenesis and tumour growth. *Int. J. Radiat. Biol.* **60**, 79–83.

Taraboletti, G., Roberts, D., Liotta, L. A., and Giarazzi, R. (1990). Platelet thrombospondin modulates endothelial cell adhesion, motility and growth: A potential angiogenesis regulating factor. *J. Cell Biol.* **111**, 765–772.

Taraboletti, G., Retolli, D., Dejana, E., Mantovuni, A., und Giovuzzi, R. (1993). Endothelial cell migration and invasiveness are induced by a soluble factor produced by murine endothelioma cells transformed by polyoma virus middle T oncogene. *Cancer Res.* **53**, 3812–3816.

Taub, M., Wang, Y., Szensny, M. T., and Kleinman, H. K. (1990). Epidermal growth factor or transforming growth factor alpha is required for kidney tubulogenesis in matrigel cultured is serum free medium. *Proc. Natl. Acad. Sci. U.S.A.* **87**, 4002–4006.

Tay, E. M. L., Ryhanen, L., and Uitto, J. (1983). Proline analogues inhibit human skin fibroblasts growth and collagen production in culture. *Invest. Dermatol.* **80**, 261–267.

Taylor, S., and Folkman, J. (1982). Protamine is an inhibitor of angiogenesis. *Nature (London)* **297**, 307–312.

Thomas, K. A., Rios-Candelore, M., Gimenez-Gallego, G., DiSalvo, J., Bennet, C., Rodkey, J., and Fitzpatrick, S. (1985). Pure brain-derived acidic fibroblast growth factor is a potent angiogenic vascular endothelial cell mitogen with sequence homology to interleukin 1. *Proc. Natl. Acad. Sci. U.S.A.* **82**, 6409–6413.

Thorpe, P. E., Derbyshire, E. J., Andrade, S. P., Press, N., Knowles, P. P., King, S., Watson, G. J., Yang, Y.-C., and Rao-Bette, M. (1993). Heparin-steroid conjugates: New angiogenesis inhibitors with anti-tumor activity in mice. *Cancer Res.* **53**, 3000–3007.

Tisher, E., Mitchell, R., Hartman, T., Silva, M., Gospodarowicz, D., Fiddes, J. C., and Abraham, J. A. (1991). The human gene for vascular endothelial growth factor. *J. Biol. Chem.* **266**, 11947–11954.

Tolsma, S. S., Volpert, O. V., Good, D. J., Frazer, W. A., Polverini, P. J., and Bouck, N. (1993). Peptides from two separate domains of the matrix molecule TSP-1 have antiangiogenic activity. *J. Cell Biol.* **122**, 497–511.

Toole, B. P. (1976). *In* "Neuronal Recognition" (S. H. Baronacles, ed.), pp. 275–329. Plenum, New York.

Tsarfaty, I., Resau, J. H., Rulong, S., Keydar, I., Faletto, D. L., and VandeWoude, G. F. (1992). The net proto-oncogene receptor and lumen formation. *Science* **257**, 1258–1261.

Uitto, J., and Prockop, D. J. (1974). Incorporation of proline analogues into collagen polypeptides. Effects on the production of extracellular pro-collagen and on the stability of the triple helical structure of the molecule. *Biochim. Biophys. Acta* **336**, 234–251.

van Mourik, J. A., Lawrence, D. A., and Loskutoff, D. J. (1984). Purification of an inhibitor of plasminogen activator (antiactivator) synthesised by endothelial cells. *J. Biol. Chem.* **259**, 14914–14921.

Vernon, R. B., Angello, J. C., Iruela-Arispe, L., Lane, L. F., and Sage, E. H. (1992). Reorganisation of basement membrane matrices by cellular traction promotes the formation of cellular networks *in vitro. Lab. Invest.* **66**, 536–547.

Vlodavsky, I., Folkman, J., Sullivan, R., Fridman, R., Ishai-Micheals, R., Sasse, J., and Klagsbrun, M. (1987). Endothelial cell-derived basic fibroblast growth factor: Synthesis and deposition into subendothelial extracellular matrix. *Proc. Natl. Acad. Sci. U.S.A.* **84,** 2292–2296.

Voss, K., Jakob, W., and Roth, K. (1984). A new image analysis method for the quantitation of neovascularisation. *Exp. Pathol.* **26,** 155–161.

Vukanović, J., Passanti, A., Hirata, T., Traystman, R. J., Hartley-Asp, B., and Isaacs, J. T. (1993). Antiangiogenic effects of the quinoline-3-carboxamide linomide. *Cancer Res.* **53,** 1833–1837.

Wahl, S. M., Hunt, D. A., Wakefield, L. M., McCartney-Frances, N., Wahl, L. M., Roberts, A. B., and Sporn, M. B. (1987). Transforming growth factor type $\beta$ induces monocyte chemotaxis and growth factor production. *Proc. Natl. Acad. Sci. U.S.A.* **84,** 5788–5792.

Walz, G., Aruffo, A., Kolanus, W., Bevilacqua, M., and Seed, B. (1990). Recognition by ELAM-1 of the sialyl-Lex determinant on myeloid and tumor cells. *Science* **250,** 1132–1135.

Weidner, N., Semple, J. P., Welch, W. R., and Folkman, J. (1991). Tumor angiogenesis and metastasis: Correlation in invasive breast carcinoma. *N. Engl. J. Med.* **324,** 1–8.

Weidner, N., Carrol, P. R., Flax, J., Blumefield, W., and Folkman, J. (1993). Tumor angiogenesis correlates with metastasis in invasive prostate carcinoma. *Am. J. Pathol.* **143,** 401–409.

Weiner, H. L., Weiner, L. H., and Swain, J. L. (1987). Tissue distribution and developmental expression of the messenger RNA encoding angiogenin. *Science* **237,** 280–282.

West, D. C., Hampson, I. N., Arnold, F., and Kumar, S. (1985). Angiogenesis induced by degradation products of hyaluronic acid. *Science* **228,** 1324–1326.

Whitham, S. E., Murphy, G., Angel, P., Rahhmsdorf, H. J., Smith, B. J., Lyons, A., Harris, T. J. R., Reynolds, J. J., Herrlich, R., and Docherty, A. S. P. (1986). Comparison of human stromelysin and collagenase by cloning and sequence analysis. *Biochem. J.* **240,** 913–916.

Wicha, M. S., Liotta, L. A., Lewko, L. A., and Kidwell, W. R. (1981). Blocking basement membrane collagen deposition inhibits the growth of 7,12-dimethylbenzanthracene-induced rat mammary tumours. *Cancer Lett.* **12,** 9–12.

Winkles, J. A., Fresel, R., Burgess, W. H., Howk, R., Mehlman, T., Weinsteiner, R., and Maciag, T. (1988). Human vascular smooth muscle cells both express and respond to heparin-binding growth factor (endothelial cell growth factor). *Proc. Natl. Acad. Sci. U.S.A.* **84,** 7124–7128.

Yamada, K. M. (1981). *In* "Cell Biology of the Extracellular Matrix" (E. D. Hays, ed.), pp. 9–114. Plenum, New York.

Yamagami, I. (1970). Electron microscopic study on the cornea. I. The mechanism of experimental new vessel formation. *Acta Soc. Ophthalmol. Jpn.* **73,** 1222–1242.

Yanagisawa-Miwa, A., Uchida, Y., Nakamura, F., Tomaru, T., Kido, H., Kamijo, T., Sugimoto, T., Kaji, K., Utsuyama, M., Kurashima, C., and Ito, H. (1992). Salvage of infarcted myocardium by angiogenic action of basic fibroblast growth factor. *Science* **257,** 1401–1403.

Yatsunami, J., Fujiki, H., Suganuma, M., Yoshizawa, S., Erikson, J. E., Olson, M. O., and Goldman, R. D. (1991). Vimentin is hyperphosphorylated in primary human fibroblasts treated with okadaic acid. *Biochem. Biophys. Res. Commun.* **177,** 1165–1170.

Yoshimura, T., Matsushima, K., Tanaka, S., Robinson, E. A., Appella, E., Oppenheim, J. J., and Leonard, E. J. (1987). Purification of a human monocyte-derived neutrophil chemotactic factor that has peptide sequence similarity to other host defense cytokines. *Proc. Natl. Acad. Sci. U.S.A.* **84,** 9233–9237.

Ziches, M., Alessandri, G., and Gullino, P. M. (1989). Gangliosides promote the angiogenic response. *Lab. Invest.* **61,** 629–643.

# Nuclear Remodeling in Response to Steroid Hormone Action

Klaus Brasch* and Robert L. Ochs[†]

*Department of Biology, California State University, San Bernardino, California 92407, and

[†]W. M. Keck Autoimmune Disease Center and Department of Molecular and Experimental Medicine, The Scripps Research Institute, La Jolla 92093, California

Steroid and similar hormones comprise the broadest class of gene regulatory agents known, spanning vertebrates through the lower animals, and even fungi. Not unexpectedly, therefore, steroid receptors belong to an evolutionarily highly conserved family of proteins. After complexing with their cognate ligands, receptors interact with hormone response elements on target genes and modulate transcription. These actions are multifaceted and only partly understood, and include large-scale changes in the structure and molecular composition of the affected cell nuclei. This chapter examines steroid hormone action and the resultant nuclear remodeling from the following perspectives: (1) Where are the receptors located? (2) Which nuclear domains are most affected? (3) Are there extended or permanent nuclear changes? (4) What is the role of coiled bodies and similar structures in this regard? To address these and related questions, information is drawn from several sources, including vertebrates, insects, and malignant tissues. Entirely new data are presented as well as a review of the literature.

**KEY WORDS:** Steroid hormones, Receptors, Nucleus, Coiled bodies, Cancer.

## I. Introduction

Steroid and other lipid hormones constitute perhaps the broadest class of gene regulatory agents so far identified. These ligands modulate transcription by coupling with specific protein receptors that bind to target genes via hormone-responsive elements (HREs). Steroid receptors are part of a superfamily of control elements that also includes those for thyroids, retinoids, and vitamin D (Beato, 1989; Carson-Jurica *et al.*, 1990). In addition, a host of other genes with sequence homology to the superfamily (orphan receptors), as well as various receptor isoforms, have also been identified (Tsai and O'Malley, 1994).

Although best understood in higher vertebrates, steroid-mediated growth and tissue differentiation have been recognized in all major animal groups and even in the fungus *Achlya* (Yamamoto, 1985; Beato, 1989; Ham and Parker, 1989; Brunt *et al.*, 1990; Gronemeyer, 1992). Moreover, in tranformation experiments with yeast and *Escherichia coli,* mammalian estradiol (ES) and glucocorticoid (G) receptors functioned in a fashion similar to that in vertebrate target tissues, both with respect to DNA binding and in terms of ligand-dependent control of transcription (Metzger *et al.*, 1988; Schema and Yamamoto, 1988; Carson-Jurica *et al.*, 1990). These experiments demonstrate that the regulatory mechanisms underlying the action of steroids and similar hormones are highly conserved and likely operate at some level in most organisms.

Although it is well established that steroid hormones act at the nuclear level through a complex sequence of receptor-mediated interactions, the multifaceted nature of these actions is only partly understood. This is clearly illustrated in the livers of roosters and male *Xenopus,* following experimental induction of vitellogenesis by estradiol (Sharpiro *et al.*, 1989). This entails not only *de novo* activation of major target genes such as those encoding the vitellogenins (Vg) and very low density lipoprotein (VLDL), but also large-scale nuclear remodeling to accommodate enhanced transcription of all major classes of RNA (Evans *et al.*, 1987; Shapiro *et al.*, 1989; Brasch, 1990a, and Section V). Moreover, even in the extended absence of hormone, once exposed to ES, hepatocytes remain hyperresponsive to subsequent stimulation. This phenomenon has been termed the "memory" effect and likely includes hormone-induced structural changes at the level of chromatin and even higher order nuclear domains (Brasch and Peters, 1985; Burch and Evans, 1986; Evans *et al.*, 1987).

This chapter focuses primarily on the large-scale aspects of steroid hormone action in the nucleus, with particulr emphasis on structural domains and remodeling of nuclear architecture. An effort is made to integrate information from mammalian through invertebrate systems, as well as normal and malignant tissue responses.

## II. Functional Domains of the Nucleus

It is becoming increasingly evident that the cell nucleus can be subdivided into descrete structural and functional domains. Although this is not a new concept (Comings, 1968; Brasch and Setterfield, 1974), only through

modern developments in antibody and nucleic acid technology have tools
become available to address this question in truly experimental terms
(Newport and Forbes, 1987; Nigg, 1988; Brasch, 1990a; Manuelidis and
Chen, 1990; Raska *et al.*, 1990b; Jackson, 1991; Raska *et al.*, 1992; Almouzni and Wolffe, 1993; Spector, 1993). An overview of the primary
nuclear domains so far recognized is summarized in Fig.1.

The interphase nucleus is dominated structurally by three major components, the nuclear envelope and pore-complex lamina, chromatin in varied
stages of condensation and decondensation, and nucleoli (Raska *et al.*,
1990b). The interchromatin areas contain a heterogeneous assortment of
less well-defined structures, including perichromatin fibrils and granules,
interchromatin granules, and various types of nuclear bodies (Brasch and
Ochs, 1992; Raska *et al.*, 1992). It is also likely that the nuclear interior
is pervaded by an underlying, integrating protein matrix (Brasch, 1990a;
Jackson, 1991; van Driel *et al.*, 1991).

Concerning structural compartmentalization in the nucleus, the classic
observations by Rabl (1985) and Boveri (1909) provided the first evidence

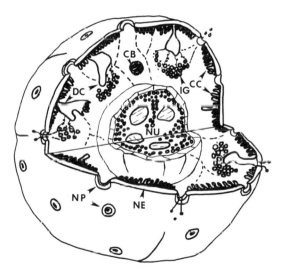

FIG.1   Model of the major functional domains of the interphase nucleus, showing the nuclear
envelope and pore-complex lamina (NE), nuclear pores (NP), condensed (CC) and decondensed (DC) chromatin, the nucleolus (NU), perichromatin fibrils and interchromatin granule
clusters (IGs), and coiled bodies (CBs). RNA polymerase I activity is confined to the NU,
which also contains fibrillar centers and granular components and may also spawn the as
yet undefined CBs. Chromosomal regions transcribed by RNA polymerase II are distributed
throughout the nucleoplasm. Pre-mRNA transcripts are associated with splicing factors,
recruited from storage and/or assembly sites (IGs). Large IG clusters may service numerous
genes in their vicinity. On completion of processing, the mature mRNAs may leave the
nucleus along directed paths. (Adapted with modifications, courtesy of Dr. D. L. Spector.)

that chromosomes are not randomly distributed but are confined to discrete regions or territories. These territories are not rigidly fixed, however, and can vary both in relation to the cell cycle and as a function of gene expression (Jackson, 1991; Spector 1993). The most compelling evidence to support these observations comes from studies with living *Drosophila* polytene nuclei stained with vital dyes (Agard and Sedat, 1983; Gruenbaum *et al.*, 1984; Hochstrasser and Sedat, 1987) that show chromosomes to be closely associated with the inner nuclear membrane mainly through telomeric and centromeric contact points. Similar chromosome positioning has also been observed in diploid nuclei from many different species (Hilliker and Appels, 1989; Gilson *et al.*, 1993; Spector, 1993). In some extreme manifestations, the overall pattern of gross nuclear morphology may actually be transmitted from one somatic cell generation to the next (Locke, 1990).

Despite strong evidence for nonrandom arrangement, interphase chromosomes are also highly dynamic. For example, chromosomes can move and regionally condense/decondense not only in terms of the cell cycle, but also in a tissue-specific manner and in relation to developmental and physiological factors (Manuelidis and Chen, 1990; Haaf and Schmid, 1991). Thus although chromosomes may be anchored to the nuclear membrane through telomeric and centromeric links, their partially decondensed arms are free to loop out and extend into the nuclear interior as needed, for transcription, replication, or transient remodeling.

This emerging picture of interphase chromosome organization, combining both order and highly dynamic characteristics, likely applies to most other nuclear structures as well, including nucleoli, ribonucleoprotein (RNP) elements, and the nuclear matrix. For example, not only are factors involved in transcription and pre-mRNA splicing highly coordinated in active nuclei (Smith, 1992; Jimenez-Garcia and Spector, 1993; O'Keefe *et al.*, 1994), specific gene transcripts appear to be channeled within the interchromatin domain (Xing *et al.*, 1993; Carter *et al.*, 1993). Such actions may also involve large-scale RNP elements (Visa *et al.*, 1993), and possibly proteins such as actin and myosin (Crowley and Brasch, 1987; Sahlas *et al.*, 1993; Milankov and De Boni, 1994). As is demonstrated below, the action of steroid hormones influences almost all structural domains in target cell nuclei.

## III. Coiled Bodies

Besides chromatin and nucleoli, the most obvious morphological elements of the nucleus are a class of structures known as nuclear bodies (Brasch and Ochs, 1992). Because of their relative infrequency and small size (0.2–1.0 $\mu$m), as well as their general lack of detectability except by electron microscopy, these structures have not been the focus of serious investigation until recently. Relative to this chapter, however, because nuclear bodies appear to proliferate after hormonal stimulation, they may represent one of the best examples of nuclear remodeling in response to steroid hormone action (Section V,C).

On the basis primarily of electron microscopic studies of pathological specimens, Bouteille et al. (1967) originally described five different types of nuclear bodies associated with cellular hyperactivity as induced by physiological, hormonal, drug, viral, or malignant conditions (Brasch and Ochs, 1992). Many such bodies occur in proximity to the nucleolus and some were observed to "bud" from the nucleolar surface (Dupuy-Coin and Bouteille, 1972). On the basis of sensitivity to nucleases and proteases, these structures were shown to be composed of complexes of RNA and protein. Nuclear bodies have also been studied in several model systems that are amenable to controlled induction of cellular hyperactivity, including concanavalin A (ConA)-stimulated mouse splenic lymphocytes (Chaly et al., 1983a,b), ES-stimulated rat uterus (Clark et al., 1978; Padykula and Clark, 1981), and after ES-induced vitellogenesis in rooster hepatocytes (Brasch and Peters, 1985; Brasch et al., 1989). In all of these systems, the numbers and complexity of nuclear bodies increase markedly after stimulation of cellular activity, implying a physiological role for these structures in as yet undefined nuclear events.

To date, the best studied type of nuclear body is the coiled body (CB) (Raska et al., 1990b; Brasch and Ochs, 1992; Spector, 1993; Lamond and Carmo-Fonseca, 1993). This is due primarily to the present availability of specific antibody probes to the CB-associated protein, p80-coilin (Raska et al., 1990a, 1991; Andrade et al., 1991, 1993). Coiled bodies were first observed at the light microscope level in 1903 by Ramón y Cajál (reviewed by Lafarga and Hervas, 1983), who referred to them as "accessory bodies" to denote their close association with nucleoli in neurons stained by silver impregnation. Coiled bodies were first named and studied in detail in the pioneering electron microscopic studies of Monneron and Bernhard (1969), who described them as round bodies 0.5–1.0 $\mu$m in diameter and composed of tightly packed coiled fibers. Staining by the EDTA-regressive method (Bernhard, 1969) revealed their ribonucleoprotein nature. Similar CBs have been observed in a variety of plant (Lafontaine, 1965; Moreno Diaz de la Espina et al., 1982a,b; Williams et al., 1983; Chamberland and Lafontaine, 1993) and animal cells (Raska et al., 1990a, 1991; Andrade et al., 1991, 1993; Lafarga et al., 1991). Coiled bodies, along with simple nuclear bodies, also predominate in many cultured animal cells (Raska et al., 1990b). Despite containing RNA and protein, CBs do not incorporate uridine (Fakan and Bernhard, 1971; Fakan et al., 1976; Moreno Diaz de la Espina et al., 1982a) and lack any detectable DNA (Monneron and Bernhard, 1969; Raska et al., 1991).

Other than marker protein p80-coilin, for which a partial cDNA sequence is known (Andrade et al., 1991), CBs have been reported to contain the U3 small nuclear ribonucleoprotein particles (snRNP)-associated nucleolar protein fibrillarin; the nucleolus organizer region (NOR) silver-staining protein; DNA topoisomerase I; the U1, U2, U4, and U6 snRNP-associated Sm proteins; and the U2 snRNP auxiliary splicing factor U2AF (Fakan et al., 1984; Eliceiri and Ryerse, 1984; Raska et al., 1990a, 1991; Zhang et al., 1992). Coiled bodies do not have detectable amounts of nucleolar proteins B23, nucleolin, RNA polymerase I, nucleolar transcription factor NOR-90/UBF, 5S rRNP, splicing factor SC35, heteronuclear RNP (hnRNP) protein L, or interchromatin granule protein 3C5 (Raska et al., 1991; Ochs et al., 1994). Using in situ hybridization with oligonucleotide antisense probes, CBs have been reported to contain U2, U4, U5, U6, and U12 snRNAs, but no detectable RNAs for U1, U3, 7SK, 5S RNA, or rRNA (Carmo-Fonseca et al., 1991a,b, 1993; Matera and Ward, 1993). Consequently, CBs remain somewhat enigmatic structures, because they contain protein and RNA molecules involved in the metabolism of both pre-rRNA and pre-mRNA (Lamond and Carmo-Fonseca, 1993).

Coiled bodies may be analogous to the snRNP-containing sphere organelles or snurposomes of amphibian oocytes, which are associated with active sites of transcription on lampbrush chromosomes and, as free bodies in the nucleoplasm, may be sites for assembly and/or storage of snRNP complexes (Gall and Callan, 1989; Gall, 1991; Wu et al., 1991; Tuma et al., 1993). Data on the presence of CBs in adipocytes of the hibernating dormouse and their disappearance on arousal from hibernation (Malatesta et al., 1994) may be taken as further evidence of a storage function for these structures.

Coiled bodies vary in size and number throughout the cell cycle and in different cell types (Andrade et al., 1991, 1993; Raska et al., 1991). Also, fewer CBs are present in contact-inhibited cells of defined passage than in immortal or transformed cells (Spector et al., 1992). In general, CBs are not detectable during mitosis even though the total amount of p80-coilin remains constant throughout all stages of the cell cycle (Andrade et al., 1993). This is presumably due to M-phase phosphorylation of coilin and subsequent disassembly of the CBs (Carmo-Fonseca et al., 1993). Coiled bodies are also sensitive to the proliferative state of the cell, because their expression can be "downregulated" in 3T3 cells by serum starvation and "upregulated" by refeeding or by addition of TSH (thyrotropin) to hormone-depleted rat thyroid FRTL-5 cells (Andrade et al., 1993).

In the only functional studies to date, inhibition or alteration of transcription by actinomycin D, $\alpha$-amanitin, 5,6-dichloro-1-$\beta$-D-ribofuranosylbenzimidazole (DRB), or heat shock, displaced snRNPs from CBs and resulted in the association of p80-coilin with the nucleolus (Raska *et al.*, 1990a; Carmo-Fonseca *et al.*, 1992). All such studies serve to demonstrate the dynamic nature of CB expression and its sensitivity to changes in cell metabolism, growth, and differentiation. Clearly the action of steroid hormones can influence all of these factors in the affected target cells.

## IV. Steroid Hormone Receptors

### A. General Characteristics

It is now well established that steroid and similar hormone receptors are proteins that exhibit the following basic characteristics (Yamamoto, 1985; Beato, 1989; Ham and Parker, 1989; Carson-Jurica *et al.*, 1990; O'Malley *et al.*, 1991; Gronemeyer, 1992; Tsai and O'Malley, 1994): (1) Each receptor has a central DNA (HRE)-binding domain that is well conserved, a ligand-binding, C-terminal domain that is moderately sequence conserved, and an extremely variable N-terminal domain; (2) receptors must complex with ligand to become functionally active, either as enhancers or repressors of transcription. Activation of receptors may involve release of inhibitory proteins such as HSP-90, dimerization, and/or other transformation steps necessary for tighter binding to the HRE sequences on DNA; (3) unbound mineralocorticoid and glucocorticoid receptors (GRs) may first reside in the cytoplasm and then translocate to the nucleus on hormone binding. This does not seem to be the case with most other steroid receptors, including those for estrogen (ERs), progesterone (PRs), and androgens (ARs). A simplified model of steroid hormone action is shown in Fig. 2.

### B. Receptor Location: Nucleus or Cytoplasm?

The exact cellular location of steroid receptors is still an issue under intense investigation (Picard *et al.*, 1990; Dauvois *et al.*, 1993). It was possible to address this question directly only after the development of specific antibodies to individual receptors (King and Greene, 1984; Press *et al.*, 1985; Perrot-Applanat *et al.*, 1985; Welshons *et al.*, 1985). Prior to that time, the classic two-step model of steroid receptor activation first suggested by Gorski *et al.*, (1968) and Jensen *et al.* (1968) was widely

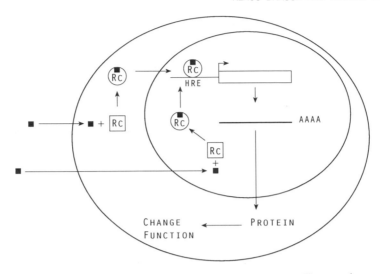

FIG. 2   A simplified model of steroid hormone action. The hormone (■) enters its target cell and binds to specific receptors (RCs). The RC exists in an inactive apoprotein form, present either in the cytoplasm or in the nucleus. Ligand binding initiates a transformation or activation that facilitates effective binding of the complex to hormone response elements (HREs) on DNA. Transcription of a *cis*-linked gene is then initiated. [Reproduced, with permission, from the *Annual Review of Biochemistry*, Volume 63, © 1994, by Annual Reviews, Inc. From Tsai and O'Malley (1994).]

accepted. Briefly, this model suggested that receptors exist in two forms: the unliganded cytoplasmic form and the hormone-bound nuclear receptor complex. A translocation step was postulated, whereby the receptor first complexed with ligand in the cytoplasm and then shifted to the nuclear compartment. This translocation was seen as an essential regulatory step in the chain of events by which steroid hormones affect cell functions.

Although this model became untenable in its original form, after immunocytochemical and related studies indicated that the ER, PR, and AR reside almost exclusively in the nucleus (Gasc and Baulieu, 1986; Savouret *et al.,* 1989; Carson-Jurica *et al.,* 1990), the situation is clearly more complex than that. For example, in the absence of ligand, the GR may first accumulate in the cytoplasm owing to masking of nuclear targeting sequences, possibly through interaction with HSPs and/or other factors (Picard and Yamamoto, 1987; Picard *et al.,* 1990; O'Malley *et al.,* 1991). In contrast, most other steroid receptors appear predominantly nuclear under normal conditions, although several studies have indicated that the ER and PR are not static but constantly shuttling between nucleus and cytoplasm (Guiochon-Mantel *et al.,* 1991; Chandran and DeFranco, 1992; Dauvois *et al.,* 1993).

For example, using antibodies against the mammalian ER, anti-estrogens, and metabolic inhibitors, Dauvois *et al.* (1993) have shown in transfected COS-1 cells that accumulation of receptors in the nucleus is (1) energy dependent, (2) blocked by pure anti-estrogens such as ICI 182780 but not by the partial anti-estrogen tamoxifen, and (3) independent of estradiol. These and similar observations with the PR suggest that although many receptors may reside predominantly in one or another subcellular compartment, they are clearly not fixed but exist in a state of dynamic equilibrium. In this context, a number of studies have focused on the roles of HSP-90, its associated immunophilin p59, and the receptor's own nuclear localization signal (Pratt, 1990; Cadepond *et al.*, 1992; Kang *et al.*, 1994). Although still incomplete, the emerging picture is that all aforementioned elements are probably involved in receptor chaperoning, activation, and nuclear–cytoplasmic shuttling. This leaves open a range of possibilities for linking of steroid receptors to a wide variety of other as yet undefined signaling pathways and functions.

## C. Location of Receptors in Nucleus

Electron microscopic studies have indicated that steroid receptors occupy several nuclear domains. Again, however, there appear to be hormone and/or tissue-related variations in this regard. Initial efforts to localize the ER by immunoelectron microscopy utilized antiserum against partially purified cytosolic receptor (Morel *et al.*, 1981). Although confirming the presence of receptors in target tissues, these results were of limited reliability owing to low signal strength. Subsequent investigations, using monoclonal and highly specific polyclonal antibodies, have provided more detailed information on the subnuclear location of both the estrogen and progesterone receptors (Press *et al.*, 1985; Isola, 1987; Savoret *et al.*, 1989; Vazquez-Nin *et al.*, 1991).

The results to date may be summarized as follows. In a wide range of both estrogen- and progesterone-responsive cells, including avian and mammalian tissues, the respective receptors are predominantly nuclear even in the absence of hormone. The trace amounts of cytoplasmic signal observed likely represent newly synthesized receptors and/or nontranslocated molecules. Within nuclei, however, the patterns or ER and PR distribution appear quite different. The ERs seem confined almost exclusively to the intechromatin (euchromatin) regions, closely associated with RNP elements (Press *et al.*, 1985; Vazquez-Nin *et al.*, 1991). Labeling of compact chromatin and nucleoli is either absent or relatively low. This situation prevailed both before and after exposure to estradiol.

In sharp contrast, the PRs were initially associated primarily with condensed chromatin but then shifted to the interchromatin domain following administration of hormone (Isola, 1987; Savouret *et al.*, 1989). This apparent relocation was attributed to partial decondensation of chromatin, which brought the PR sites into close association with RNP elements. In this way, receptor-bound chromosomal regions might be repositioned after stimulation, perhaps for enhanced transcription. The problem with this attractive notion is that, like progesterone, estrogen can also induce large-scale chromatin remodeling (Vic *et al.*, 1980; Brasch and Peters, 1985) and yet, as indicated above, the ERs seem to reside in the interchromatin regions at all times. Clearly, therefore, more information is needed to evaluate fully this apparent major difference in subnuclear distribution of these two important receptors.

## D. Receptors and the Nucleolus

It is difficult to assess from available data whether steroid receptors are abundant in nucleoli. Immunoelectron microscopy is inconclusive in this respect, because signal strength over nucleoli is usually very low, if at all present (Savouret *et al.*, 1989; Vazquez-Nin *et al.*, 1991). Although there is much evidence that enhanced transcription of pre-rRNA and changes in nucleolar morphology follow exposure to hormone in many steroid target tissues (see below), such action could well be entirely indirect and/ or involve only a minimal number of receptor sites.

## E. Receptors and the Nuclear Matrix

One nuclear domain of potentially major significance in terms of steroid hormone action is the nuclear matrix. The ''nuclear matrix'' is perhaps best defined collectively as the nonchromatin, structural elements of the nucleus. Although it is still unclear whether the matrix constitutes a true scaffold or skeleton within nuclei (Fey *et al.*, 1986; Cook, 1988), it does include an array of highly dynamic elements associated with DNA replication, RNP components, and the pore-complex lamina (Mills *et al.*, 1989; Lawrence *et al.*, 1989; Brasch, 1990a; van Driel *et al.*, 1991; Davis *et al.*, 1993; Hozak *et al.*, 1993). Not surprisingly, therefore, steroid receptors have also been detected in nuclear matrix preparations from a diversity of target tissues (Kaufmann *et al.*, 1986; Barrack, 1987; Alexander *et al.*, 1987; Carmo-Fonseca, 1988; Metzger and Korach, 1990). The results to date can be summarized as follows.

Nuclear matrix isolates, as operationally defined following sequential treatment of isolated nuclei with nonionic detergents, nucleases, and high salt concentrations (Gasser *et al.*, 1989; Stuurman *et al.*, 1992b), have been shown to contain receptors for androgens, estrogen, and glucocorticoids (Nelson *et al.*, 1986). It is noteworthy that in cell-free studies, receptor binding to matrix isolates is comparable to that of whole nuclei in terms of tissue specificity, high affinity, hormone dependence, and saturability (Barrack, 1987; Getszenberg *et al.*, 1990; Metzger and Korach, 1990). Moreover, direct evidence of matrix-bound receptors has been provided with monoclonal antibodies to the GR (Kaufmann *et al.*, 1986) and the ER (Alexander *et al.*, 1987). Association of specific hormone-induced genes with the nuclear matrix has also been demonstrated in several instances. These include not only major genes such as those encoding vitellogenin (Jost and Seldran, 1984) and ovalbumin (Robinson *et al.*, 1982; Ciejek *et al.*, 1983), but also a chimeric construct containing a viral origin of replication and a hormonally regulated transcription unit (Adom *et al.*, 1992).

Collectively, the foregoing information underscores the view that the isolated matrix constitutes a key nuclear subfraction (Gasser *et al.*, 1989; Brasch, 1990a; Berezney, 1991). It is clear that in addition to various other components, this fraction contains most elements involved in binding of steroid hormone receptors and their actively transcribed target genes. Whether these internal matrix elements, derived primarily from the interchromatin domain, also constitute a structural "skeleton" or "scaffold" *in situ* remains to be determined. This would certainly be in keeping, however, with increasingly strong evidence that the interchromatin domain houses not only the transcriptional and RNA splicing apparatus in nuclei, but may also serve to channel new transcripts to the cytoplasm (Carter *et al.*, 1993; Davis *et al.*, 1993; Jimenez Garcia and Spector, 1993).

## F. Memory Effect

One of the most intriguing aspects of steroid hormone action is the so-called "memory effect." First noted by Bergink *et al.* (1974) in vitellogenic rooster liver, this phenomenon has also been observed in *Xenopus* liver and in an ES-responsive human hepatocarcinoma cell line (Shapiro, 1982; Burch and Evans, 1986; Jost *et al.*, 1986; Tam *et al.*, 1986; Ho, 1987; Shapiro *et al.*, 1989). In the avian and amphibian systems, an extended lag period follows primary hormone induction of Vg and VLDL synthesis in immature livers and in males. A similar lag phase is manifest in the human hepatocarcinoma cells with respect to apolipoprotein synthesis.

This lag phase presumably reflects the reprogramming and restructuring of affected cells, necessary for *de novo* synthesis and export of Vg, VLDL, and other newly induced gene products (Shapiro *et al.*, 1989). Subsequent responses to ES, however, even several months after hormone withdrawal, display no such lag period and can be greater in both magnitude and duration, hence the term "memory effect." This implies that hormone-naive hepatocytes undergo extensive differentiation after primary exposure and that this memory effect is long term. The effect is also passively maintained through several rounds of cell division in embryonic and transformed hepatocytes (Burch and Evans, 1986; Evans *et al.*, 1987; Hache *et al.*, 1987).

As a potential key to understanding long-term effects of transient exposure to estrogen, the memory effect is clearly of considerable interest. Not only would this "memory" have to apply at the level of individual target genes such as that encoding Vg, but also on a broader nucleus-wide basis, because all facets of the secondary responses to ES are accelerated and enhanced relative to primary stimulation. Although a wide range of structural and molecular changes may be associated with this phenomenon, none appear as yet to account for it adequately. For example, several biochemical changes are associated with primary activation of the major vitellogenin genes. These include induction of nuclease hypersensitivity, demethylation of sites flanking the ER-binding domain, and bending of the HRE (Burch and Wientraub, 1983; Burch and Evans, 1986; Jost *et al.*, 1986; Saluz *et al.*, 1988; Nardulli and Shapiro, 1992; Sabbah *et al.*, 1992). Similar chromatin changes are associated with steroid hormone action generally and they probably affect target genes in several ways, either to enhance binding of receptors and attendant transcription factors or to inhibit transcription of unwanted genes by blocking access to them or by packaging them as heterochromatin (Tsai and O'Malley, 1994). Many of these do not appear to be permanent modifications, at least not in terms of the genes encoding Vg, and so are probably not sufficient in themselves for maintenance of the memory effect (Burch and Evans, 1986; Evans *et al.*, 1987; Beato, 1989).

Alternative explanations for the long-term effects of estrogen in hepatocytes have centered on the regulatory role of the ER in its own synthesis (Ho, 1987; Shapiro *et al.*, 1989), the role of nuclear matrix, and other related large-scale modifications of nuclear structure (Brasch and Peters, 1985; Brasch, 1990b). Clearly, the involvement of as yet poorly defined chromatin "acceptor" proteins is also a likely possibility (Feavers *et al.*, 1987; Goldberger *et al.*, 1987; Yu and Ho, 1989). Although such mechanisms may well contribute collectively to the memory effect, any explanation must account for (1) its longevity in both dividing and nondividing cells, (2) its broad-based nature, because it involves not only specific target genes but also many other coinduced genes, and (3) the rapid restructuring of cells to accommodate the estrogenic response, while simultaneously attenuating production of other regular hepatic proteins.


## V. Vitellogenesis as a Model System

### A. General Characteristics

The hormonal regulation of yolk protein production, including synthesis and transport to and incorporation by the oocyte, is collectively termed "vitellogenesis." This evolutionarily highly conserved process serves as a prime model of steroid hormone action in both egg-bearing vertebrates and insects (Tata and Smith, 1979, Ho, 1987; Tata et al., 1987; Shapiro et al., 1989; Wyatt, 1991). In vertebrates, the yolk precursor proteins, the so-called vitellogenins, are synthesized in the liver under primary control of estradiol, although other phases of the vitellogenic process appear to involve additional hormones. The vitellogenin monomers are phospholipoproteins with molecular masses in the range of 140,000–240,000 Da, depending on the species. In the chicken there are at least three, and in *Xenopus* four, primary *Vg* genes (Evans et al., 1987; Shapiro et al., 1989). The major plasma protein VLDL and apolipoproteins are also coinduced in liver, as are a number of other, functionally undefined polypeptides (Herbener et al., 1985; Holland and Wangh, 1987; Brasch, 1990b).

Several features make hepatic vitellogenesis a particularly useful model with which to study cellular aspects of steroid hormone action. First, in both chicken and *Xenopus*, the process can be induced entirely *de novo* in adult males. This provides a system that is unencumbered by the other hormonal factors normally associated with ovulation. Second, during the reprogramming of male livers, the synthesis of regularly secreted plasma proteins such as albumin is sharply attenuated, even as the overall levels of RNA and protein synthesis rise dramatically (Shapiro, 1982; Tata et al., 1987). This necessitates extensive cell restructuring to accommodate rapidly changing patterns of transcription and protein synthesis; patterns that are largely transient and revert to preinduction conditions in the absence of hormone. Last, although DNA synthesis is observed in a subpopulation of *Xenopus* hepatocytes several days after exposure to ES (Spolski et al., 1985), the primary vitellogenic response in both animal systems takes place against a background of essentially nonproliferating cells (Tata et al., 1987).

## B. Nuclear Remodeling in Estradiol-Stimulated Rooster Liver

A single intramuscular injection of estradiol induces an extensive cycle of structural and biochemical changes in rooster hepatocytes (Brasch and Peters, 1985; Brasch et al., 1989; Brasch, 1990b). Although durations and magnitudes may differ, comparable changes have been observed in amphibian hepatocytes (Nicholls et al., 1968; Bergink et al., 1977; Tseng et al., 1979), in other ES-responsive tissues (Vic et al., 1980; Padykula and Clark, 1981; Roberts et al., 1989), and in other steroid target tissues (Section VI).

Although de novo induction of genes expressing Vg and VLDL occurs within hours of primary exposure to ES, approximately 24 hr elapses before rooster hepatocytes are fully reprogrammed for peak synthesis and export of these two proteins (Shapiro, 1982; Brasch and Peters, 1985; Saluz et al., 1986; Hache et al., 1987). During this induction phase, hepatocyte nuclei enlarge, undergo major increases in total protein and RNA content, and exhibit sharp rises in [$^3$H]estradiol binding and transcription capacity in vitro (Table I and Fig. 3). These changes are paralleled by a marked proliferation of nuclear bodies. This situation persists for an additional 24 hr and then declines again in the absence of hormone. It is noteworthy, however, that even after 4 weeks not all of the above changes have returned fully to preinduction levels.

Several of these nuclear modifications merit closer scrutiny. The rapid enlargement of hepatocyte nuclei after ES administration, and the accompanying increase in total protein content, are features commonly observed in response to signals that stimulate transcription and/or reprogram the affected target cells. Examples include not only other steroid target tissue (Section VI), but also lectin-stimulated lymphocytes (Setterfield et al., 1985; Bladon et al., 1988), reactivated chick erythrocyte nuclei (Harris,

TABLE I

Volumes and Molecular Composition of Liver Nuclei Isolated from Roosters at Various Times after a Single Injection of Estradiol[a]

| Time after injection | Volume | RNA/DNA | Protein/DNA |
|---|---|---|---|
| Controls | 1.00 | 1.00 | 1.00 |
| 6–8 hr | 1.24 | 0.93 | 1.23 |
| 10–48 hr | 1.38 | 1.20 | 1.49 |
| 2–4 weeks | 1.35 | 1.00 | 1.12 |

[a] All values are expressed as ratios relative to controls; for procedural details, see Brasch (1990b). Nuclear volumes are also ratios expressed relative to controls at 1.00.

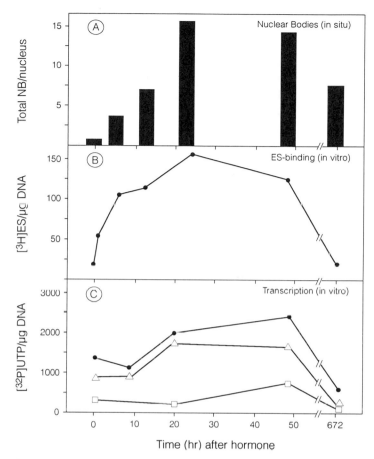

FIG. 3   Time courses of changes in nuclear body (NB) frequency, ES binding, and transcription capacity of rooster liver nuclei after a single exposure to 17β-estradiol. Animals were injected once with unlabeled 17β-estradiol (20 mg/kg) and sampled at times indicated. For procedural details, see Brasch *et al.* (1989). Transcription assays were carried out as per Marzluff and Huang (1986). (A) Total NB counts *in situ* obtained by morphometric analysis; (B) [³H]estradiol binding (cpm/μg DNA) by isolated nuclei; (C) [³²P]UTP incorporation (dpm/μg DNA) by isolated nuclei, in absence (●) and in presence of 1 μg/ml α-amarantin (△) and 10 μg/ml actinomycin D (□).

1986), and somatic nuclei transplanted into oocytes (Diberardino, 1980). In each of these instances, nuclear enlargement is accompanied by (1) rapid influx of preexisting and newly synthesized proteins, (2) partial dispersion of previously condensed chromatin, (3) enlargement of nucleoli, and (4) rapid elaboration of the interchromatin components, including RNP and nuclear matrix elements (Bergman *et al.*, 1990; Brasch, 1990a; Davis *et al.*, 1993). Comparable changes are clearly not unexpected, therefore, in rooster liver nuclei whose regular transcriptional patterns have been altered and greatly enhanced through sudden exposure to a high dose of estradiol.


## C. Coiled Bodies in Estradiol-Stimulated Rooster Liver

Nuclear bodies have been used as functional indicators of steroid hormone action in a variety of target tissues (Padykula and Clark, 1981; Brasch *et al.*, 1989; and below). Coiled bodies are the best characterized class of nuclear bodies (Section III) and appear prominently in ES-sensitive systems (Brasch and Ochs, 1992; Ochs *et al.*, 1994). This is again well illustrated in hepatocytes of ES-stimulated rooster liver (Fig. 4). The average number of CBs per nucleus rises gradually after hormone treatment, peaks about 20–48 hr later, and then declines again. Not unexpectedly, this transient pattern parallels most other manifestations of the response in animals given only a single injection of ES (Table I and Fig. 3). In animals injected on a continuous basis every second day, the CB counts were similar to those at 20–48 hr (data not shown).

What is less expected, however, is the relative heterogeneity in CB content within a given population of nuclei at all stages. For example, as shown in Fig. 4, while the mean number of CBs labeled with anti-p80-coilin is about one per nucleus in controls and 6-hr posthormone hepatocytes, a significant proportion of nuclei apparently contained none. At the other extreme, under fully induced conditions, over one-third of the nuclei contained more than three CBs.

Although it has been previously suggested that at least some CBs arise through nucleolar budding (Section III), it is not certain that they all arise this way. Because CB identification is based primarily on ultrastructure and immunological probes (Raska *et al.*, 1990b; Andrade *et al.*, 1991; Lamond and Carmo-Fonseca, 1993), these aspects could well change in terms of individual CB content or structure, the physiological status of the target cell, stage of the cell cycle, and so on. Similar developmental and numerical considerations apply to nucleoli. Because both the mechanism of formation and the functional characteristics of CBs and other nuclear bodies are still obscure (Brasch and Ochs, 1992), more information about them is clearly needed.


## D. Extended Nuclear Modifications

As indicated above, some of the large-scale modifications in rooster liver nuclei seem to persist even 4 weeks after primary stimulation with estra-

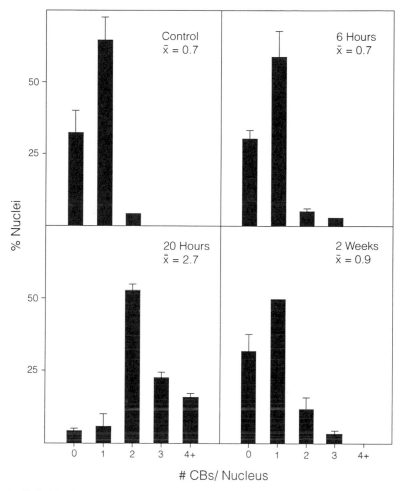

FIG. 4   Coiled body quantitation in rooster hepatocyte nuclei at various times after a single injection of estradiol. Results shown are typical for p80-coilin localization by immunofluorescence microscopy using isolated nuclei ($N = 200$ for each time point). For procedural details, see Brasch and Ochs (1992).

diol, a time when most aspects of the vitellogenic responses have subsided. These changes include, but are probably not limited to, (1) nuclear enlargement, (2) elevated total protein content, (3) persistent expansion of interchromatin elements, and (4) higher than average nuclear body levels. In addition, two-dimensional electrophoretic analyses of both nucleoplasmic and nuclear matrix proteins (Brasch and Peters, 1985; Brasch, 1990b) revealed persistent qualitative and quantitative differences among several polypeptides after 4 weeks, relative to preinduction patterns. One or more of the foregoing long-term changes could represent permanent nuclear modifications, including those contributing to the memory effect (Section IV,F). In this way, a second response to ES would be accelerated in large part because many structural elements necessary for enhanced transcription, processing, and RNA export are already in place.

## VI. Other Vertebrate Steroid Target Tissues

Extensive nuclear remodeling occurs in virtually every vertebrate tissue sensitive to the steroid/thyroid family of hormones. Not surprisingly, the best studied examples are the primary targets of the sex steroids, including both male and female reproductive organs (see, e.g., the classic ultrastructural studies by Dahl, 1976; Le Goascogne and Baulieu, 1977; Vic et al., 1980; Padykula and Clark, 1981; Arnold et al., 1983). It is also likely that comparable modifications occur in most tissues induced to hyperactivity through other kinds of hormones and/or external signals. With steroid-type hormones, however, the effects on the nucleus are probably more direct, owing to receptor-mediated interactions with many different genes simultaneously. As with rooster liver, nuclear changes in other steroid-sensitive tissues fall into a number of obvious categories. Some of these are listed in Table II. It must be stressed that this listing is neither exhaustive nor comprehensive, but merely highlights features common to a wide range of tissues and hormones.

## VII. Insect Systems

The best characterized steroid-like regulators in insects are the ecdysteroids and juvenile hormone (JH) (Wyatt, 1991; Riddiford, 1994). These ligands act at the nuclear level in manners both similar to and different from those of their vertebrate counterparts. The DNA response elements for both hormones appear to contain motifs similar to those for ES (Locke et al., 1987; Martinez et al., 1991), and the Drosophila ecdysteroid receptor is part of the steroid receptor superfamily (Koelle et al., 1991). However, unlike vertebrate steroids, neither ecdysteroids nor JH is sex specific; they exert sex-specific effects because of dimorphic gene programming in their respective target tissues (Wyatt, 1991). This is particularly well illustrated by vitellogenesis in the fat body of Locusta migratoria (Wyatt, 1988).

TABLE II

Selected Examples of Nuclear Remodeling in Vertebrate Tissues Sensitive to Steroids and Other Hormones Acting at the Level of the Nucleus

| Steroid/hormone | Nuclear expansion | Chromatin decondensation | Nucleolar expansion | Nuclear bodies | Ref. |
|---|---|---|---|---|---|
| Androgens | | | | | |
| Prostatic epithelium | × | nd[a] | nd | nd | Carmo-Fonseca (1982) |
| Seminal vesicle | × | × | × | nd | Arnold et al. (1983) |
| Epididymis | × | × | × | × | Faure et al. (1987) |
| Estrogen | | | | | |
| Endometrium | × | × | nd | nd | Vic et al. (1980) |
| Uterine epithelium | × | nd | nd | × | Padykula and Clark (1981) |
| Uterine epithelium | × | × | × | × | Lescoat et al. (1985) |
| Grandular endometrium | × | × | × | nd | Roberts et al. (1989) |
| Pituitary mammotropes | × | nd | × | nd | Poole et al. (1980) |
| Cortisol | | | | | |
| rat liver | nd | × | nd | nd | Derenzini et al. (1979) |
| Thyrotropin | | | | | |
| Thyroid cultures | nd | nd | × | × | Vagner-Capodano et al. (1982) |
| Gonadotropins | | | | | |
| Luteal cells | nd | nd | nd | × | Chegini and Rao (1988) |

[a] nd, Not determined.

Although the term "vitellogenin" (Vg) was first applied to the sex-limited, yolk precursor protein in insect hemolymph (Pan et al., 1969), the Vgs of both vertebrates and invertebrates have since been shown to comprise an evolutionary superfamily (Wahli, 1988). In most insects, the Vgs are produced in the fat body and in *Locusta* this is under the control of JH and is normally confined to adult females. Although JH probably acts at the gene level via protein receptors (Roberts and Jeffries, 1986), such action is neither as rapid nor necessarily as direct as that of ES in vertebrate liver. This is not surprising, owing to the multifaceted action of JH in locusts and similar insects. In addition to inducing Vg synthesis in females, JH also controls the overall growth and maturation of the fat body in both sexes (Wyatt, 1988, 1991). Consequently, unlike the situation in *Xenopus* and roosters, access to the *Vg* genes is blocked in adult males, but this cannot be due to the absence of JH receptors. In all likelihood, therefore, despite much overlap in structure and function, male and female fat bodies are programmed in fundamentally different ways with respect to JH.

Juvenile hormone has several marked effects on the nuclei of the locust fat body. These fall into three broad categories: (1) enlargement and restructuring, (2) DNA synthesis and polyploidization, and (3) nuclear bodies. All three aspects have been shown to be totally JH dependent, primarily on the basis of hormone replacement studies with animals in which the JH-producing glands, the corpora allata, have been surgically destroyed or chemically inactivated (Wyatt, 1988).

Fat body nuclei in both male and female locusts undergo marked enlargment and remodeling after adult emergence, but the effects are more pronounced in females (Couble *et al.,* 1979; Jensen and Brasch, 1985). These changes are directly associated with several rounds of DNA synthesis leading to extensive polyploidization (Nair *et al.,* 1981; Irvine and Brasch, 1981). In fully developed adults, the bulk of nuclei is 8C and 16C in males and females, respectively. Directly comparable, JH-dependent patterns of fat body development have been reported in other grasshoppers (Roberts and Jeffries, 1986).

Perhaps the most striking aspect of nuclear remodeling in locust fat body nuclei is the proliferation of highly complex nuclear bodies (Jensen and Brasch, 1985; Brasch and Ochs, 1992). Although morphologically different from their counterparts in vertebrate liver, locust nuclear bodies nevertheless exhibit similar dynamics with respect to hormone sensitivity. For example, they increase numerically in direct proportion to nuclear volume and ploidy levels which, in turn, are closely dependent on JH or hormone analog levels (Nair *et al.,* 1981; Wyatt, 1991). In hormone-depleted locusts, nuclear volumes do not rise, DNA synthesis is attenuated, and nuclear body numbers remain static (Jensen and Brasch, 1985). As with coiled bodies, therefore, whose exact functions are similarly obscure, nuclear body expression in the fat body is also highly dynamic and clearly sensitive to changes in cell metabolism, growth, and differentiation.

## VIII. Cancer Cells

In the United States, hormone-related cancers account for more than 20% of all newly diagnosed male and more than 40% of all newly diagnosed female malignancies (Henderson *et al.,* 1993). Hormones are thought to play a role in the etiology and pathogenesis of several human cancers, including cancers of the breast, endometrium, prostate, and ovaries.

## A. Estrogens, Breast Cancer, and Coiled Bodies

In the latest year for which figures are available, it is estimated that 176,000 women in the United States will be diagnosed with breast cancer and 46,000 will die from this disease (Harris *et al.*, 1992). Put another way, women have a one-in-eight chance of developing breast cancer during their lifetime, making this not only a leading cause of death among American women, but also a steadily rising health issue. A large part of this recent increase has been in ES-sensitive breast cancers (Harris *et al.*, 1992). This may be attributable at least in part to postmenopausal, hormone replacement therapy and to the increasing appearance of estrogenic compounds in the environment (Bulger and Kupfer, 1983; Wolff *et al.*, 1993).

Estrogen is known to have a proliferative effect on the epithelial cells lining the milk ducts, and infiltrating ductal carcinoma is by far the most prevalent malignant condition of the breast (Harris *et al.*, 1992). This information has been used to formulate an effective endocrine therapy based on the inhibitory action of the anti-estrogen tamoxifen, which competes with ES for receptor binding (Henderson *et al.*, 1993).

In a previous paper (Brasch and Ochs, 1992) and in Section V of this chapter, we reported an increase in nuclear coiled bodies in hepatocytes after injection of estradiol into young roosters. To better assess hormone-induced expression of CBs in an *in vitro* system, we have also studied their distribution in several ES-sensitive, human breast cancer cell lines (Ochs *et al.*, 1994). Antibodies to the CB-specific protein p80-coilin were used, coupled with double-label immunofluorescence, confocal microscopy, and immunoelectron microscopy. Immunofluorescence indicated not only that all cell lines exhibited prominent nucleoplasmic CBs, but also that several cell lines contained CBs within the nucleolus proper. This novel observation was confirmed by double-label immunofluorescence and confocal microscopy, showing that in addition to p80-coilin, both nucleoplasmic and nucleolar CBs also contained fibrillarin and Sm proteins. Conventional and immunoelectron microscopy further indicated that the nucleolar CBs were discrete structures and distinct from the other nucleolar domains, including the granular component, the dense fibrillar component, and the fibrillar centers. Although the significance of CBs in the nucleoli of some breast cancer lines is still unclear, this does provide the first direct evidence that both CBs and Sm determinants can be present in mammalian nucleoli.

It may not be fortuitous that nucleolar CBs were first observed in breast cancer cells, because several previous reports on nuclear bodies were actually based on ultrastructural studies of human tumor tissues (Bouteille *et al.*, 1967; Brasch and Ochs, 1992). In a study by Spector *et al.* (1992), comparing human cell lines of various degrees of transformation, the lowest percentage of cells with CBs was observed in contact-inhibited cultures of defined passage, followed by immortal cells and then by fully transformed cells with the highest percentage of CBs. A clear correlation was evident, therefore, between the degree of cell transformation and CB frequency. It is also noteworthy that the breast cancer cell lines T47D and MCF-7, shown by Ochs *et al.* (1994) to contain nucleolar CBs, has previously been shown to contain receptors for ES and PR (Vic *et al.*, 1982), vitamin $D_3$ (Sher *et al.*, 1981; Freake *et al.*, 1981), and calcitonin (Lamp *et al.*, 1981). It is clearly important now to ascertain whether or not the hormone responsiveness of such cells is related to the presence and/or nucleolar localization of CBs. This would furnish yet another example of nuclear remodeling in response to steroid/thyroid-like hormone action.

## B. Retinoic Acid and Acute Promyelocytic Leukemia

Even though not a hormone in the classic sense, retinoic acid, along with its nuclear receptors RAR and RXR (which function as ligand-inducible transcription factors), is hormone-like and may play a central role in multiple signalling pathways (Kliewer *et al.*, 1992). Clinically, retinoic acid has the remarkable ability to promote the differentiation of acute promyelocytic leukemia (APL) cells into mature granulocytes, resulting in a complete remission of the leukemic condition.

Preliminary studies by de-Thé *et al.* (1990) suggested that the RARα gene was rearranged and its expression altered in APL patients and in the APL-derived cell line NB4. It was determined that the RARα gene was translocated from chromosome 15 to 17q21, the translocation site specifically associated with APL. This translocation results in a chimeric fusion protein of 106 kDa, formed between RARα and PML, a myeloid protein encoded by a gene located on chromosome 17 (de-Thé *et al.*, 1991: Kakizuka *et al.*, 1991). Immunocytochemistry demonstrated that PML is part of a novel macromolecular domain referred to as PML oncogenic domains (PODs) (Dyck *et al.*, 1994). In APL cells, the PODs are dispersed into a microparticulate pattern following expression of the PML–RAR oncoprotein. Retinoic acid treatment of APL cells restored the PODs to normal appearance (summarized in Fig. 5, and in Dyck *et al.*, 1994; Weis *et al.*, 1994; Koken *et al.*, 1994).

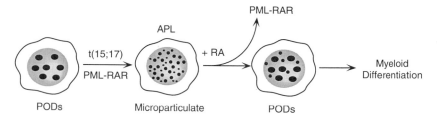

FIG. 5   Model depicting the disruption of POD structures in the presence of PML–RAR, leading to the microparticulate localization of PML and other POD-associated proteins. [Courtesy of Dyck *et al.* (1994), with permission, © 1994 Cell Press.]

In addition to the 70-kDa PML and 195-kDa PML–RAR proteins, PODs also contain a number of other nuclear proteins, including NDP55 (Ascoli and Maul, 1991), Sp100 (Szostecki *et al.*, 1990), and a component recognized by 5E10, a mouse monoclonal antibody raised against nuclear matrix (Stuurman *et al.*, 1992a), which subsequently demonstrated specificity for PML (Koken *et al.*, 1994). When immunoelectron microscopy was performed, the macromolecular complex defined by these various antibodies and proteins was a specific type of complex nuclear body 0.3–0.5 $\mu$m in diameter and typically appearing doughnut-like or ring shaped (Weis *et al.*, 1994; Koken *et al.*, 1994).

The dynamic nuclear remodeling that occurs in APL cells in response to retinoic acid is illustrated in Fig. 5 (Dyck *et al.*, 1994). It is this sort of remodeling, involving a specific type of nuclear body associated with differentiation of APL cells and a cure for leukemia, that suggests that a beneficial role of retinoic acid in this cancer lies in its ability to restore normal subnuclear organization (Weis *et al.*, 1994).

## IX. Summary and Overview

An attempt has been made in this chapter to bring together a wide range of information relating to the action of steroid and steroid-like hormones on the cell nucleus. Despite the complexity and diversity of the many systems involved, it is perhaps reassuring to note that a number of universal trends have emerged. For example, in all cases target nuclei enlarge in response to their respective hormones, presumably owing to rapid influx of newly synthesized or stored proteins from the cytoplasm. Concurrently, a partial decondensation of chromatin is evident, as well as nucleolar enlargement, elaboration of RNP and other nucleoplasmic elements, and

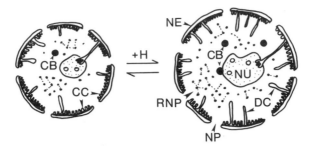

FIG. 6  Summary illustration of the large-scale modifications in the nuclei of steroid/thyroid hormone target cells, showing coiled body (CB), condensed chromatin (CC), decondensed chromatin (DC), nuclear envelope (NE), nuclear pore (NP), nucleolus (NU), and ribonucleoprotein particles (RNPs). Most changes indicated can revert in the absence of hormone.

proliferation of nuclear bodies (Fig. 6). Collectively, this remodeling reflects at least a temporary "reprogramming" of the affected nuclei, because many of these features regress or revert to preinduction levels in the absence of hormone. It is now clearly necessary to determine how specific hormones act to bring about such large-scale effects and how these interact to enhance and/or attenuate the expression of individual genes simultaneously.

## Acknowledgments

This work was funded in part by NIH Grant 1R15 AI32691-01 to K. Brasch and by NIH Grants AR-32063 and CA-56956 to Eng M. Tan. We thank Drs. B. W. O'Malley, D. L. Spector, M.-J. Tsai, and G. R. Wyatt for providing preprints and illustrations of unpublished materials, and D. Gallo, T. Stein, J. Teigen, and J. Williams for outstanding technical assistance.

## References

Adom, J. N., Gouilleux, F., and Richard-Foy, H. (1992). Interaction with the nuclear matrix of a chimeric construct containing a replication origin and a transcription unit. *Biochim. Biophys. Acta* **1171,** 187–197.

Agard, D. A., and Sedat, J. W. (1983). Three dimensional architecture of a polytene nucleus. *Nature (London)* **302,** 676–681.

Alexander, R. B., Greene, G. L., and Barrack, E. R. (1987). Estrogen receptors in the nuclear matrix: Direct demonstration using monoclonal antireceptor antibody. *Endocrinology (Baltimore)* **120,** 1851–1857.

Almouzni, G., and Wolffe, A. P. (1993). Nuclear assembly, structure and function: The use of *Xenopus in vitro* systems. *Exp. Cell Res.* **205,** 1–15.

Andrade, L. E. C., Chan, E. K. L., Raska, I., Peebles, C. L., Roos, G., and Tan, E. M. (1991). Human autoantibody to a novel protein of the nuclear coiled body: Immunological characterization and cDNA cloning of p80-coilin. *J. Exp. Med.* **173,** 1407–1419.

Andrade, L. E. C., Tan, E. M., and Chan, E. K. L. (1993). Immunocytochemical analysis of the coiled body in the cell cycle and during cell proliferation. *Proc. Natl. Acad. Sci. U.S.A.* **90,** 1947–1951.

Arnold, C., Gulbenkian, S., Carmo-Fonseca, M., and David-Ferreira, J. F. (1983). Androgen-dependent changes in nuclear ultrastructure. A steriological study of rat ventral prostate and seminal vesicle. *Biol. Cell.* **47,** 161–170.

Ascoli, C. A., and Maul, G. G. (1991). Identification of a novel nuclear domain. *J. Cell Biol.* **112,** 785–795.

Barrack, E. R. (1987). Localization of steroid hormone receptors in the nuclear matrix. *In* "Steroid Hormone Receptors" (C. R. Clark, ed.), pp. 86–127. VCH Ellis Horwood, London.

Beato, M. (1989). Gene regulation by steroid hormones. *Cell (Cambridge Mass.)* **56,** 335–344.

Berezney, R. (1991). The nuclear matrix: A heuristic model for investigating genomic organization and function in the cell nucleus. *J. Cell. Biochem.* **47,** 109–123.

Bergink, E. W., Wallace, R. A., Vande Berg, J. A., Bos, E. S., Gruber, M., and AB, G. (1974). Estrogen-induced synthesis of yolk proteins in roosters. *Am. Zool.* **14,** 1177–1193.

Bergink, E. W., Tseng, M. T., and Wittliff, J. L. (1977). Sequential changes in the structure and function of hepatocytes in oestrogen-treated *Xenopus laevis* males. *Cytobiologie* **14,** 362–377.

Bergman, M., Nyman, U., Ringertz, N., and Pettersson, I. (1990). Appearance and origin of snRNP antigens in chick erythrocyte nuclei reactivated in heterokaryons. *J. Cell Sci.* **95,** 361–370.

Bernhard, W. (1969). A new staining procedure for electron microscopical cytology. *J. Ultrastruct. Res.* **27,** 250–265.

Bladon, T., Brasch, K., Brown, D. L., and Setterfield, G. (1988). Changes in structure and protein composition of bovine lymphocyte nuclear matrix during concanavalin-A induced mitogenesis. *Biochem. Cell Biol.* **66,** 41–53.

Bouteille, M., Kalifat, S. R., and Delarue, J. (1967). Ultrastructural variations of nuclear bodies in human diseases. *J. Ultrastruct. Res.* **19,** 474–486.

Boveri, T. (1909). Die Blastomerenkeme von *Ascaris magalocephala* und die Theorie der Chromosomenindividualitat. *Arch. Exp. Zellforsch.* **3,** 181–268.

Brasch, K. (1990a). Drug and metabolite-induced perturbations in nuclear structure and function: A review. *Biochem. Cell Biol.* **68,** 408–426.

Brasch, K. (1990b). Nuclear protein modifications in vitellogenic rooster liver. *Cell. Mol. Biol.* **36,** 659–671.

Brasch, K., and Ochs, R. L. (1992). Nuclear bodies (NBs): A newly "rediscovered" organelle. *Exp. Cell Res.* **202,** 211–223.

Brasch, K., and Peters, K. E. (1985). Nuclear protein, matrix and structural changes in rooster liver after estrogenic induction of vitellogenesis. *Biol. Cell.* **54,** 109–122.

Brasch, K., and Setterfield, G. (1974). Structural organization of chromosomes in interphase nuclei. *Exp. Cell Res.* **83,** 175–185.

Brasch, K., Harrington, S., and Blake, H. (1989). Isolation and analysis of nuclear bodies from estrogen-stimulated chick liver. *Exp. Cell Res.* **182,** 425–435.

Brunt, S. A., Riehl, R., and Silver, J. C. (1990). Steroid hormone regulation of the *Achyla ambisexualis* 85-kilodalton heat shock protein, a component of the *Achyla* steroid receptor complex. *Mol. Cell. Biol.* **10,** 273–281.

Bulger, W. H., and Kupfer, D. (1983). Estrogenic action of DDT analogs. *Am. J. Ind. Med.* **4,** 163–173.

Burch, J. B. E., and Evans, M. I. (1986). Chromatin structural transitions and the phenomenon of vitellogenin gene memory in chickens. *Mol. Cell. Biol.* **6**, 1886–1893.

Burch, J. B. E., and Weintraub, H. (1983). Temporal order of chromatin structural changes associated with activation of the major chicken vitellogenin gene. *Cell (Cambridge, Mass.)* **33**, 65–76.

Cadepond, F., Gasc, J. M., Delahaye, F., Jibard, N., Schweitzer-Groyer, G., Segard-Maurel, I., Evans, R., and Baulieu, E. E. (1992). Hormonal regulation of the nuclear localization signals of the human glucocorticoid receptor. *Exp. Cell Res.* **201**, 99–108.

Carmo-Fonseca, M. (1982). Testosterone-induced changes in nuclear pore complex number of prostatic nuclei from castrated rats. *J. Ultrastruct. Res.* **80**, 243–251.

Carmo-Fonseca, M. (1988). Androgen-dependent nuclear proteins in rat ventral prostate are glycoproteins associated with the nuclear matrix. *Cell Biol. Int. Rep.* **12**, 607–620.

Carmo-Fonseca, M., Tollervey, D., Pepperkok, R., Barabino, S. M. L., Merdes, A., Brunner, C., Zamore, P. D., Green, M. R., Hurt, E., and Lamond, A. I. (1991a). Mammalian nuclei contain foci which are highly enriched in components of the pre-mRNA splicing machinery. *EMBO J.* **10**, 195–206.

Carmo-Fonseca, M., Pepperkok, R., Sproat, B. S., Ansorge, W., Swanson, M. S., and Lamond, A. I. (1991b). In vivo detection of snRNP-rich organelles in the nuclei of mammalian cells. *EMBO J.* **10**, 1863–1873.

Carmo-Fonseca, M., Pepperkok, R., Carvalho, M. T., and Lamond, A. I. (1992). Transcription-dependent colocalization of the U1, U2, U4/U6 and U5 snRNPs in coiled bodies. *J. Cell Biol.* **117**, 1–14.

Carmo-Fonseca, M., Ferreira, J., and Lamond, A. I. (1993). Assembly of snRNP-containing coiled bodies is regulated in interphase and mitosis-evidence that the coiled body is a kinetic nuclear structure. *J. Cell Biol.* **120**, 841–852.

Carson-Jurica, M. A., Schrader, W. T., and O'Malley, B. W. (1990). Steroid receptor family: Structure and functions. *Endocr. Rev.* **11**, 201–215.

Carter, K. C., Bowman, D., Carrington, W., Fogarty, K., McNeil, J. A., Fay, F. S., and Lawrence, J. B. (1993). A three-dimensional view of precursor messenger RNA metabolism within the mammalian nucleus. *Science* **259**, 1330–1326.

Chaly, N., Setterfield, G., Kaplan, J. G., and Brown, D. L. (1983a). Nuclear bodies in mouse splenic lymphocytes. I. Ultrastructural changes during stimulation by concanavalin A. *Biol. Cell.* **47**, 275–284.

Chaly, N., Setterfield, G., Kaplan, J. G., and Brown, D. L. (1983b). Nuclear bodies in mouse splenic lymphocytes. II. Cytochemistry and autoradiography during stimulation by concanavalin A. *Biol. Cell.* **49**, 35–44.

Chamberland, H., and Lafontaine, J. G. (1993). Localization of snRNP antigens in nucleolus-associated bodies: Study of plant interphase nuclei by confocal and electron microscopy. *Chromosoma* **102**, 220–226.

Chandran, U. R., and DeFranco, D. B. (1992). Internuclear migration of chicken progesterone receptor, but not simian virus-40 large tumor antigen, in transient heterokaryons. *Mol. Endocrinol.* **6**, 837–844.

Chegini, N., and Rao, C. V. (1988). Increase of nuclear bodies in bovine luteal cells after treatment with human chorionic gonadotropin. *Biol. Reprod.* **38**, 453–461.

Ciejek, E. M., Tsai, M. J., and O'Malley, B. W. (1983). Actively transcribed genes are associated with the nuclear matrix. *Nature (London)* **306**, 607–609.

Clark, J. H., Hardin, J. W., Padykula, H. A., and Cardasis, C. A. (1978). Role of estrogen receptor binding and transcriptional activity in the stimulation of hyperestrogenism and nuclear bodies. *Proc. Natl. Acad. Sci. U.S.A.* **75**, 2781–2784.

Comings, D. E. (1968). The rationale for an ordered arrangement of chromatin in the interphase nucleus. *Am. J. Hum. Genet.* **20**, 440–460.

Cook, P. R. (1988). The nucleoskeleton: Artifact, passive framework or active site? *J. Cell Sci.* **90,** 1–6.

Couble, P., Chen, T. T., and Wyatt, G. R. (1979). Juvenile hormone-controlled vitellogenia synthesis in *Locusta migratoria* fat body. Cytological development. *J. Insect Physiol.* **25,** 327–337.

Crowley, K. S., and Brasch, K. (1987). Does the interchromatin compartment contain actin? *Cell Biol. Int. Rep.* **11,** 537–546.

Dahl, E. (1976). The ultrastructure of the accessory sex organs. XI. Nuclear alterations of prostatic epithelial cells induced by castration. *Cell Tissue Res.* **171,** 285–296.

Dauvois, S., White, R., and Parker, M. G. (1993). The antiestrogen ICI 182780 disrupts estrogen receptor nucleocytoplasmic shuttling. *J. Cell Sci.* **106,** 1377–1388.

Davis, L., Cadrin, M., Brown, D. L., and Chaly, N. (1993). Reversible disassembly of transcription domains in lymphocyte nuclei during inhibition of RNA synthesis by DRB. *Biol. Cell.* **78,** 163–180.

Derenzini, M., Pession-Brizzi, A., Bonetti, E., and Novello, F. (1979). Relationship between ultrastructure and function of hepatocyte chromatin: A study with adrenalectomized rats and cortisol administration. *J. Ultrastruct. Res.* **67,** 161–179.

de-Thé, H., Chomienne, C., Lanotte, M., Degos, L., and Dejean, A. (1990). The t(15;17) translocation of acute promyelocytic leukaemia fuses the retinoic acid receptor gene to a novel transcribed locus. *Nature (London)* **347,** 558–561.

de-Thé, H., Lavau, C., Marchio, A., Chomienne, C., Degos, L., and Dejean, A. (1991). The PML–RAR fusion mRNA generated by the t(15;17) translocation in acute promyelocytic leukemia encodes a functionally altered RAR. *Cell (Cambridge, Mass.)* **66,** 675–684.

Diberardino, M. A. (1980). Genetic stability and modulation of metazoan nuclei transplanted into eggs and oocytes. *Differentiation (Berlin)* **17,** 17–30.

Dupuy-Coin, A. M., and Bouteille, M. (1972). Developmental pathway of granular and beaded nuclear bodies from nucleoli. *J. Ultrastruct. Res.* **40,** 55–67.

Dyck, J. A., Maul, G. G., Miller, W. H., Jr., Chen, J. D., Kakizuka, A., and Evans, R. M. (1994). A novel macromolecular structure is a target of the promyelocyte-retinoic acid receptor oncoprotein. *Cell (Cambridge, Mass.)* **76,** 333–343.

Eliceiri, G. L., and Ryerse, J. S. (1994). Detection of intranuclear clusters of Sm antigens with monoclonal anti-Sm antibodies by immunoelectron microscopy. *J. Cell. Physiol.* **121,** 449–451.

Evans, M. I., O'Malley, P. J., Krust, A., and Burch, J. B. E. (1987). Developmental regulation of the estrogen receptor and the estrogen responsiveness of five yolk protein genes in the avian liver. *Proc. Natl. Acad. Sci. U.S.A.* **84,** 8493–8497.

Fakan, S., and Bernhard, W. (1971). Localization of rapidly and slowly labelled nuclear RNA as visualized by high resolution autoradiography. *Exp. Cell Res.* **67,** 129–141.

Fakan, S., Puvion, E., and Spohr, G. (1976). Localization and characterization of newly synthesized nuclear RNA in isolated rat hepatocytes. *Exp. Cell Res.* **99,** 155–164.

Fakan, S., Leser, G., and Martin, T. E. (1984). Ultrastructural distribution of nuclear ribonucleoproteins as visualized by immunocytochemistry on thin sections. *J. Cell Biol.* **98,** 358–363.

Faure, J., Mesure, M., Tort, M., and Dufaure, J. P. (1987). Polyploidization and other nuclear changes during the annual cycle of an androgen-dependent organ, the lizard epididymis. *Biol. Cell.* **60,** 193–208.

Feavers, I. M. Jiricny, J., Moncharmont, B., Salvz, H. P., and Jost, J. P. (1987). Interaction of two nonhistone proteins with the estradiol response element of the avian vitellogenin gene modulates the binding of estradiol-receptor complex. *Proc. Natl. Acad. Sci. U.S.A.* **84,** 7453–7457.

Fey, E. G., Krochmalnic, G., and Penman, S. (1986). The nonchromatin substructure of the nucleus: The ribonucleoprotein (RNP)-containing and RNP-depleted matrices analyzed by sequential fractionation and resinless section electron microscopy. *J. Cell Biol.* **102,** 1654–1665.

Freake, H. C., Marcocci, C., Iwasaki, J., and MacIntyre, I. (1981). 1,25-Dihydroxyvitamin $D_3$ specifically binds to a human breast cancer cell line (T47D) and stimulates growth. Biochem. Biophys. Res. Commun. **101,** 1131–1138.

Gall, J. G. (1991). Spliceosomes and snurposomes. *Science* **252,** 1499–1500.

Gall, J. G., and Callan, H. G. (1989). The sphere organelle contains small nuclear ribonucleoproteins. *Proc. Natl. Acad. Sci. U.S.A.* **86,** 6635–6639.

Gasc, J. M., and Baulieu, E. E. (1986). Steroid hormone receptors: Intracellular distribution. *Biol. Cell.* **56,** 1–6.

Gasser, S. M., Amati, B. B., Cardenas, M. E., and Hofmann, J. F. X. (1989). Studies on scaffold attachment sites and their relation to genome function. *Int. Rev. Cytol.* **119,** 57–96.

Getzenberg, R. H., Pienta, K. J., and Coffey, D. S. (1990). The tissue matrix: Cell dynamics and hormone action. *Endocr. Rev.* **11,** 399–417.

Gilson, E., Laroche, T., and Gasser, S. M. (1993). Telomeres and the functional architecture of the nucleus. *Trends Cell Biol.* **3,** 128–134.

Goldberger, A., Horton M., Katzmann, J., and Spelsberg, T. C. (1987). Characterization of the chromatin acceptor sites for the avian oviduct progesterone receptor using monoclonal antibodies. *Biochemistry* **26,** 5811–5816.

Gorski, J., Toft, D., Shyamala, G., Smith, D., and Notides, A. (1968). Hormone receptors: Studies on the interaction of estrogen with the uterus. *Recent Prog. Horm. Res.* **24,** 45–81.

Gronemeyer, H. (1992). Control of transcription activation by steroid hormone receptors. *FASEB J.* **6,** 2524–2529.

Gruenbaum, Y., Hochstrasser, M., Mathog, D., Saumweber, H., Agard, D. A., and Sedat, J. W. (1984). Spatial organization of the *Drosophila* nucleus: A three-dimensional cytogenetic study. *J. Cell Sci. Suppl.* **1,** 223–234.

Guiochon-Mantel, A., Lescop, P., Christin-Maitre, S., Loosfelt, H., Perrot-Applanat, M., and Milgrom, E. (1991). Nucleocytoplasmic shuttling of the progesterone receptor. *EMBO J.* **10,** 3851–3859.

Haaf, T., and Schmid, M. (1991). Chromosome topology in mammalian interphase nuclei. *Exp. Cell Res.* **192,** 325–332.

Hache, R. J. G., Tam, S. P., Cochrane, A., Nesheim, and Deeley, R. G. (1987). Long-term effects of estrogen on avian liver: Estrogen-inducible switch in expression of nuclear, hormone-binding proteins. *Mol. Cell. Biol.* **7,** 3538–3547.

Ham, J., and Parker, M. G. (1989). Regulation of gene expression by nuclear hormone receptors. *Curr. Opin. Cell Biol.* **1,** 503–511.

Harris, J. R. (1986). Blood cell nuclei: The structure and function of lymphoid and erythroid nuclei. *Int. Rev. Cytol.* **102,** 53–168.

Harris, J. R., Lippman, M. E., Veronesi, U., and Willett, W. (1992). Breast cancer. *N. Engl. J. Med.* **327,** 319–380.

Henderson, B. E., Ross, R. K., and Pike, M. C. (1993). Hormonal chemoprevention of cancer in women. *Science* **259,** 633–638.

Herbener, G. H., Bendayan, M., and Feldhoff, R. C. (1985). Immunocytochemical localization of a non-vitellogenin, estrogen-induced plasma protein in the hepatocyte of the American bullfrog, *Rana catesbeiana. Eur. J. Cell Biol.* **39,** 142–146.

Hilliker, A. J., and Appels, R. (1989). The arrangement of interphase chromosomes: Structural and functional aspects. *Exp. Cell Res.* **185,** 297–318.

Ho, S. M. (1987). Endocrinology of vitellogenesis. *In* "Hormones and Reproduction in Fishes, Amphibians and Reptiles" (D. O. Norris and R. E. Jones, eds.), pp. 145–169. Plenum, New York.

Hochstrasser, M., and Sedat, J. W. (1987). Three-dimensional organization of *Drosophila melanogaster* interphase nuclei. II. *J. Cell Biol.* **104,** 1455–1470.

Holland, L. J., and Wangh, L. J. (1987). Estrogen induction of a 45 kDa secreted protein coordinately with vitellogenin in *Xenopus* liver. *Mol. Cell. Endocrinol.* **49,** 63–73.

Hozak, P., Hassan, A. B., Jackson, D. A., and Cook P. R. (1993). Visualization of replication factories attached to a nucleoskeleton. *Cell (Cambridge, Mass.)* **73,** 361–373.

Irvine, D. J., and Brasch, K. (1981). The influence of juvenile hormone on polyploidy and vitellogenesis in the fat body of *Locusta migratoria*. *Gen. Comp. Endocrinol.* **45,** 91–99.

Isola, J. J. (1987). The effect of progesterone on the localization of progesterone receptors in the nuclei of chick oviduct cells. *Cell Tissue Res.* **249,** 317–323.

Jackson, D. A. (1991). Structure-function relationships in eukaryotic nuclei. *BioEssays* **13,** 1–10.

Jensen, A. L., and Brasch, K. (1985). Nuclear development in locust fat body: The influence of juvenile hormone on inclusion bodies and the nuclear matrix. *Tissue Cell* **17,** 117–130.

Jensen, E. V., Suzuki, T., Kawashima, T., Stumpf, W. E., Jungblat, P. W., and Desombre, E. R. (1968). A two step mechanism for the interaction of oestradiol with the rat uterus. *Proc. Natl. Acad. Sci. U.S.A.* **59,** 632–638.

Jimenez-Garcia, L. F., and Spector, D. L. (1993). *In vivo* evidence that transcription and splicing are coordinated by a recruiting mechanism. *Cell (Cambridge, Mass.)* **73,** 47–59.

Jost, J. P., and Seldran, M. (1984). Association of transcriptionally active vitellogenin II gene with the nuclear matrix of chicken liver. *EMBO J.* **3,** 2005–2008.

Jost, J. P., Moncharmont, B., Jiricny, J., Salvz, H., and Hertner, T. (1986). *In vitro* secondary activation (memory effect) of avian vitellogenin II gene in isolated liver nuclei. *Proc. Natl. Acad. Sci. U.S.A.* **83,** 43–47.

Kakizuka, A., Miller, W. H., Jr., Umesono, K., Warrell, R. P., Jr., Frankel, S. R., Murty, V. V. V. S., Dmitrovsky, E., and Evans, R. M. (1991). Chromosomal translocation t(15;17) in human acute promyelocytic leukemia fuses RAR with a novel putative transcription factor, PML. *Cell (Cambridge, Mass.)* **66,** 663–674.

Kang, K. I., Devin, J., Cadepond, F., Jibard, N., Guiochon-Mantel, A., Baulieu, E. E., and Catelli, M. G. (1994). *In vivo* functional protein-protein interaction: Nuclear targeted hsp 90 shifts cytoplasmic steroid receptor mutants into the nucleus. *Proc. Natl. Acad. Sci. U.S.A.* **91,** 340–344.

Kaufmann, S. H., Okret, S., Wikström, A. C., Gustafsson, J. A., and Shaper, J. H. (1986). Binding of the glucocorticoid receptor to the rat liver nuclear matrix. *J. Biol. Chem.* **261,** 11962–11967.

King, W. J., and Greene, G. L. (1984). Monoclonal antibodies localize oestrogen receptor in the nuclei of target cells. *Nature (London)* **307,** 745–747.

Kliewer, S. A., Umesono, K., Mangelsdorf, D. J., and Evans, R. M. (1992). Retinoid X receptor interacts with nuclear receptors in retinoic acid, thyroid hormone and vitamin $D_3$ signalling. *Nature (London)* **355,** 446–449.

Koelle, M. R., Talbot, W. S., Segraves, W. A., Bender, M. T., Cherbas, P., and Hogness, D. S. (1991). The *Drosophila* EcR gene encodes an ecdysone receptor, a new member of the steroid receptor superfamily. *Cell (Cambridge, Mass.)* **67,** 59–77.

Koken, M. H. M., Puvion-Dutilleul, F., Guillemin, M. C., Viron, A., Linares-Cruz, G., Stuurman, N., de Jong, L., Szostecki, C., Calvo, F., Chomienne, C., Degos, L., Puvion, E., and de-Thé, H. (1994). The t(15;17) translocation alters a nuclear body in a retinoic acid-reversible fashion. *EMBO J.* **13,** 1073–1083.

Lafarga, M., and Hervas, J. P. (1983). Light and electron microscopic characterization of the "accessory body" of Cajál in the neuronal nucleus. *In* "Ramón y Cajál's Contribution to the Neurosciences" (Grisola, S., Guerri, C., Sampson, F., Norton, S., and Reinoso-Suorez, F., eds.), pp. 91–100. Elsevier, Amsterdam.

Lafarga, M., Andres, M. A., Berciano, M. T., and Maquiera, E. (1991). Organization of nucleoli and nuclear bodies in osmotically stimulated supraoptic neurons of the rat. *J. Comp. Neurol.* **308,** 329–339.

Lafontaine, J. G. (1965). A light and electron microscope study of small, spherical nuclear bodies in meristematic cells of *Allium cepa, Vicia faba,* and *Raphanus sativus. J. Cell Biol.* **26,** 1–17.

Lamond, A. I., and Carmo-Fonseca, M. (1993). The coiled body. *Trends Cell Biol.* **3,** 198–204.

Lamp, S. J., Findlay, D. M., Moseley, J. M., and Martin, T. J. (1981). Calcitonin induction of a persistent activated state of adenylate cyclase in human breast cancer cells (T47D). *J. Biol. Chem.* **256,** 12269–12274.

Lawrence, J. B., Singer, R. H., and Marselle, L. M. (1989). Highly localized tracks of specific transcripts within interphase nuclei visualized by *in situ* hybridization. *Cell (Cambridge, Mass.)* **57,** 493–502.

Le Goascogne, C., and Baulieu, E. E. (1977). Hormonally controlled "nuclear bodies" during development of the prepuberal rat uterus. *Biol. Cell.* **30,** 195–206.

Lescoat, D., Saboureau, M., Castaing, L., and Chambon, Y. (1985). The hedgehog uterus. *Acta Anat.* **122,** 29–34.

Locke, J., White, B. N., and Wyatt, G. R. (1987). Cloning and 5' end nucleotide sequence of two juvenile hormone-inducible vitellogenin genes of the African migratory locust. *DNA* **6,** 331–342.

Locke, M. (1990). Is there somatic inheritance of intracellular patterns? *J. Cell Sci.* **96,** 563–567.

Malatesta, M., Zancanaro, C., Martin, T. E., Chan, E. K. L., Amalric, F., Luhrmann, R., and Fakan, S. (1994). Is the coiled body involved in nucleolar functions? *Exp. Cell Res.* **211,** 415–419.

Manuelidis, L., and Chen, T. L. (1990). A unified model of eukaryotic chromosomes. *Cytometry* **11,** 8–25.

Martinez, E., Givel, F., and Wahli, W. (1991). A common ancestor DNA motif for invertebrate and vertebrate hormone response elements. *EMBO J.* **10,** 263–268.

Marzluff, W. F., and Huang, R. C. C. (1986). Transcription of RNA in isolated nuclei. *In* "Transcription and Translation: A Practical Approach" (B. D. Hames and S. J. Higgins, eds.), pp. 89–129. IRL Press, Oxford.

Matera, A. G., and Ward, D. C. (1993). Nucleoplasmic organization of small nuclear ribonucleoproteins in cultured human cells. *J. Cell Biol.* **121,** 715–727.

Metzger, D. A., and Korach, K. S. (1990). Cell-free interaction of the estrogen receptor with mouse uterine nuclear matrix: Evidence of saturability, specificity, and resistance to KCE extraction. *Endocrinology (Baltimore)* **126,** 2190–2195.

Metzger, D. A., White, J. H., and Chambon, P. (1988). The human oestrogen receptor functions in yeast. *Nature (London)* **334,** 31–36.

Milankov, K., and De Boni, U. (1993). Cytochemical localization of actin and myosin aggregates in interphase nuclei *in situ. Exp. Cell Res.* **209,** 189–199.

Mills, A. D., Blow, J. J., White, W. B. A., Wilcock, D., and Laskey, R. A. (1989). Replication occurs at discrete foci spaced throughout nuclei replicating *in vitro. J. Cell Sci.* **94,** 471–477.

Monneron, A., and Bernhard, W. (1969). Fine structural organization of the interphase cell nucleus of some mammalian cells. *J. Ultrastruct. Res.* **27,** 266–288.

Morel, G., Dubois, P., Benassayag, C., Nunez, E., Radanyl, C., Redeuilh, G., Richard-Foy, H., and Baulieu, E. E. (1981). Ultrastructural evidence of oestradiol receptor by immunochemistry. *Exp. Cell Res.* **132**, 249–257.

Moreno Diaz de la Espina, S., Risueno, M. C., and Medina, F. J. (1982a). Ultrastructural, cytochemical and autoradiographic characterization of coiled bodies in the plant cell nucleus. *Biol. Cell.* **44**, 229–238.

Moreno Diaz de la Espina, S., Sanchez-Pina, M. A., and Risueno, M. C. (1982b). Localization of acid phosphatase activity, phosphate ions and inorganic cations in plant nuclear coiled bodies. *Cell Biol. Int. Rep.* **6**, 601–607.

Nair, K. K., Chen, T. T., and Wyatt, G. R. (1981). Juvenile hormone-stimulated polyploidy in adult locust fat body. *Dev. Biol.* **81**, 356–360.

Nardulli, A. M., and Shapiro, D. J. (1992). Binding of the estrogen receptor DNA-binding domain to the estrogen response element induces DNA bending. *Mol. Cell. Biol.* **12**, 2037–2042.

Nelson, W. G., Pienta, K. J., Barrack, E. R., and Coffey, D. S. (1986). The role of the nuclear matrix in the organization and function of DNA. *Annu. Rev. Biophys. Biophys. Chem.* **15**, 457–475.

Newport, J. W., and Forbes, D. J. (1987). The nucleus: Structure, function and dynamics. *Annu. Rev. Biochem.* **15**, 535–565.

Nicholls, T. J., Follett, B. K., and Evennett, P. J. (1968). The effects of oestrogens and other steroid hormones on the ultrastructure of the liver of *Xenopus laevis* Daubin. *Z. Zellforsch. Mikrosk. Anat.* **90**, 19–27.

Nigg, E. A. (1988) Nuclear function and organization: The Potential of immunochemical approaches. *Int. Rev. Cytol.* **110**, 27–92.

Ochs, R. L., Stein, T. W., Jr., and Tan, E. M. (1994). Coiled bodies in the nucleolus of breast cancer cells. *J. Cell Sci.* **107**, 385–399.

O'Keefe, R., Mayeda, A., Sadowski, C. L., Krainer, A. R., and Spector, D. L. (1994). Disruption of pre-mRNA splicing *in vivo* results in reorganization of splicing factors. *J. Cell Biol.* **124**, 249–260.

O'Malley, B. W., Tsai, S. Y., Bagchi, M., Weigel, N. L., Schrader, W. T., and Tsai, M. T. (1991). Molecular mechanism of action of a steroid hormone receptor. *Recent Prog. Horm. Res.* **47**, 1–26.

Padykula, H. A., and Clark, J. H. (1981). Nuclear bodies as functional indicators in the target cells of sex steroid hormones. *In* "The Cell Nucleus" (H. Busch, ed.), Vol. 9, pp. 309–339. Academic Press, New York.

Pan, M. L., Bell, W. J., and Telfer, W. H. (1969). Vitellogenin blood protein synthesis in insect fat body. *Science* **165**, 393–394.

Perrot-Applanat, M., Logeat, F., Groyer-Picard, M. T., and Milgrom. E. (1985). Immunocytochemical study of mammalian progesterone receptor using monoclonal antibodies. *Endocrinology* **116**, 1473–1484.

Picard, D., and Yamamoto, K. R. (1987). Two signals mediate hormone dependent nuclear localization of the glucocorticoid receptor. *Embo J.* **6**, 3333–3340.

Picard, D., Kumar, V., Chambon, P., and Yamamoto, K. R. (1990). Signal transduction by steroid hormones: Nuclear localization is differentially regulated in estrogen and glucocorticoid receptors. *Cell Regul.* **1**, 291–299.

Poole, M. C., Mahesh, V. B., and Costoff, A. (1980). Intracellular dynamics in pituitary mammotropes throughout the rat estrous cycle. *Am. J. Anat.* **158**, 3–13.

Pratt, W. B. (1990). Interaction of hsp 90 with steroid receptors: Organizing some diverse observations and presenting the newest concepts. *Mol. Cell. Endocrinol.* **74**, C69–C76.

Press, M. F., Nousek-Goebl, A., and Greene, G. L. (1985). Immunoelectron microscopic localization of estrogen receptor with monoclonal estrophilin antibodies. *J. Histochem. Cytochem.* **33**, 915–924.

Rabl, C. (1885). Uber zellteilung. *Morphol. Jahrb.* **10**, 214–330.

Raska, I., Ochs, R. L., Andrade, L. E. C., Chan, E. K. L., Burlingame, R., Peebles, C., Gruol, D., and Tan, E. M. (1990a). Association between the nucleolus and the coiled body. *J. Struct. Biol.* **104**, 120–127.

Raska, I., Ochs, R. L., and Salamin-Michel, L. (1990b). Immunocytochemistry of the cell nucleus. *Electron Microsc. Rev.* **3**, 301–353.

Raska, I., Andrade, L. E. C., Ochs, R. L., Chan, E. K. L., Chang, C.-M., Roos, G., and Tan, E. M. (1991). Immunological and ultrastructural studies of the nuclear coiled body with autoimmune antibodies. *Exp. Cell Res.* **195**, 27–37.

Raska, I., Dundr, M., and Koberna, K. (1992). Structure-function subcompartments of the mammalian cell nucleus as revealed by the electron microscopic affinity cytochemistry. *Cell Biol. Int. Rep.* **16**, 771–789.

Riddiford, L. M. (1994). Cellular and molecular actions of juvenile hormone. I. general considerations and premetamorphic actions. *Adv. Insect Physiol.* **24**, 213–274.

Roberts, D. K., Lavia, L. A., Horbelt, D. L., and Walker, N. J. (1989). Changes in nuclear and nucleolar areas of endometrial glandular cells throughout the menstrual cycle. *Int. J. Gynecol. Pathol.* **8**, 36–45.

Roberts, P. E., and Jeffries, L. S. (1986). Grasshopper as a model system for the analysis of juvenile hormone delivery to chromatin acceptor sites. *Arch. Insect Biochem. Physiol.* Suppl. 1, 7–23.

Robinson, S. I., Nelkin, B. D., and Volgelstein, B. (1982). The ovalbumin gene is associated with the nuclear matrix of chicken oviduct cells. *Cell (Cambridge, Mass.)* **28**, 99–106.

Sabbah, M., LeRicousse, S., Redeuilh, G., and Baulieu, E. E. (1992). Estrogen receptor-induced bending of the *Xenopus* vitellogenin A2 gene hormone response element. *Biochem. Biophys. Res. Commun.* **185**, 944–952.

Sahlas, D. J., Milankov, K., Park, P. C., and DeBoni, U. (1993). Distribution of snRNPs, splicing factor SC-35 and actin in interphase nuclei: Immunocytochemical evidence for differential distribution during changes in functional states. *J. Cell Sci.* **105**, 347–357.

Saluz, H. P., Jiricny, J., and Jost, J. P. (1986). Genomic sequencing reveals a positive correlation between the kinetics of strand-specific DNA demethylation of the overlapping estradiol/glucocorticoid receptor binding sites and the rate of avian vitellogenin mRNA synthesis. *Proc. Natl. Acad. Sci. U.S.A.* **83**, 7167–7171.

Saluz, H. P. Feavers, I. M., Jiricny, J., and Jost, J. P. (1988). Genomic sequencing and *in vivo* footprinting of an expression specific DNase 1-hypersensitive sited avian vitellogenin II promoter reveal a demethylation of amCpG and a change in specific interactions of proteins with DNA. *Proc. Natl. Acad. Sci. U.S.A.* **85**, 6697–6700.

Savouret, J. F., Misrahi, M., Loosfelt, H., Atger, M., Bailly, A., Perrot-Applanat, M., Vultai, M. T., Guiochon-Mantel, A., Jolivet, A., Lorenzo, F., Logeat, F., Pichon, M. F., Bouchard, P., and Milgrom, E. (1989). Molecular and cellular biology of mammalian progesterone receptor. *Recent Prog. Horm. Res.* **45**, 65–120.

Schema, M., and Yamamoto, K. R. (1988). Mammalian glucocorticoid receptor derivatives enhance transcription in yeast. *Science* **241**, 965–968.

Setterfield, G., Bladon, T., Chaly, N., Hall, R., Brasch, K., Jones-Villeneuve, E., and Brown, D. L. (1985). Extrachromatin nuclear components and structural changes in nuclei. *UCLA Symp. Mol. Cell Biol., New Ser.* **26**, 63–86.

Shapiro, D. J. (1982). Steroid hormone regulation of vitellogenin gene expression. *CRC Crit. Rev. Biochem.* **12**, 187–203.

Shapiro, D. J., Barton, M. C., McKearn, D. M., Chang, T.-C., Lew, D., Blume, J., Nielsen,

D. A., and Gould, L. (1989). Estrogen regulation by steroid hormones. *Recent Prog. Horm. Res.* **45**, 29–64.

Sher, E., Eisman, J. A., Moseley, J. M., and Martin, T. J. (1981). Whole-cell uptake and nuclear localization of 1,25-dihydroxycholecalciferol by breast cancer cells (T47D) in culture. *Biochem. J.* **200**, 315–320.

Smith, H. C. (1992). Organization of RNA splicing in the cell nucleus. *Curr. Top. Cell Regul.* **33**, 145–166.

Spector, D. L. (1993). Macromolecular domains within the cell nucleus. *Annu. Rev. Cell Biol.* **9**, 265–315.

Spector, D. L., Lark, G., and Huang, S. (1992). Differences in snRNP localization between transformed and nontransformed cells. *Mol. Biol. Cell.* **3**, 555–569.

Spolski, R. J., Schneider, W., and Wangh, L. T. (1985). Estrogen-dependent DNA synthesis and parenchymal cell proliferation in the liver of adult male *Xenopus* frogs. *Dev. Biol.* **108**, 332–340.

Stuurman, N., De Graaf, A., Floore, A., Josso, A., Humbel, B., de Jong, L., and van Driel, R. (1992a). A monoclonal antibody recognizing nuclear matrix-associated nuclear bodies. *J. Cell Sci.* **101**, 773–784.

Stuurman, N., de Jong, L., and van Driel, R. (1992b). Nuclear frameworks: Concepts and operational definitions. *Cell Biol. Int. Rep.* **16**, 837–852.

Szostecki, C., Guldner, H. H., Netter, H. J., and Will, H. (1990). Isolation and characterization of cDNA encoding a human nuclear antigen predominantly recognized by autoantibodies from patients with primary biliary cirrhosis. *J. Immunol.* **145**, 4338–4347.

Tam, S. P., Hache, R. J. G., and Deeley, R. G. (1986). Estrogen memory effect in human hepatocytes during repeated cell division without hormone. *Science* **234**, 1234–1237.

Tata, J. R., and Smith, D. F. (1979). Vitellogenesis. A versatile model for hormonal regulation of gene expression. *Recent Prog. Horm. Res.* **35**, 47–95.

Tata, J. R., Ng, W. C., Perlman, A. J., and Wolffe, A. P. (1987). Activation and regulation of the vitellogenin gene family. *In* "Gene Regulation by Steroid Hormones" (A. K. Roy and J. H. Clark, eds.), Vol. 3, pp. 205–233. Springer-Verlag, New York.

Tsai, M. J., and O'Malley, B. W. (1994). Molecular mechanisms of action of steroid/thyroid receptor superfamily members. *Annu. Rev. Biochem.* **63**, 451–486.

Tseng, M. T., Bergink, E. W., and Wittliff, J. L. (1979). Influence of CI-628 (CN-55, 945-27) on estrogen-induced vitellogenin synthesis and on the ultrastructure of hepatocytes in male *Xenopus laevis. Eur. J. Cell Biol.* **20**, 143–149.

Tuma, R. S., Stolk, J. A., and Roth, M. B. (1993). Identification and characterization of a sphere organelle protein. *J. Cell Biol.* **122**, 767–773.

Vagner-Capodano, A. M., Bouteille, M., Stahl, A., and Lissitzky, S. (1982). Nucleolar ribonucleoprotein release in the nucleoplasm as nuclear bodies in cultured thyrotropin-stimulated thyroid cells: Autoradiographic kinetics. *J. Ultrastruct. Res.* **78**, 13–25.

van Driel, R., Humbel, B., and de Jong, L. (1991). The nucleus: A black box being opened. *J. Cell. Biochem.* **47**, 311–316.

Vazquez-Nin, G. H., Echeverria, O. M., Fakan, S., Traish, A. M., Wotiz, H. H., and Martin T. E. (1991). Immunoelectron microscopic localization of estrogen receptor on pre-mRNA containing constituents of rat uterine cell nuclei. *Exp. Cell Res.* **192**, 396–404.

Vic, P., Garcia, M., Humeau, C., and Rochefort, H. (1980). Early effects of estrogen on chromatin ultrastructure in endometrial nuclei. *Mol. Cell. Endocrinol.* **19**, 79–92.

Vic, P., Vignon, F., Derocq, D., and Rochefort, H. (1982). Effect of estradiol on the ultrastructure of the MCF7 human breast cancer cells in culture. *Cancer Res.* **42**, 667–673.

Visa, N., Puvion-Dutilleul, F., Harper, F., Bachellerie, J. P., and Puvion, E. (1993). Intranuclear distribution of poly(A) RNA determined by electron microscopy *in situ* hybridization. *Exp. Cell Res.* **208,** 19–34.

Wahli, W. (1988). Evolution and expression of vitellogenin genes. *Trends Genet.* **4,** 227–232.

Weis, K., Rambaud, S., Lavau, C., Jansen, J., Carvalho, T., Carmo-Fonseca, M., Lamond, A., and Dejean, A. (1994). Retinoic acid regulates aberrant nuclear localization of PML-RAR in acute promyelocytic leukemia cells. *Cell (Cambridge, Mass.)* **76,** 345–356.

Welshons, W. V., Krummel, B. M., and Gorski, J. (1985). Nuclear localization of unoccupied receptors for glucocorticoids, estrogens and progesterone in GH3 cells. *Endocrinology (Baltimore)* **117,** 2140–2147.

Williams, L. M., Jordan, E. G., and Barlow, P. W. (1983). The ultrastructure of nuclear bodies in interphase plant cell nuclei. *Protoplasma* **118,** 95–103.

Wolff, M. S., Toniolo, P. G., Lee, E. W., Rivera, M., and Dubin, N. (1993). Blood levels of organochlorine residues and risk of breast cancer. *J. Natl. Cancer Inst.* **85,** 648–652.

Wu, Z., Murphy, C., Callan, H. G., and Gall, J. G. (1991). Small nuclear ribonucleoproteins and heterogeneous nuclear ribonucleoproteins in the amphibian germinal vesicle: Loops, spheres, and snurposomes. *J. Cell Biol.* **113,** 465–483.

Wyatt, G. R. (1988). Vitellogenin synthesis and the analysis of juvenile hormone action in locust fat body. *Can. J. Zool.* **66,** 2600–2610.

Wyatt, G. R. (1991). Gene regulation in insect reproduction. *Invertebr. Reprod. Dev.* **20,** 1–35.

Xing, Y., Johnson, C. V., Dobner, P. R., and Lawrence, J. B. (1993). Higher level organization of individual gene transcription and RNA splicing. *Science* **259,** 1326–1330.

Yamamoto, K. R. (1985). Steroid receptor regulated transcription of specific genes and gene networks. *Annu. Rev. Genet.* **19,** 209–252.

Yu, M. S., and Ho, S. M. (1989). Nuclear acceptor sites for estrogen-receptor complexes in the liver of the turtle *Chrysemyspicta*. *Mol. Cell. Endocrinol.* **61,** 37–48.

Zhang, M., Zamore, P. D., Carmo-Fonseca, M., Lamond, A. I., and Green, M. R. (1992). Cloning and intracellular localization of the U2 small nuclear ribonucleoprotein auxiliary factor small subunit. *Proc. Natl. Acad. Sci. U.S.A.* **89,** 8769–8773.

# Effects of Axotomy, Deafferentation, and Reinnervation on Sympathetic Ganglionic Synapses: A Comparative Study

Jacques Taxi and Daniel Eugène
Institut des Neurosciences, C. N. R. S., Université Pierre et Marie Curie, 75252 Paris, France

The main physiological and morphological features of the synapses in the superior cervical ganglia of mammals and the last two abdominal ganglia of the frog sympathetic chain are summarized. The effects of axotomy on structure and function of ganglionic synapses are then reviewed, as well as various changes in neuronal metabolism in mammals and in the frog, in which the parallel between electrophysiological and morphological data leads to the conclusion that a certain amount of synaptic transmission occurs at "simple contacts." The effects of deafferentation on synaptic transmission and ultrastructure in the mammalian ganglia are reviewed: most synapses disappear, but a number of postsynaptic thickenings remain unchanged. Moreover, intrinsic synapses persist after total deafferentation and their number is strongly increased if axotomy is added to deafferentation. In the frog ganglia, the physiological and morphological evolution of synaptic areas is comparable to that of mammals, but no intrinsic synapses are observed. The reinnervation of deafferented sympathetic ganglia by foreign nerves, motor or sensory, is reported in mammals, with different degrees of efficiency. In the frog, the reinnervation of sympathetic ganglia with somatic motor nerve fibers is obtained in only 20% of the operated animals. The possible reasons for the high specificity of ganglionic connections in the frog are discussed.

**KEY WORDS:** Ganglionic synapse, Sympathetic ganglion, Synaptic ultrastructure, Synaptic transmission, Axotomy, Deafferentation, Preganglionic reinnervation, Nerve fibers.

195

## I. Introduction

All neurons form networks within which they connect to each other by synapses or to target organs by junctions. It follows that any injury to a neuron may affect both of these connections. Likewise, any alteration of the connections may affect the structure and activity of the corresponding neurons.

Because of the general morphology of nerve cells, the most frequent injury experienced by a peripheral neuron is the section of its axon, that is, axotomy. Another form of injury for a neuron is the section of the afferent fibers making synapses with it, that is, deafferentation.

The experimental study of neuronal reactions to injuries such as axotomy or deafferentation requires models that must meet certain criteria, especially the possibility of easy and selective section of all the fibers giving rise to afferent synapses, or all the axons of an identified group of neurons. In fact, only some models fulfill these conditions, each having its own peculiarities, such as the distance from the perikaryon at which the axon can be severed or the physiological characteristics of the afferent fibers. This is why comparing results obtained from different models can be difficult and generalizations from one model must be made with caution.

In the central nervous system (CNS), neurons generally receive numerous afferences from several origins, not always well known and accessible for selective cutting. This is a serious limitation for experimental studies. On the other hand, certain sympathetic ganglia, whose neurons are much more accessible than those of the CNS, are good models for studies of axotomy or deafferentation of "extrinsic" neurons, that is, neurons entirely or partly located outside the CNS. Their behavior under many experimental conditions is different from that of "intrinsic" neurons, entirely confined within the CNS (Barron, 1983).

The sympathetic ganglia contain a relatively homogeneous population of extrinsic neurons, although not as simply organized as was thought up to the 1960s. Among these ganglia, the superior cervical ganglion (SCG) of mammals is a classic material for experimentation because it is easy to divide selectively all of its preganglionic or postganglionic fibers. This is not the case for the other ganglia, especially the prevertebral ones, whose afferences have two origins, one from the CNS and the other from the periphery, the latter being associated with postganglionic fibers in the same nerve.

However, another preparation exists that exhibits interesting properties, making it in many respects complementary to the preceding one: the last two abdominal ganglia of the sympathetic chain of the frog. The anatomy of the frog nervous system is well known from the monographies of Gaupp (1899) and Pick (1970). The main histological features of the sympathetic ganglia were described by Smirnow (1890), whose interpretations were revised by Huber (1900). The origin and pathways of the preganglionic fibers were analyzed by Langley and Orbeli (1910) (for reviews, see Taxi, 1965, 1976). Briefly, the interest in the frog abdominal ganglia, already emphasized by Kelly et al. (1989b), comes from the following.

The special arrangement and accessibility of the preganglionic and postganglionic fibers of the last two abdominal ganglia [8th and 9th, according to the direct numbering of Langley and Orbeli (1910) used by us, or 9th and 10th, according to Ecker and Wiedersheim's numbering (Gaupp, 1899)], which permits selective separation of one or the other category of fibers (Fig. 1)

The relative length of the rami communicantes, and the limited amount of connective tissue around the ganglia (compared to the development of this tissue in the toad, for instance), an advantage for electrophysiological recordings

The complete absence of interneurons, in contrast with the SCG of certain mammals, for instance the rat

The disposition of ganglionic synapses, practically all axosomatic in these ganglia, which is a great advantage for the search and counting of synapses by electron microscopy

The fact that the innervation of each neuron is usually realized by only one preganglionic nerve fiber (Weitsen and Weight, 1977)

The capacity of regeneration in amphibians, generally admitted to be superior to that of mammals

The metabolism of amphibians, which is lower than that of mammals; thus certain processes can be spread out in time and rendered easier to analyze

The vicinity of the last abdominal ganglia to the last thoracic motor nerves and lumbosacral plexus, which permits substitution of motor nerves for preganglionic ones for the purpose of heterogenic reinnervation after sectioning

The existence of two types of neurons, B and C, on the basis of electrophysiological criteria (Nishi et al., 1965). They receive their preganglionic fibers by separated pathways, which makes it possible to study the specificity of the innervation of two close neuronal types

In spite of all these advantageous properties, only a limited number of experimental studies has been devoted to the sympathetic ganglia of amphibians, except those of the group of Gordon, Kelly, and Smith, some of whose results were fairly surprising (Gordon et al., 1987). This situation prompted us to reinvestigate the properties of neurons and especially of ganglionic synapses after axotomy, deafferentation, and reinnervation either by normal preganglionic fibers or by somatic motor ones.

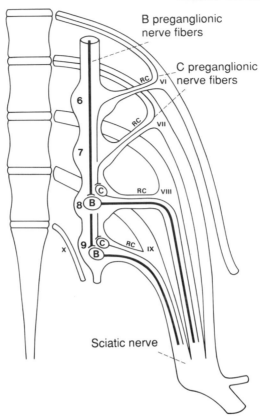

FIG. 1    Scheme of the connections of the 8th and 9th abdominal ganglia in the frog *Rana esculenta*. Ganglia are numbered in arabic numerals; roman numerals are used for the spinal nerves and the spinal roots of the lumbosacral plexus. The 10th ganglion, which is inconstant, is not represented. RC, Ramus communicans. In the 8th and 9th ganglia, the two types of sympathetic neurons are represented by B and C. Both the myelinated pre- and postganglionic axons are represented by thick lines, whereas the unmyelinated ones are represented by thin lines.

Our approach was to use the same ganglia in parallel, first for electrophysiological recordings at the cellular level, then for ultrastructural observations, including counting of synaptic areas.

In this article, in order to make comparisons, the main data concerning the same types of experiments in other vertebrates, particularly mammals, are reported. Data concerning the parasympathetic ganglia, whose neurons are cholinergic, are taken into consideration only occasionally.

## II. Synaptology in Normal Sympathetic Ganglia

Here we summarize only the main features of the ganglionic synapses of the preparations used in experimental studies, that is, the SCG of "laboratory mammals" and the last two abdominal ganglia of anuran amphibians.

### A. Physiology of Ganglionic Transmission

Electrical stimulation of preganglionic nerve fibers evokes complex postsynaptic responses. These results have been reviewed in a book edited by Karczmar *et al.* (1986). Acetylcholine (ACh) is the principal neurotransmitter released by preganglionic endings. By activation of nicotinic receptors, ACh induces in all the neurons fast excitatory postsynaptic potentials (EPSPs) with short synaptic delays (2 10 msec), whereas the activation of muscarinic receptors can induce slow EPSPs or inhibitory postsynaptic potentials (IPSPs) with longer synaptic delays (100–400 msec). The peak amplitude of the fast EPSP is generally large enough to trigger an action potential in the sympathetic neuron, thus ensuring effective ganglionic transmission. In addition, peptides released after repetitive preganglionic stimulation can generate late slow EPSPs (1- to 2-sec delay). The functional significance of this peptidergic excitatory transmission is not clear, but it might modulate cholinergic neurotransmission (Adams *et al.*, 1986). Peptides may also play a trophic role (see Hökfelt, 1991).

### 1. Mammals

Many neurons of the guinea pig SCG can be activated by electrical stimulation of several ventral roots. These neurons receive afferences from about 12 preganglionic fibers originating from different levels of the spinal cord (approximately 4 levels) and differing in their conduction velocities (Njå and Purves, 1977a; Purves and Litchman, 1985). Moreover, although the cell bodies of neurons with identical peripheral targets are not highly localized within the same areas, generally, there is a rostrocaudal organization of neurons within a ganglion: it has been shown in the rat SCG, both electrophysiologically (Kiraly *et al.*, 1989) and morphologically (Flett and Bell, 1991), that the neurons are preferentially located near the exit sites of their axon from the ganglion. However, no fundamental differences could be detected between the electrical characteristics of neurons projecting into either of the major postganglionic trunks (Kiraly *et al.*, 1989).

Consequently, in the SCG, a clear classification into two or more types of sympathetic neurons based on different synaptic inputs or electrical characteristics is not as easy as in the amphibian abdominal ganglia (see Section II,A,2). Nevertheless, this situation is not the same in prevertebral ganglia, because in the guinea pig inferior mesenteric or celiac ganglia, two or even three classes (respectively) of sympathetic neurons have been electrophysiologically distinguished (Cassel *et al.*, 1986; McLachlan and Meckler, 1989; Jänig and McLachlan, 1992).

The three cholinergic postsynaptic potentials (fast EPSP and slow EPSP and IPSP) were first recorded in the SCG of the cat (Eccles, 1952) and later in the SCG of the rat (Dunant and Dolivo, 1967; Dun and Karczmar, 1980). The fast EPSP due to nicotinic receptor activation is sensitive to *d*-tubocurarine and hexamethonium whereas the muscarinic slow potentials are both antagonized by atropine (Eccles and Libet, 1961). The slow EPSP is mediated by $M_1$ muscarinic receptors and the slow IPSP by $M_2$ receptors (Newberry and Connolly, 1989). However, there is considerable controversy over the synaptic mediation of slow IPSPs in mammalian ganglia (see Koketsu, 1986). Eccles and Libet (1961) have suggested the existence of an adrenergic step in the synaptic pathway of the slow IPSP: chromaffin-like cells of the ganglia would be stimulated by ACh via muscarinic receptors and would release adrenergic substances. The latter would in turn activate the adrenergic postganglionic receptor and thereby generate the slow IPSP of these neurons. But there are several weaknesses in this hypothesis of disynaptic mediation of the slow IPSP. Cole and Shinnick-Gallager (1980, 1984) presented direct evidence indicating a monosynaptic mediation. Therefore, it may be only speculated that in mammalian ganglia the slow IPSP is generated by combined direct actions of adrenergic substances and ACh. The alteration of one or the other slow potential modifies the excitability of the ganglionic neurons and subsequent synaptic transmission through nicotinic receptors (Yarosh *et al.*, 1988).

Noncholinergic excitatory transmission was first demonstrated in the SCG of the dog (Chen, 1971), then in the cat (Alkadhi and McIssac, 1973) and later in the rabbit (Ashe and Libet, 1981). On the other hand, networks of peptide-positive fibers [enkephalin, vasoactive intestinal polypeptide (VIP), and substance P] were observed by immunocytochemistry as surrounding the principal neurons in the rat and guinea pig ganglia, but the plexus of varicose fibers were less dense in the SCG than in the prevertebral ganglia (Hökfelt *et al.*, 1977a,b; Schultzberg *et al.*, 1979; Matthews *et al.*, 1987; Masuko and Chiba, 1988). Moreover, strong evidence was given for the coexistence of ACh and enkephalin in preganglionic nerve endings (Kondo *et al.*, 1985). It was therefore suggested that such peptides may serve as transmitters generating late slow EPSPs, but the pertinent demonstration is available only in respect to the prevertebral ganglia

(Simmons, 1985; Love and Szurszewski, 1987; Griffith *et al.*, 1988). In addition, fast noncholinergic EPSPs due to the activation of ionotropic γ-aminobutyric acid (GABA) receptors were also evidenced in the rat SCG (Eugène, 1987) and GABAergic-positive fibers were seen forming basket-like networks around some principal neurons (Eugène and Taxi, 1988; Kása *et al.*, 1988; Wolff *et al.*, 1992). Thus, GABA release induced by preganglionic fiber excitations could also modulate ganglionic neurotransmission.

## 2. Amphibia

Since the work of Langley and Orbeli (1910) showing that the preganglionic and postganglionic nerve fibers of the last two abdominal ganglia go through different rami communicantes, these two ganglia have been extensively studied. In the toad *Bufo vulgaris*, the sympathetic neurons have been first classified into two types (B and C) on the basis of electrophysiological criteria: the B neuron has a myelinated axon with a fast conduction velocity (3–4 m/sec) whereas the C neuron has an unmyelinated axon with an 8 to 10 times slower conduction velocity (Nishi *et al.*, 1965; Honma, 1970). Moreover, these two types are innervated by B (myelinated) and C (unmyelinated) preganglionic fibers, respectively (Skok, 1964; Libet *et al.*, 1968; Francini and Urbani, 1973; Eugène and Taxi, 1990, 1991). Therefore at the level of the last two abdominal ganglia, each type of neuron can be distinguished by both conduction velocity evoked by antidromic stimulation and by their synaptic input. However, in the big frog *Rana catesbeiana*, a third type of neuron has also been identified (Dodd and Horn, 1983a; Tokimasa, 1984). It is innervated by B myelinated preganglionic fibers but has an unmyelinated axon: it was named the slow-B type. This third type was not found in the small frog *Rana esculenta* (Eugène and Taxi, 1991). By selective horseradish peroxidase labeling, perikarya of B and C preganglionic neurons have been located in separated segments of the spinal cord (Horn and Stofer, 1988) and in axons of B and C postganglionic neurons in different peripheral nerves (Horn *et al.*, 1988). Thus in contrast with mammals, the sympathetic nervous system of amphibians is clearly organized into two different pathways (see also Horn, 1992).

Electrical stimulation of amphibian preganglionic fibers releases ACh (Nishi *et al.*, 1967) but also a luteinizing hormone-releasing hormone (LHRH)-like peptide that is localized only in the C preganglionic endings (L. Y. Jan *et al.*, 1980; Branton *et al.*, 1986). However, both the B and C neurons have LHRH receptors because late slow EPSPs, owing to this peptide, can be recorded in the two types of neurons (Jan and Jan, 1982). Repetitive stimulations are more effective than single ones in releasing the LHRH-like peptide (Peng and Horn, 1991). Pharmacologically, the fast EPSP provoked by nicotinic ACh receptor activation is similar to the end plate potential recorded at the neuromuscular junction (Lipscombe and Rang, 1988). But the kinetic properties of the nicotinic receptor–ion channel are different in B and C neurons: the mean open time of the channel is slower in C than in B neurons (Marshall, 1986; Minota et al., 1989). Moreover, the slow EPSP is recorded in B neurons only in response to repetitive stimulations of B preganglionic fibers whereas the slow IPSP is recorded in C neurons after C preganglionic fiber stimulation (Tosaka et al., 1968; Dodd and Horn, 1983b; Smith and Weight, 1986; however, see Koketsu, 1986). As in mammalian ganglia, the slow EPSP in the frog is mediated by $M_1$ muscarinic receptors whereas the $M_2$ receptor activation would be responsible for the slow IPSP (Kuba et al., 1989). But in contrast with mammals, a disynaptic mediation of the slow IPSP cannot exist in amphibians, first because the chromaffin-like cells are few in number and they do not exhibit morphological characteristics of interneurons (Weight and Weitsen, 1977), and second because adrenergic antagonists do not inhibit the slow IPSP (Yavari and Weight, 1988). All these differences in ganglionic transmission of amphibian B and C neurons reinforce the distinction with the mammalian paravertebral ganglia, for which a clear anatomical or physiological separation between several pathways has never been found.

## B. Morphology of Synaptic Areas

### 1. Mammals

The morphological studies of synapses in sympathetic ganglia of mammals have been reviewed and illustrated by Matthews (1974, 1983) and Smolen (1988). All the ganglia contain two types of cells involved in synaptology. By far the most numerous population is constituted by the "principal neurons" (Matthews and Raisman, 1969); the second type, at most a few percent, is represented by the small intensely fluorescent (SIF) cells (Eränkö and Härkönen, 1965), also called chromaffin cells (Siegrist et al., 1968), small granule-containing (SG) cells (Matthews and Raisman, 1969), or chromaffin-like cells (Taxi, 1979b). The bulk of ganglionic synapses is similar to those described by Palay (1956) in the CNS and are called here type I. Each type I synapse is an asymmetrical unit formed by the apposition of a presynaptic and a postsynaptic profile, separated by a cleft of about 15–20 nm, and is characterized by the "synaptic complex" (Palay, 1958), also called "active zone" (Couteaux, 1961) (Fig. 2). At this level, the presynaptic profile contains a cluster of clear synaptic vesicles, about

40 nm in diameter, associated with more or less distinct dense patches attached to the presynaptic membrane (Gray, 1963). In contrast to this, the postsynaptic profile exhibits a "membrane thickening" or "postsynaptic differentiation," which can survive presynaptic degeneration, giving a "vacated postsynaptic differentiation" (VPD); VPDs are encountered rarely in normal ganglia (Raisman et al., 1974). The presynaptic endings, whose diameters are on an order of magnitude of 1 $\mu$m, may also contain large dense-cored vesicles (LDVs) (65–100 nm in diameter), mitochondria, and, more rarely, inclusions of autophagic vacuole type (Raisman et al., 1974).

Another type of synapse, type II, first described by Williams (1967), is rare. The only difference from type I synapses is that at least some of their vesicles contain a dense granule and for this reason can be considered as adrenergic (Wolfe et al., 1962); also the size of vesicles is extended from that of classical synaptic vesicles to that of LDVs. Type II synapses are restricted to a few percent of all the synaptic population in the rat and the cat SCG according to Tamarind and Quilliam (1971). In the rabbit SCG, the proportion of type II synapses reaches 25% (Tamarind and Quilliam, 1971; Dail and Evan, 1978). According to Williams (1967), they correspond to endings of SIF cell processes. This was also the opinion of Ramsay and Matthews (1985), who improved the detection of the dense granule in the vesicles by a treatment with the false transmitter 5-hydroxy-dopamine (Tranzer and Thoenen, 1967) and concluded that, in normal conditions, type II synapses in the rat SCG only originate from SIF cells. However, some authors considered that they are endings of axon collaterals of ganglionic neurons; this hypothesis is especially attractive for the rabbit SCG, in which there is a dramatic disproportion between the number of SIF cells and that of type II (adrenergic) synapses (Dail and Evan, 1978; Williams et al., 1976).

The principal neurons are multipolar, provided with large and long dendrites. The axodendritic position of the synapses was soon noticed in the cat SCG (Elfvin, 1963; Ceccarelli, 1968) as well as in the rat SCG (Taxi, 1965). In the latter, the synapses are practically all axodendritic, on dendritic shafts or preferentially (about 80%) on spinelike protrusions (Raisman et al., 1974), the ratio of dendritic to spine synapses having been estimated to be 1.33 : 1 by Ramsay and Matthews (1985). Axosomatic synapses were estimated to be 1% by Tamarind and Quilliam (1971), but only 4 of 1060 by Matthews and Nelson (1975) and 2 or 3 of 650 by Ramsay and Matthews (1985), some of them being type II. In the mouse SCG, Yokota and Yamauchi (1974) found that axon terminals occupied 0.6% of the perikaryon surface, of which 0.03% corresponds to synapses, whereas 0.07% is occupied by postganglionic, adrenergic elements that could be recurrent axon collaterals of ganglionic neurons or vesiculated segments

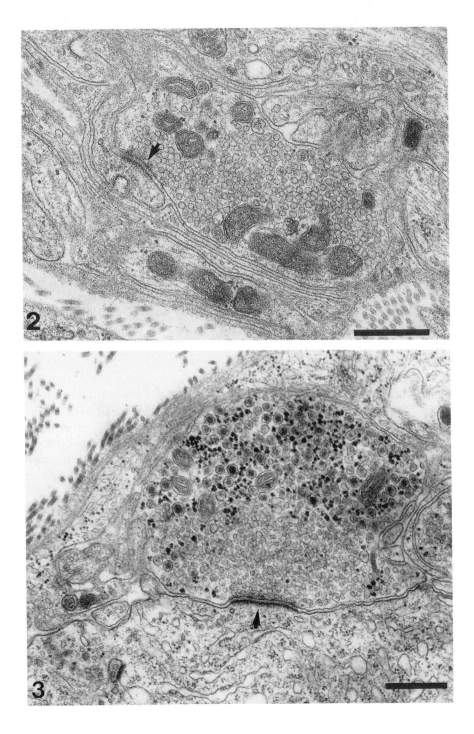

of dendrites. These elements are usually involved in asymmetric synapses. More numerous were symmetrical attachment plaques between dendrites, between a dendrite and a perikaryon, or between a dendrite and a ganglionic axon collateral, but the identification criteria for these are ambiguous. As for the guinea pig SCG, there is a surprising discrepancy between McLachlan (1974), according to whom 26% of synapses are axosomatic, and Purves (1975), who found only two axosomatic synapses of 1957 observed. Perhaps the discrepancy comes from the fact that short endocapsular dendrites could have been interpreted as dendrites by one author and not by the other?

## 2. Amphibia

A thorough description and illustration of frog sympathetic ganglia was given by Taxi (1965) and their synapses reviewed by Taxi (1976) and Watanabe (1983). Briefly, the neurons are all unipolar. Generally, the ganglia also contain one or several clusters of SIF cells, but none of them was reported to form synapses with a principal neuron.

The main features of synapses are the same as in mammalian ganglia, but they are as a rule axosomatic (Fig. 3), each neuron being innervated by one preganglionic branchlet. Only some synapses are located on short endocapsular dendrites. Synapses are especially numerous in the region of origin of the unique process and on the process itself up to 0.5 $\mu$m from the perikaryon (Murata et al., 1989). The size of the presynaptic ending sections is usually several micrometers. Thus, one synaptic contact can exhibit two or even more synaptic complexes. In addition, a variable proportion of synaptic complexes is provided with a dense synaptic apparatus or "subsynaptic bar" (Taxi, 1961b, 1965).

The two cell types electrophysiologically defined as B and C neurons have synapses exhibiting different transmission properties (see Section II,A,2). Morphologically, C neurons are smaller in diameter (<30 $\mu$m) than B neurons (>40 $\mu$m), but this criterion of size is not absolute, because both populations overlap between 30 and 40 $\mu$m, at least in the European frog R. esculenta (Lascar and Taxi, 1992). More reliable is the presence

---

FIG. 2    An axodendritic synapse in a rat superior cervical ganglion. The fairly large presynaptic ending, which contains numerous synaptic vesicles, mitochondria, and some large dense-cored vesicles, is associated with a dendritic spine. The arrow indicates the synaptic complex. Bar: 0.5 $\mu$m.

FIG. 3    An axosomatic synapse in a frog abdominal sympathetic ganglion. The synapse is identified by its synaptic complex (arrow). The presynaptic ending contains numerous synaptic vesicles and, in the region remote from the synaptic complex, large dense-cored vesicles, glycogen granules, and mitochondria. Bar: 0.5 $\mu$m.

of neuropeptide Y selectively in C neurons (Horn *et al.,* 1987). On the other hand, Y. N. Jan *et al.* (1979, 1980) demonstrated that the synapses of C neurons release, in addition to ACh, an LHRH-like peptide, which is located within the nerve endings (L. Y. Jan *et al.,* 1980), more precisely in the LDVs of the presynaptic endings (Lascar *et al.,* 1982). In the B preganglionic fibers, the LDVs contain calcitonin gene-related peptide (CGRP) (Kuramoto and Fujita, 1986; Horn and Stofer, 1989). According to Watanabe and Burnstock (1978), the subsynaptic bar is restricted to C neurons in the frog *Limnodynastes dumerili.* B neurons are provided with "junctional subsurface organs" (Watanabe and Burnstock, 1976), which are flat cisternae of endoplasmic reticulum, but not generally associated with synaptic differentiations.

## 3. Quantitative Data

In mammals, the quantitation of synapses has been expressed by the number of synapses, defined by the synaptic complex (Palay, 1958), per surface unit explored, assuming that the distribution of synapses is random within the ganglia (Östberg *et al.,* 1976) or defined by their total number (Siklós *et al.,* 1990).

In the frog, the distribution of neurons is heterogeneous; the perikarya are concentrated at the surface of the ganglia, the center being mainly occupied by fiber bundles. Moreover, the synapses, all axosomatic, are located especially in the region of origin of the unique process. Thus, it is improper to relate the number of synapses to a surface unit, and we have defined and used a "synaptic index," corresponding to the ratio of the number of synapses, or more precisely the number of synaptic complexes, to that of the neuronal sections studied, irrespective of their size. At least 30, and usually more, sections of perikaryon were examined for each ganglion. This type of index is justified by the fact, demonstrated by Sargent (1983) on similar unipolar neurons in the parasympathetic ganglion of the mudpuppy heart, that the number of synapses is roughly proportional to the cell size. All details and discussion concerning this index were given in Eugène and Taxi (1991). Its mean value in the frog *R. esculenta,* calculated from nine pairs of abdominal ganglia (8th and 9th) taken from nine frogs, was found to be 1.45 ± 0.10. In the same way, an "index of simple contacts" has been determined: it takes into account the number of vesicle-containing profiles (VCPs) in contact with the surface of the neuron, but which are devoid of a synaptic complex (Section III,B,2,b). Its value was found to be 0.96 ± 0.10. Additionally, the proportion of synapses endowed with a subsynaptic bar was also calculated, equaling 20 ± 10 percent of synapses, the high standard error being due to large variations from one ganglion to another. For instance, in one animal, the percentage was 16 in the last two right-hand lumbar ganglia and 40 in the left ones.

## III. Axotomy of Sympathetic Neurons

### A. Introduction

The term *axotomy* is used for any process resulting in the interruption of the axonal flow, such as axon section, crushing, or ligation. In our own work, we have used almost exclusively axon section.

The modern study of the effects of axotomy on nerve cells was initiated by Nissl (1892a,b), thanks to the use of his original staining method, which revealed a dramatic loss of staining by methylene blue in the perikaryon of motor neurons in the rabbit facial nerve. This is because basophil material, normally concentrated mainly around the cell nucleus, is displaced to the cell periphery. This chromatolytic reaction, or chromatolysis, is the most conspicuous event of several changes occurring in axotomized neurons, called "axon reaction" or "retrograde reaction." It includes a migration of the nucleus to the periphery of the perikaryon and a general swelling of the cell. Early descriptions of these changes were reviewed in Ramón y Cajal's treatise (1911); then reviews by Nicholson (1924) and Geist (1933) marked the way to the histochemical study by Bodian and Mellors (1945). More recently, all the data, including those of the first 15 years of electron microscopy, were thoroughly reviewed by Lieberman (1971). An account of the influence of various factors, such as animal species and age, nature of injury, distance of the lesion from the perikaryon, and prevention of the reestablishment of connections with the target, on neuronal changes after axotomy was given by Barron (1983). For instance, the age of animals is of the utmost importance: in newborns or very young animals, axotomy results in a marked atrophy of the ganglionic neurons, whereas they undergo only chromatolysis and a certain swelling in adults (Hendry, 1975b). Only the latter are considered in this article.

Axotomy causes metabolic alterations that induce changes in neuronal structure and also in membrane properties and spike electrogenesis. Titmus and Faber (1990) have reviewed and attempted to summarize the effects of nerve injury on the electrophysiological properties of different types of mature neurons. First, the largest myelinated axons generally show a decline in conduction velocity following nerve injury, whereas the unmyelinated axons react differently to axotomy. However, the conduction velocity of frog B (myelinated) sympathetic neurons does not change after axotomy and that of C (unmyelinated) neurons increases (Shapiro *et al.*, 1987). Second, passive membrane properties remain relatively stable in the majority of axotomized neurons, but excitability in the soma and proximal dendrites is augmented whereas it is diminished in the initial segment–axon hillock region. Thus, there is no change in input resistance or resting potential of axotomized sympathetic neurons (Purves, 1975; Gordon *et al.*, 1987), although in mammalian motoneurons, axotomy enhanced both input resistance and excitability (Eccles *et al.*, 1958; Kuno and Llinas, 1970a,b; Gustafsson, 1979; Laiwand *et al.*, 1988). Last, the action potential waveforms and discharge patterns of many axotomized neurons are altered. For example, both the afterhyperpolarization amplitude and duration of action potentials of axotomized sympathetic neurons are decreased. These reductions have been attributed to a loss of calcium-dependent potassium efflux (Kelly *et al.*, 1986) due to a slowing of calcium dynamics (Sanchez-Vives *et al.*, 1993; Jassar *et al.*, 1993). Consequently, the electrophysiological changes provoked by nerve injury vary not only from one neuronal type to another, but also within similar neuronal populations.

## B. Mammalian Ganglia

Although the cytological modifications affecting the perikarya are not within the scope of this article, it is necessary to recall some fundamental data.

### 1. Perikarya

An account on the axotomized sympathetic neurons in the cat SCG was given by de Castro (1930), to whom we refer for review of previous studies. de Castro reported a marked axonal reaction starting 2–3 days after axon injury and emphasized the changes in the neurofibrillar system of the perikaryon and dendrites. Later, the retrograde reaction of sympathetic neurons to axotomy was revisited by Bianchine *et al.* (1964), who reported an enlargement of cells in the cat SCG and an early increase in nuclear size, followed by a decrease. Acheson and Schwarzacher (1956) investigated changes induced by axotomy in Nissl bodies and nuclear position in the cat inferior mesenteric ganglion. The displacement of the nucleus from its normal central position to the periphery of the perikaryon was noted. They considered the position of the nucleus as a clue to the degree of cellular abnormality compatible with normal synaptic transmission. Then, during the phase of reversal of nuclear distortion, there was an increase in density of the nuclear rim, which was maximal during the second postoperative week. These authors also found a slight shrinkage of the neurons at early stages (1–3 days), reinforced from 7 days onward. More recently, Yawo (1987) described changes in dendritic geometry in axotomized mouse SCG, which recovered with the reinnervation of target organs.

A thorough electron microscopy study of the rat SCG was carried out by Matthews and Raisman (1972). After ligation of both of the carotid nerves 1 to 2 mm away from the SCG, they observed the onset of chromatolysis as soon as 6 hr after the operation. At 12–13 hr, 60–70% of neurons were chromatolytic, the maximum (80–88%) occurring at 24 hr. The recovery began at 3–6 days, the proportion of chromatolytic neurons falling to 60–50% between 7 and 21 days. However, 20% were still chromatolytic at 143 days. One of the main features of the reaction to axotomy is already clear from these results, that is, large individual variations between neurons. A spectacular change concerned a large increase in the number of autophagic vacuoles of highly varied morphology, because all types of cytoplasmic organelles could be sequestrated and transformed within them. They rapidly evolved into dense bodies containing stacks of lamellae and increasing in size by coalescence. This was considered a partial involution of the neurons. Concerning the Golgi apparatus, their conclusions are in agreement with those of Holtzmann et al. (1967) about the neurons of the nodose ganglion, in that the Golgi elements became concentrated in the center of the cell, instead of the Nissl stacks, and that a reduction in size of Golgi profiles was observed starting 4–5 days after axotomy. Between 7 and 14 days, many neurons recovered their normal appearance, although they may sometimes keep an unusually high number of dense bodies. A sprouting reaction was observed at the surface of the perikarya. This reaction led to processes containing an organelle-free cytoplasm, often organized in several thin layers at the neuronal surface. It started to develop 3 days after axotomy, reached its maximum at 10 days, and then progressively disappeared. An extension of sheath cells was also noted. They can penetrate the neurons and sequestrate portions of neuronal cytoplasm, which are digested in glial autophagic vacuoles. Beyond 14 days, a large heterogeneity was visible among neurons. Many were much smaller than normal, with signs of nuclear and cytoplasmic atrophy. By contrast, others were hypertrophic, with massive Nissl clumps and large nucleoli.

A number of cellular activities are modified after axotomy. These include, for instance, an increase in glucose utilization and reduction in phosphorus incorporation in phospholipids (Nagata et al., 1973), an induction of neurofilament epitope phosphorylation in perikarya (Shaw et al., 1988), a decrease in specific activity of glycolytic enzymes as well as in RNA and protein concentration (Härkönen, 1964; Härkönen and Kauffman, 1973, 1974), as well as other activities involved in synapse functioning (see Section III,B,2,b).

There are some discrepancies concerning the survival of ganglionic neurons after axotomy. First, Levinsohn (1903) claimed that the bulk of ganglionic neurons died after axotomy. The same conclusion was also reached by Acheson and Schwarzacher (1956), who reported that about 50% of the neurons disappeared 2 weeks after axotomy, confirming the electrophysiological data of Acheson and Remolina (1955). Similar results were reported in the guinea pig SCG by Purves (1975), according to whom the critical condition for neuronal survival is the reestablishment of connections with a target organ. When this is prevented, the number of surviving neurons falls dramatically. Matthews and Raisman (1972) and Matthews and Nelson (1975) did not mention such important loss of neurons, although they described neuronal alterations.

## 2. Synapses

Concerning synapses, Barr (1940) found no noticeable changes in the number, size, or appearance of afferent endings to the perikaryon of axotomized spinal motor neurons. Thus, Blinzinger and Kreutzberg (1968), using electron microscopy, were the first to mention a fall in the number of synapses in the axotomized motor neurons, as well as an active participation of microglial cells in synaptic disruption. These conclusions were confirmed in several papers in the following years, especially in the quantitative study of Sumner and Sutherland (1973), and then analyzed in the electrophysiological studies of "reactive" deafferentation (reviewed by Titmus and Faber, 1990).

*a. Electrophysiology*   The main physiological consequence of deafferentation is a depression of synaptic transmission. The first electrophysiological data concerning synapses were published in 1954 by Brown and Pascoe, who reported a complete failure of ganglionic transmission in the stellate ganglion of the cat, from 3 to 12 weeks after sectioning postganglionic fibers. This situation could not be relieved by posttetanic facilitation or pharmacological treatments, such as local injection of ACh, whose output from the preganglionic nerves remained normal. Thus, it appeared that it was the sensitivity of ACh of axotomized neurons that was decreased, while the total acetylcholinesterase (AChE) content determined in rat SCG was reduced (McLennan, 1954). However, about 5% of normal transmission remained. Acheson and Remolina (1955) explored the temporal course of the effects of axotomy in the inferior mesenteric ganglion of the cat. Stimulation at high frequency was already less effective at 2 days and the response diminished 5 days after axotomy. At 11 days, only small and rapidly decreasing responses were elicited. During the third week, no response was obtained. A slow recovery began at 4 weeks. The sensitivity to ganglion-blocking agents was strongly increased 1 week after axotomy. Moreover, in the guinea pig SCG, Purves (1975) has recorded a 70% reduction in evoked EPSP amplitude paralleled by a 65–70% loss of synapses, from 4 to 7 days after axotomy, and also a reduction in frequency as well as in amplitude of spontaneous EPSPs. The decrease in ACh sensitivity following axotomy could therefore be attributed to a smaller number of synaptic nicotinic ACh receptors (Jacob and Berg, 1987, 1988). Futhermore, Ramcharan and Matthews (1991) found that both muscarinic binding and the muscarinic responsiveness of sympathetic neurons of the rat SCG were reduced after axotomy.

*b. Morphology* In their statistical study of the effects of axotomy on rat SCG synapses, Matthews and Nelson (1975) distinguished "synaptic profiles" and VCPs; after 1 week, they observed a limited fall in VCPs, whereas synapses fell dramatically from 100–180 to 30 in a reference area. Expressed in terms of percentage of synapses per total number of VCPs, the minimum value (about 25%) was reached between 3 and 7 days. The first affected synapses were those on dendritic spines, and then those on dendritic shafts, in parallel with increasing chromatolysis. At the same time, they observed VCPs exhibiting the typical features of preganglionic endings, especially the "dense projections" attached to the membrane. They were interpreted as presynaptic endings separated from their target (dendritic spine or shaft) (Fig. 4). The highest level of these detached endings was evaluated as around 7% on 2690 VCPs at 3–7 days after axotomy, the period of highest synaptic depression. Postsynaptic differentiations were no longer visible as VPDs, as they are after deafferentation (see Section IV,B,1,b), and persisting desmosome-like structures were noted. However, a large decrease in membrane differentiations of the desmosome type, which are frequent between dendritic profiles in normal ganglia, was noted. The few type II synapses, originating from SIF cells, are also presumed to be disrupted (Matthews, 1976). The gradual recovery of synapse number began after 7 days; it was almost complete at 42 days, whereas about half the neurons were still exhibiting chromatolysis.

Similar results were observed by Purves (1975) on guinea pig SCG. In this case, 30 to 35% of the normal number of synapses per surface unit were found at the time of maximal depression of synaptic transmission. Even 3 months after axotomy, the number of synapses per surface unit did not return to its normal value. However, this is paralleled by the loss of about half the neurons in the axotomized ganglion, compared to the unoperated side. Also between 1 and 4 weeks, large profiles characterized by tubular and vesicular organelles appeared in the axotomized ganglia. This presumably represents terminal varicosities of dendrites, in line with

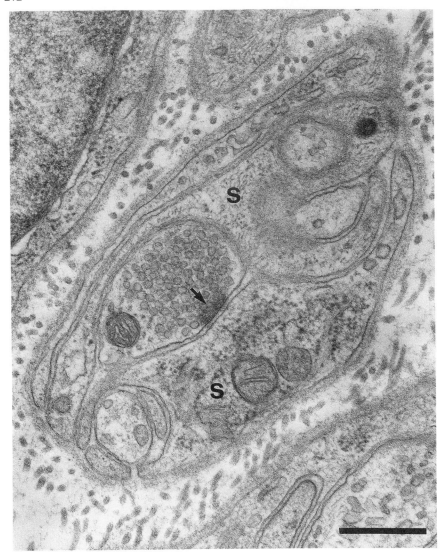

FIG. 4   Axotomized rat superior cervical ganglion, 5 days after operation. A presumably detached presynaptic profile, surrounded by Schwann cell profiles (s), exhibits a presynaptic specialization (arrow). Bar: 0.5 μm. [From Matthews, M. R., and Nelson, V. H. (1975). *J. Physiol.* (*London*) **245,** 91–135. Reprinted with permission.]

the observations of de Castro (1930), although the alternative, that some of these widenings were axonal, could not be eliminated.

An active role for satellite cells has been postulated in the separation of presynaptic endings from postsynaptic profiles having already lost their "postsynaptic thickenings." The activity of satellite cells in axotomized ganglia was also attested to in other ways (Matthews and Nelson, 1975): they form protrusions into the neurons, which can pinch off and isolate small bits of neuronal cytoplasm and give rise to lamellar extensions that wrap around the neurons and preganglionic endings. Profiles recognizable as degenerating nerve profiles or lipid droplets derived from them were extremely rare, except during the first 2 days. These reactions are similar to those noticed in the CNS by Blinzinger and Kreutzberg (1968) and Kirkpatrick (1968). Nacimiento and Kreutzberg (1990) suggested that the increase in 5'-nucleotidase activity in the plasma membrane of satellite cells was related to their role in synapse disruption.

The retraction of preganglionic endings and interposition of glial layer(s) probably cut their inductive influence onto the ganglionic neurons by way of substances released from the endings. Among other things, the transmitter synthesis could be reduced.

*c. General Considerations*  Changes after axotomy of various cell components more or less involved in synapse functioning have been reported: an initial fall in catecholamine content of the perikarya of rat SCG (Härkönen, 1964, Boyle and Gillepsie, 1970; Cheah and Geffen, 1973), a dramatic fall in perikaryon AChE activity of rat and cat SCG (McLennan, 1954; Taxi, 1961a; Härkönen, 1964), in neuropeptide content (Nagata *et al.*, 1989; Rao *et al.*, 1993) and in receptor expression (Sinicropi *et al.*, 1979), and a transient decrease in transported proteins as well as transsynaptic effects on preganglionic endings (Koenig and Droz, 1971). In all these studies, cytochemical data in electron microscopy are lacking and the actual changes in synaptic areas *sensu stricto* would need new investigations. Hendry (1975b) paid special attention to the effects of axotomy in the newborn, in which a critical period of 21 days exists, after which the response is similar to that of adults. During the critical period, axotomy resulted in degeneration of ganglionic neurons; this effect can be antagonized by nerve growth factor (NGF) (Hendry, 1975a). Some neuronal components seem under the control of factors originating from target tissues and transported by retrograde transport (Kessler and Black, 1979); others, such as calmodulin (Seto-Oshima *et al.*, 1987) or substance P (Kessler and Black, 1982), seem minimally or not at all affected by axotomy, whereas VIP immunoreactivity is strongly increased (Hyatt-Sachs *et al.*, 1993). These observations on the SCG, as well as others on other preparations, have led to the generally accepted idea that axotomy results in an increase in synthesis of molecules necessary for axonal regeneration at the expense of materials related to other functions, such as synaptic transmission and maintenance of the cytological organization, including special adhesion structures of the synapses (Watson, 1974).

Biochemical changes and electrophysiological properties after axotomy were interpreted by many authors as a return to a kind of embryonic state. This could also explain the observations of Ramsay and Matthews (1985), who found that the proportion of type II ganglionic synapses, which appeared in chronically deafferented ganglia, was increased 3.6-fold in both deafferented and axotomized ganglia. This is an expression of the higher plasticity of the axotomized neuron. These observations are in agreement with those made in tissue culture of adult SCG neurons, where such adrenergic synapses were observed (Johnson *et al.*, 1980).

According to Purves (1975), a close relationship exists between the reestablishment of synapses and the survival of neurons: both are largely dependent on the reestablishment of connections between the neurons and a target organ. The long delay for recovery would be related to this condition, with added difficulties for the axons in finding their way to their original target or to find an alternative target. Under normal conditions of axonal regeneration in the guinea pig SCG, about half the neurons disappeared between 1 and 3 months after the operation; this proportion reached more than 80% if the connection with target organs was prevented by ligation of the postganglionic nerve. It is assumed that the surviving neurons were only those able to find a new target; if they are unable to do so, they are presumed to undergo degeneration leading to death.

Another question raised by the reestablishment of synapses is that of specificity. In the normal ganglion, it has been demonstrated that the innervation of each sympathetic neuron involved several preganglionic axon terminals originating from different spinal segments, one of them being predominant (Njå and Purves, 1977a). Some months after axotomy, this selective pattern was reestablished. On the other hand, the response of end organs was abnormal, because the new connections made by the regenerating sympathetic axons were not specific to the original target (Purves and Thompson, 1979).

A puzzling fact is the persistence, at least in certain animals, of a certain number of synapses not affected by the synaptic disruption, during the maximum depression of synaptic transmission. Except for the hypothesis according to which synapses survive at random as a consequence of the level of depression of the synthesis of specific molecules (which can vary from one neuron to another), the surviving synapses could represent a particular class of terminals, whose other features remain to be established. Alternatively, they may represent a peculiar stage of a functional cycle of synapses, as suggested by the instability of synapse location on parasympathetic neurons described by Purves and Lichtman (1987).

Last, there is evidence that the signal for synaptic disruption could be the loss of some peripheral factor rather than the axon injury per se: the effect of colchicine, which mimicks axotomy in interrupting the anterograde and retrograde axonal flows, was interpreted in this way by Pilar and Landmesser (1972) in the parasympathetic ciliary ganglion of the chick and by Purves (1976a) in mammalian sympathetic ganglia. This particular topic is considered in detail below, in comparison with the frog (see Section III,D).

## C. Frog Ganglia

There are few studies that deal with the effects of axotomy of the sympathetic neurons in anuran amphibians, and none at all in urodelans, in which studies on sympathetic ganglia were discouraged by the small size and bad accessibility of these organs. In anurans, the views of different authors are conflicting. The pioneers Hunt and Riker (1966) found that axotomy in *Rana pipiens* produced a failure in synaptic transmission, whereas Gordon *et al.* (1987) reported that axotomy did not significantly affect synaptic transmission and morphology in *R. catesbeiana*. However, axotomy promotes distinct and reproducible changes in the electrical properties of the membrane, but not in perikaryon morphology (Gordon *et al.*, 1987; Kelly *et al.*, 1988), although axonal sprouting was described in B cells by Kelly *et al.* (1989a). These results are highly conflicting with the general views of Lieberman (1971), who wrote that the "absence of chromatolysis in axotomized rabbit, mouse, rat and amphibian neurons are almost certainly wrong and suggest either inadequate histologic technique or inexpert observer judgments." At variance with the results of Gordon *et al.* (1987), we have found dramatic changes in synaptic transmission and ultrastructure of ganglionic neurons of *R. esculenta* (Eugène and Taxi, 1990, 1991).

### 1. Perikarya

As far as we know, there is no comprehensive study in the literature of the chromatolysis and related events following axotomy of frog sympathetic neurons. Although we have not focused on this point, from our observations on the frog *R. esculenta* it seems that the general features known in mammalian ganglia, such as the fall in basophilia by disruption of Nissl bodies and migration of their remnants to the periphery of the perikaryon, the migration of the Golgi stacks, and the development of autophagic processes, are also true in the frog, but with a different chronology. For instance, the first sightings of chromatolysis were made only some days after axotomy. There are also changes concerning the shape of the perikarya, especially the appearance of deep infoldings and short winding dendrites (Taxi and Eugène, 1991).

## 2. Synapses

*a. Electrophysiology*   From 7 to 47 days after axotomy, Gordon *et al.* (1987) showed that EPSP amplitude in axotomized B neurons was somewhat larger than in the control. This result is contradictory with the data of Hunt and Riker (1966) and ours (Eugène and Taxi, 1990, 1991). In response to 10-Hz stimulations of the B or C preganglionic fibers innervating, respectively, the B or C neurons, we detected the first effect on synaptic transmission at 4 days after axotomy. At 1 week 19% of neurons showed subthreshold EPSPs (Fig. 5), and at 2 weeks this proportion reached 63%. At 1 month the proportion was the same, while the EPSP amplitude was at its smallest and the EPSP latency was at its largest, indicating a strong depression of synaptic transmission. At 2 months there was some recovery toward normal transmission, which was not, however, reached by 4 months.

*b. Morphology*   The beginning of changes in the synaptic index was found at 4 days after axotomy (Eugène and Taxi, 1990, 1991). The minimum value was reached after about 2 weeks, with large individual variations of the minimum value of the synaptic index, because no normal synapses remained in certain animals, whereas the index reached 0.23 for others (normal value, 1.45). Rare VPDs were observed at various times soon after axotomy, coexisting with remaining normal synapses. This suggests that VPDs could survive briefly or disappear at the time of disruption of synaptic contacts. The place of presynaptic endings was immediately occupied by glial cells surrounding the neurons (amphicytes). It is not known whether glia play an active role in the disruption of synapses, as it has been shown in the rat CNS (Blinzinger and Kreutzberg, 1968). In the sympathetic ganglia, amphicytes showed cytoplasmic features of increased activity after axotomy (reactive glia; Taxi and Eugène, 1991). Between 1 and 3 weeks after axotomy, unusual "free endings" were found in frog ganglia: these were fibers containing synaptic vesicles with some dense projections attached to the membrane, surrounded by at least one glial layer (Fig. 6). They are similar to the structures described in rat SCG by Matthews and Nelson (1975) and interpreted as presynaptic endings detached from their target. These structures were never numerous in the frog ganglia and, although no counting was performed, they represented only a few percent of normal synapses. This implies that, as in mammals, the disappearance of presynaptic dense projections is delayed after synapse disruption in certain endings. The other presynaptic endings probably quickly lose their dense projections and become members of the numerous populations of VCP present in the axotomized ganglia. Hypertrophic preganglionic fibers and/or endings, with different cytological contents, were

**B NEURON**                                    **C NEURON**

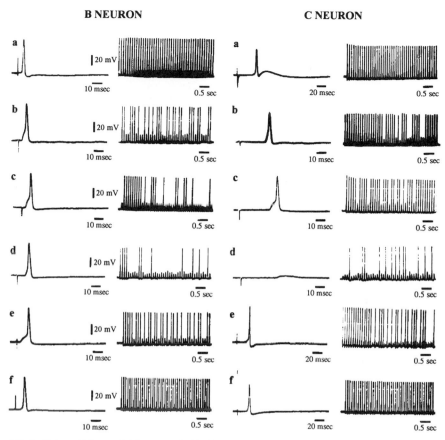

FIG. 5   Intracellular activities recorded in frog B and C sympathetic neurons from a normal abdominal ganglion (a) and from ganglia at 8 days (b), 13 days (c), 27 days (d), 52 days (e), and 114 days (f) following axotomy. For each B and C neuron, the recording on the left is a response to a single stimulation of its respective preganglionic nerve fibers, whereas the recording on the right is a response to 10-Hz stimulations. Under these repetitive stimulation conditions, axotomized B and C neurons showed subthreshold excitatory postsynaptic potentials, the number of which was maximum at about 1 month, suggesting a strong depression of synaptic transmission. [From Eugène, D., and Taxi, J. (1990). *C. R. Séances Acad. Sci.* **310,** 599–606. Reprinted with permission.]

also observed, once more with large individual variations in number (Taxi and Eugène, 1991).

The index of simple contacts, after an initial phase of stability or even increase, was also reduced, but kept a value of about half of the normal value. The initial increase is probably due to dedifferentiation of synapses in simple contacts, of which some rare intermediary stages were observed (Eugène and Taxi, 1991). Both indices rose gradually and slowly from 3

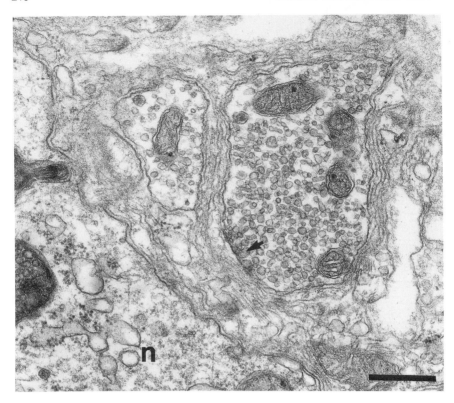

FIG. 6    Axotomized frog abdominal ganglion, 8 days after operation. A presumably detached presynaptic profile is filled with synaptic vesicles and keeps its presynaptic dense projections (arrow); several glial layers seperate it from a ganglionic neuron (n). Bar: 0.5 μm. [From Eugène, D., and Taxi, J. (1990). *C. R. Séances Acad. Sci.* **310**, 599–606. Reprinted with permission.]

to 4 weeks, the synaptic index reaching significantly higher values only 108–131 days after the operation. Synapses provided with a subsynaptic bar were rare. During the first period of synapse reestablishment, axodendritic synapses were fairly frequent in relation to the transient outgrowth of short dendrites, one of the aspects of the changes in perikaryon shape reported in Section III,C,1. These changes are probably related to changes in skeletal proteins involved in the maintenance of cell shape. Later, the situation turned progressively back to normal, the dendritic processes being probably incorporated in the perikaryon, which regained its smooth surface.

According to many authors (see Lieberman, 1971; Barron, 1983), the distance from the perikaryon at which the axon section is made may have important effects on the events affecting the perikaryon and thus may account for the discrepancy between the results of Gordon *et al.* (1987) and ours. In our experiments, we usually cut the rami communicantes just at the point where they entered the corresponding root of the lumbar plexus. In other words, the cut was made about 0.5 to 2 mm from the ganglion. To make experiments more similar to those of Gordon *et al.* (1987), the axon section was performed in some animals at about 10 mm or more from the ganglia, on the corresponding branches of the lumbosacral plexus (see Fig. 1). No significant differences were noted in both the synaptic and simple contact indices at 1–2 weeks after the operation; thus it does not seem that the discrepancy between results of Gordon *et al.* and ours can be explained by differences in the position of axon section.

**c. Parallel between Electrophysiological and Morphological Data**   A correlation between synaptic and simple contact indices and synaptic transmission appears when data are plotted on the same graph (Fig. 7). The general correlation is consistent between the minimum value of the

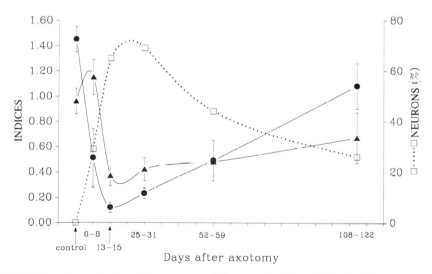

FIG. 7   Variation with time after axotomy of the synaptic index (filled circles), the simple contact index (filled triangles), and the number of B and C neurons that initiated subthreshold excitatory postsynaptic potentials (EPSPs; open squares). For each ganglion analyzed, both indices were calculated as the ratio of the number of morphological contacts between pre- and postsynaptic elements, in ultrathin sections of neurons examined by electron microscopy. In the graph, the means of each index (±SEM) are reported at different times. The number of neurons with subthreshold EPSPs is expressed as a percentage of the total number of electrophysiologically investigated neurons. [From Eugène, D., and Taxi, J. (1991). *J. Neurocytol.* **20**, 404–419. Reprinted with permission.]

indices and the maximum number of neurons exhibiting subthreshold EPSPs. However, some discrepancies can be noticed when the data are examined in detail. There is already a drop in the synaptic index whereas the electrophysiological recordings remain normal in most neurons. Thus transmission remains possible, at least for a short while, after the disappearance of many synapses. On the other hand, the efficacy of synaptic transmission remains low 4 weeks after axon section, whereas the synaptic index has already increased. This suggests that the newly formed synapses were not immediately functional. Moreover, in all the animals studied, a certain number of neurons kept normal synaptic transmission, presumably those with the remaining normal synapses. As cholinergic transmission is considered impossible when nerve endings are detached and separated from the postsynaptic neuron by one or more glial layers, it is likely that a certain amount of transmission can occur at simple contacts. In fact, the population of simple contacts is heterogeneous, formed by those existing in normal ganglia and those representing "dedifferentiated synapses." In this latter group, postsynaptic membranes might keep a certain concentration of nicotinic receptors in spite of having lost their "postsynaptic differentiation." Thus the correlation between electrophysiological and morphological data is not strict: a certain amount of transmission can occur with a limited number of identifiable synapses; on the other hand, reappearance of synaptic complexes does not mean that they are immediately functional.

## D. Comparison between Mammals and Frog

After axotomy of sympathetic neurons, the most obvious reaction of synaptic areas is similar in mammalian and frog ganglionic synapses, that is, detachment of presynaptic endings from postsynaptic sites and, simultaneously or rapidly afterward, disappearance of postsynaptic differentiations (Matthews and Nelson, 1975; Eugène and Taxi, 1990, 1991). But in both classes of vertebrates, a certain percentage of normal synapses persisted, with large individual variations.

In the frog, the parallel between electrophysiological and morphological data has led to the conclusion that a certain amount of transmission after axotomy is possible at "simple contacts" (Eugène and Taxi, 1990, 1991). This suggests that there is no direct relationship between the disappearance or drastic decrease in molecules responsible for the postsynaptic differentiations and synaptic efficacy. Similar studies in mammals have to date not yet been performed.

In the sympathetic ganglia of the frog, reactive deafferentation appears some days after axon section but lasts several months (Eugène and Taxi,

1990, 1991). In rat SCG, a total recovery was observed 42 days after axotomy (Matthews and Nelson, 1975). This discrepancy is certainly attributable to the different metabolisms of the frog (a cold-blooded animal) and the rat (a warm-blooded animal).

Axotomy also modifies some electrophysiological properties of sympathetic neuron membranes similarly in mammals (Purves, 1975) as well as in the frog (Hunt and Riker, 1966; Gordon *et al.*, 1987; Eugène and Taxi, 1991). However, Purves and Njå (1976) have shown that the application of NGF on guinea pig SCG prevented the biochemical, morphological, and electrophysiological changes due to axotomy. Moreover, reactive deafferentation similar to axotomy and alterations of electrical properties could be provoked in normal SCG by colchicine application (Purves, 1976a) or by an anti-NGF antibody (Njå and Purves, 1978a). In the frog, some of the altered electrical properties are restored after axonal regeneration (Kelly *et al.*, 1988) and we have also applied murine NGF on axotomized sympathetic neurons and found a smaller loss of synaptic contacts (J. Taxi and D. Eugène, unpublished data). Moreover, the effects of murine NGF have been tested on the electrical properties of frog sympathetic neurons in culture (Traynor *et al.*, 1992): the results confirm that some properties altered by target deprivation were restored by NGF. Consequently, both the reduction in synapse number and the modification of electrical properties of frog sympathetic neurons could be attributed to a loss of trophic factors that would again be supplied to the neurons during axonal regeneration (Smith *et al.*, 1988). Thus in the frog, like in mammals, a NGF-like trophic factor released by the peripheral targets of sympathetic neurons and retrogradely carried in the sympathetic axon could participate in the maintenance of the ganglionic synapses and of some electrical properties.

In mammals, it seems that the regeneration of contacts with the target organs is fundamental for the survival of neurons and the regeneration of ganglionic synapses (Purves and Lichtman, 1978). This point seems less important in the frog, at least within the limits of duration of our experiments, in which the axons were not prevented from regenerating to their target organs. Indeed, the large number of frog neurons surviving in culture without NGF (Kelly *et al.*, 1989b) or in adult ganglia autografts performed in the intercostal muscles (see Section V,B) seems to indicate that the relationships with definite targets are not necessary for frog sympathetic neuron survival or that frog neurons are more able to find new targets. By all means, it is well known that the relationships of nerve terminals with the vessels, one of the main targets of the sympathetic fibers, are not very close.

The disruption of synapses in both cases is clearly related to modifications of the current program of protein synthesis of the neuron, which

has important consequences for proteins taking part in the synaptic specializations and/or in the shape of the perikaryon.


## IV. Deafferentation and Reinnervation of Ganglia by Preganglionic Nerve Fibers

### A. Introduction

The sympathetic ganglia and especially the SCG of mammals appeared
for a long time to be appropriate preparations with which to examine
questions raised by deafferentation and/or reafferentation studies of neurons, such as the chronology of failure and recovery of synaptic transmission in parallel with the disappearance and reappearance of synaptic structures and their maturation as well as the specificity of the newly formed
connections. This last problem can be studied from two angles: the specificity of the connections between preganglionic fibers and neurons acting
on a definite target organ, and the possibility of substituting other nerve
fibers for preganglionic ones. This last topic, called *heterogenic reinnervation,* is examined in Section V.


### B. Mammals

#### 1. Total Deafferentation

As all preganglionic fibers enter the SCG via the sympathetic trunk through
the posterior end of the ganglion, complete deafferentation is easy to carry
out by sectioning, crushing, or freezing of the trunk more or less near the
caudal pole of the ganglion.

*a. Physiology of Transmission*   Cannon and Rosenblueth (1936) were
the first to show that chronic preganglionic denervation increased the
sensitivity of sympathetic ganglia to ACh. This phenomenon occurs in
all peripheral receptors and was called "denervation supersensitivity"
(Cannon and Rosenblueth, 1949). In the deafferented SCG of the cat,
this supersensitivity is mainly the result of a reduction in AChE activity
(Brown, 1969). Previously, the effects of denervation on AChE activity
in sympathetic ganglia had been explored histochemically by Koelle (1951,
1955) in the cat SCG and then by Taxi (1961a) in the rat SCG. The activity
disappeared in the structures occupying the interneuronal spaces and
thus seemed localized in preganglionic fibers, including the presynaptic

endings. Later, Gautron and Gisiger (1976) and Gisiger *et al.* (1978) reported an initial fall in AChE activity after sectioning preganglionic fibers in the rat SCG, but there was a secondary increase some days later, which appeared as a reaction to denervation. Moreover, in the rabbit SCG, it has also been shown that ACh sensitivity of the neurons is different according to the two types of cholinergic receptors: the nicotinic response is decreased whereas the muscarinic response is increased (Dun *et al.*, 1976a). On the other hand, passive and active membrane properties of the neurons are not affected by preganglionic denervation (McLachlan, 1974; Dun *et al.*, 1976b).

After degeneration following injury, the preganglionic fibers spontaneously regenerate within the ganglion. At the end of the last century, Langley (1897), using the physiological methods available at that time on the cat SCG, showed the ability of the preganglionic fibers to restore synaptic transmission, usually very efficiently. However, considerable deficit was observed in some animals. Langley had already established that the regeneration of preganglionic fibers after sectioning is selective, in the sense that the fibers of each ramus communicans form new connections with the same class of neurons as determined by their target organ (blood vessels, hairs, skin glands, etc.). These new connections were established by unmyelinated fibers, instead of myelinated ones as in normal ganglion. On the other hand, Langley also observed that postganglionic axons were able to innervate a target of a different type than the normal one. Preganglionic axons regenerating within the ganglia always divided into several sprouts. Moreover, the chronology of ganglion reinnervation was highly dependent on the existence of a scar: if there was no scar and in the absence of retraction of the two nerve stumps, new perineuronal arborizations began to appear about 12 days after sectioning (de Castro, 1930, 1951). At this time, the physiological effects of reinnervation were still limited and became more obvious only after 15 to 21 days. The return to anatomical and functional normality did not take place before 30–40 days. In case of a large discontinuity, such as the ablation of 1 cm of nerve, the beginning of recovery was delayed to 35–40 days, and sometimes there was no recovery at all. The new distribution of connections between preganglionic fibers and ganglionic neurons is evidenced by the difference in the effects of stimulation of rami communicantes between normal and reinnervated animals. Langley (1897) noted that these differences tended to diminish with time, but it is not known whether this is due to reorganization of synapses according to specific relations, or to an increase in the number of connections with time, which allows the activation of more neurons.

In the guinea pig SCG, McLachlan (1974) found that the normally low frequency of miniature EPSPs was reduced following preganglionic regeneration and that, after 3 months, there was no sign of functional transmission in about 20% of neurons, despite the normal amplitude of the postganglionic action potential. These data were confirmed by Purves (1976b), who recorded 13% of subthreshold evoked EPSPs after 2–3 months and only 2% after 6–7 months. Nevertheless, Purves considered reinnervation to be incomplete even after 15 months, because the amplitude distribution of evoked EPSPs remained in the low normal or subnormal range and the number of synaptic steps was also lower than normal.

Njå and Purves (1977a), using intracellular recordings, added that each neuron of the guinea pig SCG receives inputs from several axons arising generally from different contiguous segments of the spinal cord, one of them being predominant for each group of neurons. After denervation, although the number of axons reinnervating each neuron did not exceed 50–60% of the normal level, the functional schedule is reestablished (Njå and Purves, 1977b). They also observed that some synaptic contacts were reestablished as early as 8–11 days after freezing the preganglionic fibers (Njå and Purves, 1978b). After this, the proportion of reinnervated neurons increases rapidly up to 3 months, but the innervation remains weaker than normal even after 6 months. Physiological data suggest that the reinnervation occurs by progressive addition of synapses that are each immediately appropriate, and not by initially inappropriate synapses that would be then submitted to a remodeling process with elimination of unsuitable contacts (Njå and Purves, 1978b).

Finally, in mammals the pattern of activation of end organs after preganglionic reinnervation is indistinguishable from normal (Langley, 1897; de Castro, 1930; Njå and Purves, 1977b, 1978b; Östberg and Vrbová, 1982), which implies that the reestablishment of ganglionic synapses is selective (Liestøl et al., 1986). However, preganglionically reinnervated sympathetic neurons may show some electrophysiological abnormalities.

**b. Morphology of Synapses**   The studies of the beginning of the century were reviewed by Lawrentjew (1925) and de Castro (1930), who revised the chronology of the events following the section of preganglionic fibers in young cats, as revealed by silver nitrate staining. Hypertrophy and alteration of terminal rings of neurofibrils began between 6 and 12 hr after the operation; the peak of this process was reached 24 hr after sectioning, and the degeneration was complete after 43–56 hr, except for some fibers that were still hypertrophic at this time. All degenerative fibers had disappeared by 6–7 days. This chronology was controlled by stimulation of the central tip of the preganglionic nerve.

In electron microscopy, pioneer attempts were made by Causey and Hoffman (1956) and Causey and Barton (1958), but the limitations of their technique did not permit accurate observations. Hamori et al. (1968)

and Ceccarelli (1968) gave a precise chronology of the first modifications affecting synaptic vesicle numbers and clumping 6–12 hr after decentralization of the cat SCG; these alterations became general, affecting all presynaptic organelles by 24 hr, although with important variations from one synapse to another. This results in formation of dense autophagic vacuoles and, as early as 12 hr, removal of axonal cytolysomes by ganglionic glia is visible. The endocytotic vacuoles soon lose their double membrane and become glial cytolysomes. The number of endings showing degenerative features decreased dramatically in the postoperative 72 hr, and became rare after 1 week. This was confirmed in the rat by Quilliam and Tamarind (1972) and Raisman and Matthews (1972). The speed of degeneration was deduced from the scarcity of degenerating structures (Ramsay and Matthews, 1985). After the disappearance of synapses, a number of VPDs persisted, now covered by a glial layer. All afferent synapses, concerning either principal neurons or SIF cells, degenerate according to Taxi et al. (1969) and Matthews and Östberg (1973). However, Quilliam and Tamarind (1972) mentioned some persisting synapses, whose number increased after the third day and reached about 10% of the normal amount of ganglionic synapses after 1 week. They are mainly of type II (adrenergic) and are thought to be formed by axon collaterals of ganglionic neurons. However, they may originate from SIF cells (Matthews, 1976).

Raisman et al. (1974) and Östberg et al. (1976) confirmed and extended the observations of Quilliam and Tamarind (1972). They counted the VPDs, which were found to be about 35% of the normal number of synapses per surface unit, with a possible slow decrease as time went on. However, VPDs were not found in deafferented ganglia of neonates (Smolen and Raisman, 1980). Synapses reappeared about 25 days after deafferentation by gentle freezing the preganglionic chain; this delay was longer than that mentioned by de Castro (1930) in the cat SCG. Thereafter, the number of synapses increased until about 75 days, when it reached values in the lowest range for normal animals. After this, no further changes were observed. Even much later, the synaptic density never exceeded that of normal animals and often did not reach normal values. This is also the conclusion of Purves (1976b), who considers that reinnervation is achieved within 6 months after surgery. This was taken by Raisman et al. (1974) as strong evidence that synapses can be formed only at sites where synapses existed before deafferentation or, in other words, that neurons are not able to evolve new "postsynaptic differentiations." There is, however, a problem with this interpretation: the number of VPDs per unit area being about 35% of that of normal synapses, the authors considered that the discrepancy is explained by the fact that VPDs are more difficult to identify than complete synapses. Although this is true, is it sufficient to explain that identified VPDs represent only one-third of normal synapses? These results were to a certain extent challenged by those of Ramsay and Matthews (1985), who evaluated the loss of postsynaptic thickenings as 74%, 2 days after deafferentation, the decrease continuing later on at a slow rate. Experiments involving a second section of the preganglionic chain led, after 3 months, to the disappearance of all synapses and thus eliminated the possibility that newly formed synapses were produced by axonal sprouting of intraganglionic neurons. This is a fundamental difference with the phenomenon of reinnervation by sprouting of neighbor fibers in the CNS observed by Raisman *et al.* (1974).

The persistence of a few "intrinsic" synapses after sectioning presumably all preganglionic fibers and at postoperative times too short to allow regeneration, first mentioned by de Castro (1930), was confirmed in the cat SCG by Lakos (1970) and in the rat SCG by Quilliam and Tamarind (1972), McLachlan (1974), Raisman *et al.* (1974), and Östberg *et al.* (1976). Ramsay and Matthews (1985) later described their uneven distribution. These synapses were attributed either to preganglionic fibers following unexpected routes to the ganglion and thus having escaped surgery, or to axon collaterals sprouting from the few ganglionic cholinergic neurons present in the SCG (Hamberger *et al.*, 1965).

In chronically denervated ganglia, Purves (1976b) found that about 2–3% of the normal number of synapses, mainly of type I, and rare VPDs survive when observed a short time after surgery. But after 4–5 months, Purves observed a ninefold increase in the number of synapses. Various physiological experiments led Purves to the conclusion that these synapses, although provided with clear vesicles, were formed by axon collaterals of principal ganglionic neurons, but no pharmacological tests concerning the nature of the transmission were performed. Ramsay and Matthews (1985), using the labeling of presumably catecholamine-containing vesicles by 5-hydroxydopamine, observed that only 15% of the synapses were type I and the others were type II (adrenergic). This reinforced the interpretation that they are formed by endings of postganglionic axon collaterals. A similar result was previously obtained by Yokota and Burnstock (1983) in the deafferented pelvic ganglia of the guinea pig. If the ganglia were both deafferented and axotomized, a threefold increase in type II synapses was observed (Fig. 8); this means that the type II synapses represent 34% of the number of type I synapses in normal ganglia (Ramsay and Matthews, 1985). Still more surprisingly, they found the same number of VPDs in axotomized–deafferented ganglia as in only chronically deafferented ones, whereas Matthews and Nelson (1975) reported a complete disappearance of such structures after axotomy.

McLachlan (1974), using intracellular injections of Procion Yellow in the guinea pig SCG, showed that the geometry of denervated neurons was not modified. The recovery of synaptic transmission in the ganglion

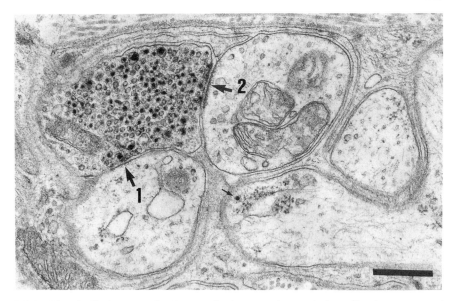

FIG. 8  Chronically denervated and axotomized rat superior cervical ganglion that was treated with 5-hydroxydopamine before fixation. A profile containing numerous small dense-cored vesicles forms two synapses, whose synaptic complexes are indicated by the arrows. Synapse 1 was considered to be symmetrical, on a dendritic profile; synapse 2 is asymmetrical, on a not characteristic structure, which could be a dendrite containing some clear vesicles. Bar: 0.5 μm. [Reprinted from Ramsay, D. A., and Matthews, M. R (1985). *Neuroscience* **16**, 997–1026. Copyright (1985), with kind permission from Elsevier Science Ltd., The Boulevard, Langford Lane, Kidlington OX5 1GB, UK.]

started 4 weeks after nerve section; at this time, only a few fibers forming synaptic contacts were found in the ganglia, and these synapses exhibited few synaptic vesicles. The number of synapses then increased regularly and the author concluded that after 3 months, it is difficult to distinguish reinnervated from normal ganglia, but quantitative data were lacking in this study. In their thorough comparative study, Östberg *et al.* (1976) found that denervation caused a loss of 93% of ganglionic synapses and VPDs were less than half the number of normal synapses. In the case of denervation by freezing the preganglionic chain, regeneration restored up to 85% of the normal synapses. If the preganglionic chain was cut off and then sutured, there were large individual variations in the number of regenerated synapses, reaching at best 60% of the normal number, and the number of VPDs varied in inverse proportion to that of the number of synapses. This does not coincide with the view expressed by Purves and Lichtman (1978), according to which the mode of denervation (i.e., sectioning, crushing, or freezing) has practically no effect on regeneration.

## 2. Partial Deafferentation

As preganglionic fibers to the SCG originate at different levels of the spinal cord and join the sympathetic trunk through the rami communicantes of the first to fourth thoracic segments (Njå and Purves, 1977a; Purves and Lichtman, 1985), partial deafferentation can be obtained by sectioning rami communicantes and the degree of specificity of ganglionic connections can be evaluated after regeneration. This type of investigation was initiated long ago by Langley (1897), who observed that the stimulation of different thoracic roots produced responses specific to different target organs: for instance, the first thoracic root in the cat controls pupillary and palpebral muscles, while the fourth controls ear blood vessels and piloerection of the face and neck. This implies that preganglionic fibers do not send their endings equally to all neurons. Murray and Thompson (1957) observed that, after sectioning about 90% of preganglionic fibers of the cat SCG, an intense sprouting of the remaining axons occurred that made contacts with many neurons; normal ganglionic transmission was reestablished within 4 to 8 weeks, although the amount of released ACh was only 60% of normal. However, stimulation of residual preganglionic axons can elicit abnormal peripheral sympathetic effects (Guth and Bailey, 1961; Guth and Bernstein, 1961) indicating a possible loss of specificity after sprouting, which would certainly be due to the procedure of partial deafferentation, that is, the complete removal of preganglionic axons arising from particular spinal cord segments.

More recently, Mæhlen and Njå (1981, 1982, 1984), Henningsen et al. (1985), and Liestøl et al. (1986, 1987) reinvestigated the problem of selective reestablishment of synaptic connections in response to a different type of partial deafferentation: they crushed the cervical trunk of the guinea pig SCG at the level of the subclavian artery. Following this operation, deafferented neurons were also reinnervated by sprouting of intact preganglionic axons and this reinnervation was relatively selective because deafferented neurons were reinnervated only by certain preganglionic axons arising from different spinal cord segments. But surprisingly, the sprouting axons abandoned most of their synaptic connections in the stellate ganglion (Mæhlen and Njå, 1984). Moreover, when the severed preganglionic axons regenerated within the SCG, the synapses formed by the sprouting axons were not totally eliminated. Consequently, these experiments indicate that the pattern of synaptic connections in normal ganglia is maintained thanks to a competition between preganglionic fibers. The formation and maintenance of synapses certainly result from both selective recognition mechanisms between the preganglionic axons and the sympathetic neuron and the axon availability. A model of selective synapse formation has been proposed by Liestøl et al. (1992).

## 3. General Considerations

The Wallerian degeneration of preganglionic fibers first provokes a failure in ganglionic transmission, even before visible morphological alterations of preganglionic endings can be observed. This failure is faster the closer the preganglionic injury is to axonal endings. Beside the loss of molecules and functions directly related to the disappearance of preganglionic fibers, such as ACh content or choline acetyltransferase (CAT) activity (Raisman *et al.*, 1974), ganglionic deafferentation also causes metabolic and functional modifications related to neurotransmission or other cellular functions of sympathetic neurons.

For example, in the deafferented SCG of the rat, there is an increase of substance P-like (Kessler and Black, 1982), VIP-like (Hyatt-Sachs *et al.*, 1993), and protein kinase C-like (Roivainen, 1991) immunoreactivities, a decrease in tyrosine hydroxylase (TH) activity (Raisman *et al.*, 1974), and a reserpine-induced increase in its messenger RNA that was prevented by deafferentation (Black *et al.*, 1985). Previously, Black *et al.* (1971, 1972), using ganglionic denervation experiments, demonstrated that the ontogeny of catecholamines synthesizing enzymes and their level in adults are regulated by synaptic activity. Similarly, the level of neurofilament phosphorylation increases after deafferentation (Shaw *et al.*, 1988), whereas the atrial natriuretic peptide content decreases (Nagata *et al.*, 1989). This peptide, known as a humoral factor in regulating body fluid volume and blood pressure, occurs in large amounts in the adult rat SCG. All these results indicate that transsynaptic factors certainly play important roles in the regulation of protein synthesis in sympathetic neurons. One of these factors could be glycyl-L-glutamine, because its application on the preganglionically deafferented SCG of the cat maintains the AChE content (Koelle *et al.*, 1988; Koelle, 1988).

Moreover, the concentration of certain molecules involved in transmission has been directly related to synaptic activity because it is modified after preganglionic denervation or transmission block. Thus, Greif (1986) described a transient increase in a vesicle-specific membrane protein after deafferentation, followed by a drop below the control level after 2 weeks, whereas chlorisondamine, a blocking agent of ganglionic transmission, mimicked to a certain extent this effect. This suggests that the regulation of synthesis of this protein as well as that of the synaptic vesicle pool of the sympathetic neuron was under the control of synaptic transmission. The effects of synaptic transmission on the properties of a 65-kDa protein located in the postsynaptic density of the ganglionic synapse, similar to that of CNS synapses, were studied by Wu and Black (1987, 1988), using its ability to bind calmodulin. The normal calmodulin-binding capacities dropped dramatically after denervation, whereas the protein content remained unchanged. Reserpine, which induces an increase in synaptic transmission, also elicits a 90% increase in calmodulin binding. All these data suggest that the presynaptic innervation certainly regulates the concentration of specific molecules of the postsynaptic neuron.

## C. Frog

### 1. Total Deafferentation

Total deafferentation of both the 8th and 9th ganglia was obtained by section of the sympathetic trunk 1 to 2 mm before the 8th ganglion (see Fig. 1), according to the origin of preganglionic fibers established by Langley and Orbeli (1910).

*a. Physiology of Transmission*   Degeneration of presynaptic endings is obvious after 24–36 hr according to Hunt and Nelson (1965), who recorded at this time a disappearance of miniature EPSPs. The degeneration is more evident after 3–4 days according to Streichert and Sargent (1990) and also according to our own experience. After deafferentation, Hunt and Nelson (1965) also reported an increase in input resistance with no change in the resting membrane potential. However, for Dunn and Marshall (1985) this increase is not significant and the cholinergic supersensitivity studied electrophysiologically would be due to a decrease in AChE as found in mammals, and not to a modification in the number or distribution of ACh receptors. On the other hand, histochemical studies demonstrated that in normal ganglia, AChE activity is associated with pre- and postsynaptic elements (Taxi, 1965; Brzin et al., 1966; Weitsen and Weight, 1977) and that after deafferentation this activity decreased extracellularly, but not within the sympathetic neuron (Streichert and Sargent, 1990).

From 2 weeks after sectioning the preganglionic sympathetic trunk, B but also C neurons were reinnervated by regenerating B preganglionic fibers (Feldman, 1988; Eugène and Taxi, 1994). It is certainly because myelinated fibers regenerate faster than unmyelinated fibers (Schmidt and Stefani, 1976). However, when C preganglionic fibers were fully regenerated, the inappropiate synapses were eliminated as a result of competition between the two groups of preganglionic fibers (Feldman, 1988). Consequently, the reinnervation of frog ganglia is highly selective. In addition, the comparison between the synaptic indices of the 8th and 9th ganglia suggests that certain regenerating preganglionic fibers forming contacts with sympathetic neurons of the 8th ganglion would limit or stop their growth toward the 9th ganglion. That could result from a decrease in the vigor of axonal growth in adult animals (Fawcett, 1992).

*b. Morphology of Synapses*   Both aspects of degenerating fibers, the so-called dark and clear degenerations, were described by Taxi (1962, 1965, 1979a), Hunt and Nelson (1965), Sotelo (1968), and Russel and Riker (1977). As early as 1 to 3 days, degenerating fibers were engulfed in glial cells, lost their own membrane to turn into dense bodies, and disappeared within 10 days. Synapses were disrupted with interposition of a glial layer as soon as the degenerating aspects were visible. A number of VPDs survived, a certain proportion of them being provided with a subsynaptic bar. The number of VPDs surviving synaptic disruption was variable from one frog to another. An index of VPDs was calculated on the same basis as the synaptic index; its mean value is about 0.4 ± 0.1, for 16 frogs taken from a few days up to 150 days after deafferentation; in the case of a period longer than 2 weeks, regeneration was prevented. The disappearance of postsynaptic thickenings seems to occur mainly by an "all or nothing" process at the time of synapse disruption, because countings made at different postoperative times showed that there is no obvious modification of the VPD index. For instance, for 3 frogs kept from 75 to 150 days after sectioning with prevented regeneration, the VPD index was 0.47, compared with the mean value of 0.4 mentioned above. Thus, it appears that postsynaptic differentiations are of two kinds: the labile and the stable ones; unfortunately the origin and nature of this difference are not yet understood. Another fact is that the proportion of VPDs provided with a subsynaptic apparatus is significantly higher (30 ± 17% from 17 animals) than that of normal synapses (20 ± 10%, see Section II,B,3), in other words, that such VPDs are statistically more stable than the others.

In certain animals, there is a small number of persisting synapses at times when normally all synapses have degenerated (from 5 days to 2 weeks) and regeneration has not yet started, a state of things that has also been reported in mammals. As frogs are cold-blooded animals, the possibility cannot be eliminated that some synapses are exceptionally resistant to degeneration, as it is well documented in certain CNS regions of cold-blooded animals (Repérant *et al.*, 1991); however, the most probable interpretation is that these synapses correspond to endings of preganglionic fibers that followed abnormal pathways. This would be a supplementary aspect of the plasticity of the sympathetic nervous system, including variations in size, position, and even in the number of the ganglia, the 8th and 9th forming only one ganglion in some animals or one of them being divided in two in others.

The reappearance of synapses began about 1 month after sectioning the sympathetic trunk, as always with large individual variations: the nerve fibers within the ganglion were numerous as a result of the sprouting of the regenerating axons. It seems obvious that the VPDs are privileged sites for the reestablishment of synapses because a rapid fall in number of VPDs occurred after reinnervation has begun. However, there were also newly formed synapses, because the final synaptic index often reached or was near normal values, whereas the VPD index was only about 35% of that. The question remains, whether the new postsynaptic differentiations are formed *de novo* or whether they appear at the same places where synapses were located before deafferentation; in this case, these places would have kept a kind of "imprinting" of their previous differentiation. This imprinting could be due to the permanence of ACh-binding sites, as it was observed in the parasympathetic neurons of the denervated cardiac ganglion of the frog (Sargent *et al.*, 1991). However, the ACh receptor-like clusters at synaptic sites appeared subject to variations in size and distribution in the cell membrane after denervation (Sargent and Pang, 1988).

The permanence of some VPDs a long time after deafferentation, when the synaptic index already recovered normal values, indicated that the sprouting of preganglionic fibers to more than one neuron is a transient phenomenon, as it was observed electrophysiologically in various regenerative processes, and did not concern all neurons. Thus, VPDs do not seem to exert a powerful attraction on regenerating axons. Possible intermediary stages of synaptic formation were extremely rare. This indicates that the formation of a new synapse is also a rapid process, and could be described as an "all or nothing" process.

When reinnervation was prevented for several months, formation of intrinsic synapses was never observed in sympathetic ganglia, at variance with what happened in the parasympathetic cardiac ganglia where new, preferentially axoaxonic synapses were reported by Sargent and Dennis (1981). Moreover, in chronically denervated ganglia, the organization of the glial sheath around the unique process, which is fairly complex but typical (Taxi, 1965), is not modified, and this suggests that the glial pathways leading to each neuron persist and might be followed by the regenerating fibers provided that they are able to enter one of these pathways.

## 2. Selective Deafferentation of B or C Neurons

One interesting feature of frog ganglia is that they contain two types of neurons (B and C) characterized electrophysiologically by Nishi *et al.* (1965) (see Section II,A,2), who also indicated that B neurons are larger than the C ones. It was demonstrated by Skok (1964) and confirmed by Libet *et al.* (1968) that B and C neurons can be selectively denervated; the preganglionic myelinated fibers to B neurons of 8th and 9th ganglia come from spinal segments anterior to the 6th ganglion via the sympathetic trunk, whereas the preganglionic unmyelinated fibers to C neurons of the same ganglia come via the rami communicantes of the 6th and 7th ganglia (see Fig. 1).

The consequences of partial denervation in *R. pipiens* and *R. catesbei-*

*ana* sympathetic ganglia were explored electrophysiologically by Feldman (1988), with regeneration of cut fibers being prevented. After selective sectioning of B preganglionic fibers, C preganglionic fibers sprout and innervate B neurons. Previously, Marshall (1985) had shown that the kinetic properties of ACh receptor channels of B neurons reinnervated by C fibers were similar to those of normally innervated C neurons, but not to those of normal B neurons (Marshall, 1986). Therefore, the innervation appears to control the kinetic properties of postsynaptic receptors. Later on, if B fibers regenerate, they reinnervate only B neurons, in such a way that synapses originating from sprouting of C fibers completely disappear in a few weeks (Feldman, 1988).

We also have undertaken a study of the effects of selective section of B or C fibers according to Fig. 1 (D. Eugène and J. Taxi, unpublished). What follows is a description of our preliminary results. At various times after sectioning, electrophysiological recordings and gathering of morphological data, including synaptic and VPD indices calculated according to Section II,B,3, were carried out.

*a. Section of B Preganglionic Fibers*   At short times (up to 1 week) after sectioning, we found that synaptic transmission in B neurons of both 8th and 9th ganglia stopped, and the synaptic index fell to less than 50% of its normal value. After one month, synaptic transmission recovered in 20 to 50% of B neurons in response to stimulation of C preganglionic fibers, but the synaptic index remained low. Thus, a number of B neurons appear to be reinnervated by sprouting C preganglionic fibers. The permanence of low synaptic indices can be explained by a compensatory mechanism similar to that described in mammals by Mæhlen and Njå (1984); the C fibers sprouting onto B neurons could lose a certain number of their previous synaptic contacts with C neurons; however, this hypothesis still needs confirmation. After 2 months, no more B neurons were reinnervated whereas slightly higher synaptic indices were found; however, normal values were generally not obtained, even up to 5 months after the operation. Large individual variations were observed, as has been noticed already in other cases of regeneration after sectioning nerve fibers. In the present case, these variations may have been exacerbated by sampling problems, because B and C neurons are not randomly distributed in the limited areas of the sections used for electron microscopy, as was observed in the selective identification of C neurons by staining of their pericellular apparatus with antibodies to LHRH (Lascar *et al.,* 1982). Finally, it appears that the correlation between electrophysiological and morphological data is acceptable, because not all but a significant proportion of B neurons was functionally reinnervated by C preganglionic fibers, whereas the synaptic index remained below the normal value.

It was also noticed that the proportion of synapses provided with a subsynaptic bar is low after sectioning B preganglionic fibers. On the other hand, the proportion of VPDs provided with a subsynaptic bar remains within normal values. This suggests that the synapses with such a bar are mainly located on B neurons.

***b. Section of C Preganglionic Fibers***    Five days after sectioning, it was no longer possible to record synaptic events from C neurons of the 8th and 9th ganglia. At the same time, the synaptic index dropped to about 50% of its normal value. This result is unexpected, if one remembers that the number of C neurons, identified by the presence of a positive LHRH pericellular apparatus, is about double that of B neurons (Jan and Jan, 1982; Lascar and Taxi, 1992). However, this result means that B neurons are more richly provided with synapses than C neurons, probably in relation to their larger size (see Sargent, 1983). After 1 month, synaptic responses to preganglionic trunk stimulation were evoked in all the C neurons recorded. However, the synaptic indices remained at the same mean value as after 5 days, with considerable individual variations, the index being normal in certain animals while low in others. As there was no anatomical evidence that C preganglionic fibers could rapidly regenerate after total ablation of the 6th and 7th rami communicantes, the value of the synaptic index is presumed to be due to synapses of sprouting B fibers onto C neurons and also to normal synapses of B neurons, a certain number of which might, however, be lost as a consequence of the compensatory mechanism described in mammals (Mæhlen and Njå, 1984). After 2 months, the synaptic index reached normal values, confirming the electrophysiological results that all neurons were reinnervated.

After C fiber sectioning, the proportion of subsynaptic bar-bearing synapses is at least equal to and often higher than that of normal ganglia; on the other hand, the proportion of VPDs provided with a subsynaptic bar is very low or zero. This indicates that subsynaptic bars are preferentially located in B neurons, in complete agreement with the conclusions drawn from our sectioning experiments of B preganglionic fibers.

***c. Discussion***    The selectivity of sectioning was checked by immunocytochemistry in the case of C preganglionic fibers, taking advantage of the fact that the endings of preganglionic C fibers contain a LHRH-like substance (L. Y. Jan *et al.,* 1980). It was therefore observed that, 10 days after surgery, only a few endings reacted with an anti-LHRH serum (G. Lascar and J. Taxi, unpublished).

The results obtained after selective sectioning of B or C preganglionic fibers were not symmetrical. At 1 month after sectioning (and perhaps even less), all C neurons were reinnervated, whereas less than 50% of B neurons were, and this proportion did not increase even after 5 months.

This strongly suggests that C neurons could be rapidly reinnervated by sprouting B fibers, whereas the sprouting of C fibers toward B neurons is much less efficient. Nevertheless, the connections seemed stable, at least in the absence of competition with regenerating B fibers. One possible explanation is that there is a specificity factor in the synapses between B preganglionic fibers and B neurons that is not present in the synapses of C neurons. That explains the relative inability of C fibers to make synaptic connections with B neurons. This factor could be related to the myelinated nature of B preganglionic fibers. An alternative hypothesis cannot be discarded, namely, that C fibers have a more limited intrinsic sprouting power. The limited possibilities of each preganglionic fiber in forming synapses in mammalian SCG were first suggested by Östberg *et al.* (1976), and Purves and Litchman (1978) added that different types of nerve fibers probably have different sprouting potential.

Another aspect is the distribution of subsynaptic bars between the two types of neurons. The two types of sections gave convergent results, suggesting that the subsynaptic bars are preferentially, although not exclusively, located in B neurons. Thus, it does not seem possible to relate the presence of this organelle with a specific function on B or C neurons. This conclusion is exactly the opposite of that reached by Watanabe and Burnstock (1978) in another frog species, *L. dumerili* (see Section II,B,2). The discrepancy seems difficult to explain convincingly by species differences. It seems more likely that the unique criterion of neuronal section size used by Watanabe and Burnstock for the discrimination between B and C neurons is inadequate.

## D. Comparison between Mammals and Frog

Experiments involving deafferentation and reinnervation by preganglionic fibers, as carried out in mammals and the frog, lead to the same conclusions: in sympathetic ganglia, the reestablishment of synaptic connections is selective and there is a competition between the different types of preganglionic fibers to reestablish normal connections. In the frog, however, because the synaptic connections are clearly organized in two different and separated pathways, selectivity and competition are certainly higher. Nevertheless, the existence of several functional types of ganglionic neurons in mammals, according to the nature of the neuropeptide(s) they contain, has been established (reviewed by Schultzberg *et al.,* 1983, and Smolen, 1988), but the eventual incidence of these types on the specificity of synaptic regeneration is not known. In the two classes of vertebrates, experiments involving partial deafferentation also show that, according to the types of neurons, recognition mechanisms between preganglionic axons and sympathetic neurons can present a certain degree of plasticity, because preganglionic fibers innervating neurons of one type can form synapses by sprouting onto neurons of another type with various degrees of efficiency.

## V. Reinnervation of Ganglia with Foreign (Nonpreganglionic) Nerve Fibers

Replacing a nerve by another and studying the functional effects of this substitution is an old idea (for history, see Langley and Anderson, 1904). One interesting aspect of this type of experiment, called heterogeneous reinnervation, is to test the plasticity or, *a contrario,* the specificity of interneuronal connections, which may be different according to the type of transmitter or environmental factors. The possible role of many of these was not even suspected when the first experiments of this type were performed.

### A. Mammals

Attempts to replace the preganglionic fibers by nerve fibers of another origin began with the contributions of Langley (1898) and Langley and Anderson (1904). Different connection pathways of the central vagus ending with the SCG in young cats were studied by Langley, the vagus nerve being cut beyond the nodose ganglion. The effects were examined 40 to 120 days later. All experiments gave similar results, that is, a response of the main target organs of the SCG (nictitating membrane, pupil, hairs, etc.) on stimulation of the vagus nerve. Local application of nicotine to the SCG abolished the responses. Thus, although histological controls were not made, Langley concluded that the vagus nerve formed synapses with the ganglionic neurons similar to those normally formed by preganglionic fibers.

This type of experiment was revised and extended by de Castro (1934, 1937), who replaced the preganglionic nerve fibers by those of the vagus or the hypoglossal nerves. Using the same physiological and pharmacological tests as Langley, de Castro concluded that heterologous ganglionic synapses were created either with autonomic motor nerve fibers of the vagus (isomorphous synapses) or with somatic fibers of the hypoglossal nerve (heteromorphous synapses). de Castro added a detailed morphological study of the regenerated fibers in the ganglion by silver impregnation, especially their mode of branching and the distribution of the "boutons terminaux" (de Castro, 1934). In the case of the hypoglossal nerve, de Castro noticed that the fibers were of a larger diameter, often myelinated, and their diameter remained larger than that of normal preganglionic fibers up to the terminal ramifications.

Ceccarelli (1968) and Ceccarelli *et al.* (1971) performed an electron microscopic study of the new synapses formed in the cat SCG after suturing the vagus nerve, cut beyond the nodose ganglion, to the preganglionic trunk cut 2 cm caudally. The synapses were similar in structure and size of those of the normal ganglia. It was concluded that the vagal fibers, which normally give rise to large motor endings, were also functional in small axodendritic areas. However, it should be noticed that vagus nerve fibers are not all motor, and some of them are preganglionic fibers normally ending on intramural neurons of the gut. The same type of experiment was performed by Taxi and Babmindra (1972), who sutured the central end of the phrenic nerve, a purely motor nerve, to the posterior pole of the SCG in the dog. The general aspect of nerve terminals after silver impregnation was not different from that of normal ganglia, and large fibers and endings were not found in the vicinity of perikarya. As a rule, the newly formed synapses were axodendritic and their ultrastructure was similar to that of normal ganglia. In one operated animal, fibers and presynaptic endings containing synaptic vesicles with a dense core (i.e, vesicles of the noradrenergic type) were found, suggesting sprouting of sympathetic axons.

In 1974, McLachlan was the first to produce electrophysiological evidence of transmission by intracellular recordings of sympathetic neurons reinnervated by nonpreganglionic axons. Following anastomosis of the guinea pig SCG to the sternohyoid muscle nerve, ganglionic transmission was detected after 2 months in 13 of 18 animals operated, but small EPSPs were evoked only in about 5% of impaled neurons. Moreover, only six typical synaptic complexes were found by electron microscopy. Therefore, it is obvious that in this case, the neoformation of synapses was exceptional.

A thorough comparative electron microscopy study of the reinnervation of the rat SCG by preganglionic or somatic fibers was carried out by Östberg *et al.* (1976), who anastomosed the central end of the SCG with the vagus or the hypoglossal nerve at midcervical level and compared the number of synapses in ganglia reinnervated with these nerves with that of ganglia preganglionically reinnervated. The vagus nerve was able to give rise to about half the normal number of synapses observed in guinea pig SCG reinnervated by preganglionic fibers (Purves, 1976b). Among those synapses originating from vagus nerve fibers, Östberg *et al.* (1976) distinguished two types (Figs. 9 and 10): in one type, the presynaptic endings contained a large number of LDVs mixed with synaptic vesicles; they were assumed to belong to preganglionic axons of the vagus nerve. In the second type, a smaller population of presynaptic endings contained

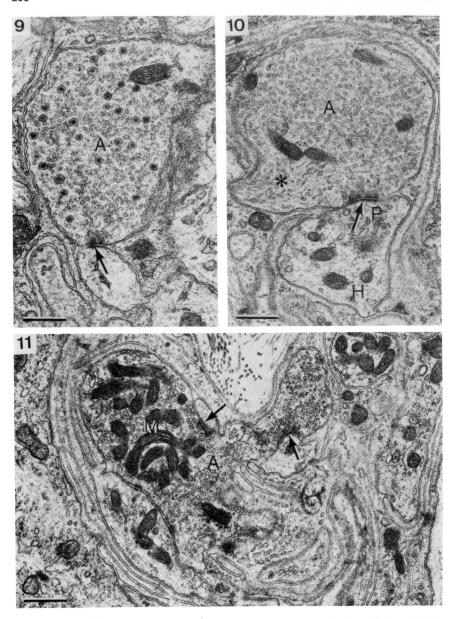

FIGS. 9–11   Figures 9 and 10 were obtained from rat superior cervical ganglia in which the vagus nerve was substituted to the preganglionic trunk for reinnervation, whereas in Fig. 11 the hypoglossal nerve was substituted to the preganglionic trunk. Bars: 0.5 μm. [From Östberg, A. J. C., Raisman, G., Field, P. M., Iversen, L. L., and Zigmond, R. E. (1976). *Brain Res.* **107**, 445–470. Reprinted with permission.]

almost exclusively synaptic vesicles. A progressive increase in the number of synapses was noted for up to 100 postoperative days. With the hypoglossal nerve, less than 20% of the normal number of synapses and a limited number of large synaptic terminals reappeared (Fig. 11), as already indicated by de Castro (1937) and Hillarp (1946) on the basis of silver impregnation. Östberg et al. (1976) hypotesized that these large synapses might be due to a special group of neurons of the hypoglossal nucleus, having some properties in common with the neurons of the dorsal vagal nucleus, but considered that the bulk of the hypoglossal neurons are unable to form synapses with ganglionic neurons. It is possible that the small number of synapses would be located only on the presumably small number of cholinergic neurons present in the SCG, although the presence of such neurons was well documented only in the cat SCG (Hamberger et al., 1965). The fact that even after long survival periods, VPDs could still be observed, indicates that the synaptogenic power of regenerating fibers is limited and suggests that VPDs do not play a decisive role in the reestablishment of synapses. As a general conclusion, these authors consider that most motor skeleton axons (except some special ones), sensory axons, and postganglionic axons are unable to form ganglionic synapses. This is in total contradiction with the previous results of de Castro (1937, 1942) and Hillarp (1946).

Competitive reinnervation in the rat SCG by vagus and hypoglossal nerves was also studied physiologically and electron microscopically by Östberg and Vrbová (1982). They noted that both nerves reinnervated preferentially a population of neurons located at the cranial pole of the ganglion and innervating the nictitating membrane. But the proportion of newly formed synapses did not exceed one-fifth of the normal value in the case of reinnervation by the hypoglossal nerve. They also showed

---

FIG. 9   A ganglionic synapse of the commoner type, presumably formed with a vagal preganglionic axon, 117 days after vagal anastomosis. The synapse is formed by an axon terminal (A) containing synaptic vesicles and numerous large dense-cored vesicles, associated with a small dendritic spine; the synaptic complex is pointed out by the arrow.

FIG. 10   An axodendritic synapse of a less common type, found after vagal anastomosis (112-day survival). The axon terminal (A) is crowded with synaptic vesicles and some tubules (asterisk). The arrow indicates the synaptic complex, on a protrusion (P) of the dendritic shaft (H).

FIG. 11   A ganglionic synapse of the type characteristic of the hypoglossal nerve, 87 days after the anastomosis. The large, highly irregular axon terminal (A) has a central cluster of mitochondria (M), numerous synaptic vesicles, and practically no large dense-cored vesicles; it makes two synaptic contacts with two spinelike profiles, whose arrows indicate the synaptic complexes.

that the vagus nerve is much more efficient in reinnervation than the hypoglossal nerve, the reinnervation efficacy being measured by the contraction of the nictitating membrane. When both nerves were connected to the SCG, only reinnervation by the vagus nerve could be detected. The results reported by Purves (1976b) in the guinea pig SCG were fairly different. In the competition between preganglionic and vagal fibers, 25% of the SCG neurons were reinnervated by preganglionic fibers, 25% by vagal fibers, and 50% by both these nerves. This double reinnervation was stable, because it was still present 1 year later. Thus, Purves (1976b) concluded that reinnervation by both preganglionic fibers and vagus nerve have an additive effect more than a competitive effect. Concerning double reinnervation, unexpected results were reported after chronic application of GABA (Wolff *et al.*, 1978) or sodium bromide (Joó *et al.*, 1980): in the intact rat SCG, both substances were shown to increase the number of VPDs. In addition, when the hypoglossal or vagus nerve was surgically implanted onto this ganglion, foreign axons developed functional synapses with sympathetic neurons (Dames *et al.*, 1985; Toldi *et al.*, 1986). The effect of GABA and sodium bromide is probably due to a decrease in cholinergic transmission (Kása *et al.*, 1987).

The most ambitious experiments and impressive results in heterogeneous reinnervation of the SCG were obtained by de Castro (1942) with the reinnervation of the cat SCG by sensory fibers growing from the central pole of the nodose ganglion after appropriate sectioning of the vagus nerve, a type of experiment in which neither Langley and Anderson (1904) nor their predecessors succeeded. The functional reinnervation was controlled by ocular, pilomotor, or vascular reactions, and complementary operations were carried out some days before sacrificing the operated animals, in order to avoid interference of other nerves in regeneration of ganglionic synapses. Thus, 45–60 days after surgey, more or less intense signs of recovery were observed in about 30% of the operated animals under stimulation of the vagal sensory pathways, the activity of SCG being coupled, for instance, with the ingestion of solid food. Histologically, the regenerated fibers exhibited some peculiar features, such as a large diameter and a special mode of ramification.

De Castro did not observe any potentiation of the ganglionic transmission by eserine and concluded that it was not cholinergic. However more complete pharmacological controls of the nature of this ganglionic transmission were revised by Matsumura and Koelle (1961). They obtained reinnervation of the SCG with sensory fibers in 17 of 44 operated cats. Selective blockade of ganglionic transmission by tetraethylammonium and potentiation by eserine led them to the conclusion that the transmission was cholinergic. On the other hand, in such reinnervated SCG, CAT and AChE activities were higher than in preganglionically denervated ganglia, although lower than in normally innervated ganglia (Fujiwara and Kura-hashi, 1976; Fujiwara et al., 1978), whereas the ACh content was also higher (Tsubomura et al., 1988). Among several hypotheses concerning the origin of this cholinergic innervation, the most attractive is that a small population of the neurons of the nodose ganglion is cholinergic. This interpretation received support from Ternaux et al., (1989), who demon-strated CAT activity in the nodose ganglion of several mammals, the maximum activity having been measured in the cat and an immunocyto-chemical control made in the rabbit (see also Fujiwara et al., 1978). More-over, cholinergic neurons participating in the SCG reinnervation might also be present through the entire length of the vagus nerve (Okamoto et al., 1988). Acetylcholine might coexist with substance P in those neurons, although Hayashi et al. (1983) reported the same type of experiments in cat SCG, and showed that regenerating fibers immunopositive to anti-substance P pass through the ganglion without forming synapses. What-ever the real nature and mechanism of the transmission, it seems well established that fibers originating from sensory ganglia are able to form synapses with sympathetic neurons in mammals.

Another type of heterogeneous reinnervation was reported by Ceccarelli et al. (1974), who obtained 50% success in reinnervating cat left SCG by postganglionic fibers of the right SCG, the reinnervation being controlled by physiological as well as morphological tests. As the physiological re-sponses were abolished by hexamethonium and electron microscopy re-vealed that endings contained normal (clear) synaptic vesicles and LDVs, they concluded that reinnervation was realized by cholinergic neurons from the right SCG (see Hamberger et al., 1965).


## B. Frog

As far as we know, there is only one study in the frog of reinnervation of autonomic neurons by somatic axons, but it concerns parasympathetic neurons. This study was performed by Proctor et al. (1982) in deafferented cardiac ganglia by suturing the hypoglossus nerve. They found that about 70% of the neurons were reinnervated and that the reinnervation lasted as long as 60 weeks. However, in response to stimulation of the hypoglossus nerve, the EPSP amplitude was much smaller than that induced by the native nerve. Well-defined synapses revealed by electron microscopy were formed between hypoglossal terminals and axons of parasympathetic neu-rons, whereas the perikaryal surface remained devoid of terminals. More-over, inappropriate synapses were totally eliminated when the normal preganglionic innervation regenerated into the ganglia (Proctor and Roper, 1982).

We have carried out such experiments with the sympathetic ganglia (Eugène and Taxi, 1993, 1994). In a first group of experiments (31 frogs), after sectioning the sympathetic chain between the 7th and the 8th ganglion, the posterior stump of the chain was anastomosed with the central stump of the cut 6th spinal motor nerve, which is slightly larger in diameter than the sympathetic chain. The reinnervation by regenerating preganglionic fibers was tentatively prevented by excision of rami communicantes and of the sympathetic chain to a level right before the 6th ganglion. However, because of the difficulty in preventing the regeneration of preganglionic fibers in this way, a complementary set of experiments (11 frogs) was carried out, in which the cut sympathetic chain was directly inserted in place of a 1-cm segment of the 8th root of the lumbar plexus, taking care not to injure the rami communicantes of the 8th ganglion. In this way, a large excess of motor fibers was available for growing within the sympathetic chain and the reinnervation by preganglionic fibers was prevented by suturing the central stump of the chain to the lung. However, in both procedures, the ganglia of about 80% of the operated frogs were not reinnervated or were reinnervated only with regenerating preganglionic axons. For the other frogs, the ganglia were functionally reinnervated with somatic axons. However, many neurons were not reinnervated at all and some ganglia were further reinnervated with preganglionic axons, certain neurons thus receiving double reinnervation. In all these experiments, functional reinnervation was checked by intracellular recordings of synaptic activity in response to electrical stimulation of the regenerating nerves. In both B and C neurons, somatic reinnervation was found only 3 months after anastomosis (Fig. 12) whereas preganglionic reinnervation could be observed starting at 25 postoperative days.

In those ganglia reinnervated only with somatic axons, the synaptic index was far from its normal value as well as from its value in ganglia reinnervated with regenerating preganglionic fibers, even when regeneration was counteracted as described above. As for the VPD index, it is fairly close to that of denervated ganglia. This is in full agreement with electrophysiological data showing poor reinnervation. In addition to normal synapses, some interesting observations were made: in Fig.13, three VPDs are visible close to a simple contact of a preganglionic fiber with the same neuron. This suggested that VPDs did not exert an attraction on the somatic terminals, and correlated well with the almost unchanged VPD index after somatic regeneration. Thus, somatic endings do not seem necessarily to use VPDs to make synapses, but seem to be able to induce postsynaptic differentiations. The heterogeneous synapses had the usual characteristics, with some exhibiting a subsynaptic bar, as in Fig. 14. It was also noticeable that, although no quantitative study was carried out, LDVs could be present, in relatively large numbers, in presynaptic areas

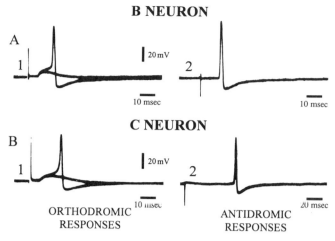

**B NEURON**

A

1        | 20 mV        2

—
10 msec                              —
                                    10 msec

**C NEURON**

B

1        | 20 mV        2

—
10 msec                              —
                                    20 msec

ORTHODROMIC                    ANTIDROMIC
RESPONSES                      RESPONSES

FIG. 12 Orthodromic and antidromic electrical responses of frog B and C sympathetic neurons from a deafferented abdominal ganglion reinnervated with somatic axons of the 8th spinal root of the lumbosacral plexus, 90 days after operation. The orthodromic responses ($A_1$ and $B_1$) were obtained after stimulations of the spinal root and had identical latencies, suggesting that the B and C neurons were reinnervated by similar somatic axons. Moreover, the subthreshold excitatory postsynaptic potentials evoked by juxtathreshold stimulations indicate that the heterogeneous reinnervation was polyaxonal. The antidromic responses ($A_2$ and $B_2$) were provoked by stimulations of the sciatic nerve and recorded from the same neurons, in order to identify the type of reinnervated neurons: in myelinated B neurons, antidromic conduction velocities were more than 2 m sec$^{-1}$, whereas in unmyelinated C neurons, they were less than 1 m sec$^{-1}$.

regenerated by motor nerve fibers, whereas they are rare in endings of neuromuscular junctions. This would indicate a retrograde transsynaptic regulation of LDVs.

In the ganglia doubly reinnervated with both somatic and preganglionic axons, the values of both morphological indices exhibited large variations, and only for one (the 8th ganglion) was the synaptic index close to normal. Normal synapses coexisted with VPDs, although never observed on the same neuronal profile. Unfortunately, morphology per se does not allow the identification of the origin of nerve endings. Dark endings, generally presumed to be in the course of degeneration, were observed in certain animals (Fig. 15). We considered them to be somatic fibers that were in competition with preganglionic fibers and then were pushed aside and underwent a remodeling or definitive degeneration process. However, nothing else is known about the end of this process.

Another type of experiment was performed in order to obtain the addition of two effects presumably favorable to the formation of new synapses (Eugène and Taxi, 1994). For the reinnervation of the ganglion, no regener-

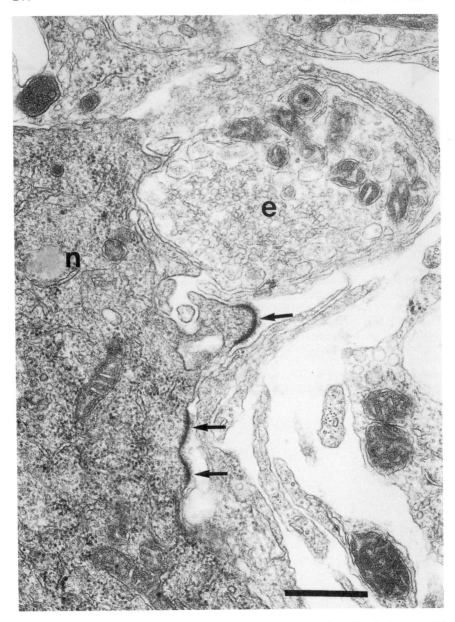

FIG. 13  Deafferented frog abdominal ganglion reinervated with the 6th spinal nerve, 124 days after operation. A (presumably) motor nerve ending (e) forms a "simple contact" with a sympathetic neuron (n), whereas three vacated postsynaptic membrane thickenings (arrows) near the ending are not reoccupied. Bar: 0.5 μm.

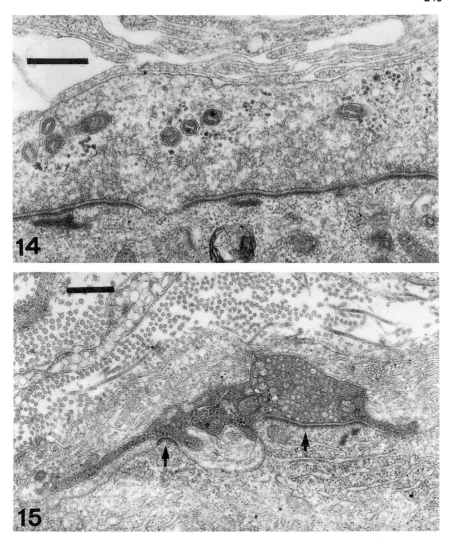

FIGS. 14 AND 15   Deafferented frog abdominal ganglia reinnervated with axons of the 8th spinal root of the lumbosacral plexus.

FIG. 14   Ninety days after operation. A large terminal bears three synaptic complexes, two of which are provided with a subsynaptic bar. Only one large dense-cored vesicle is visible on the left upper part of the presynaptic area. Bar: 0.5 μm.

FIG. 15   Ninety one days after operation. A (presumably) motor ending, having formed a synapse with two postsynaptic differentiations (arrows), is in the course of "dense degeneration." Bar: 0.5 μm.

ating somatic axons but axonal sprouts of motor nerves were used. Following the protocols reported by Diaz and Pécot-Dechavassine (1990), sprouting was induced by inserting a degenerating piece of nerve of the same animal in the vicinity of a motor nerve. After a sufficient delay for obtaining significant sprouting (2 months), the 8th and 9th ganglia were removed and grafted in the vicinity of the sprouting nerve. Such ganglia are, of course, axotomized, a circumstance that might be favorable for the formation of new synapses, if the results observed in mammals (Ramsay and Matthews, 1985; Matthews and Clowry, 1987) are applicable to amphibians. Electrophysiological recordings and morphological observations did not reveal any synapses, although numerous neurons were in good health, as judged by their morphology, and vesicle-containing fibers were seen inserted between the glial sheath and the perikaryon. The origin of these fibers can obviously not be ascertained. However, it seems reasonable to consider them of somatic origin, because no sprouting capacity of ganglionic neurons was observed in the denervated frog ganglia, even after several months, at variance with what is known for mammals (Purves, 1976b; Ramsay and Matthews, 1985).

Last, we have tested the eventual reinnervation of ganglia with postganglionic axons. In some animals, the rami communicantes of the 9th ganglion were cut at their entrance to the root of the lumbosacral plexus and joined with the peripheral stump of the left chain divided just before the 8th ganglion. In three animals examined after several months by electron microscopy, we were unable to find any synapse.

## C. Comparison between Mammals and Frog

Although amphibians are generally considered to be more capable of regenerating processes than mammals, it is difficult and even exceptional to obtain the reestablishment of sympathetic ganglionic synapses from somatic motor axons. These results were completely different from those obtained in parasympathetic cardiac ganglia by Proctor et al. (1982), who succeeded in 100% of the experimented animals, although the EPSPs were large enough to generate action potentials in only some neurons. In contrast with this, we have recorded a small number of reinnervated sympathetic neurons, but practically all these neurons generated action potentials in response to supramaximal presynaptic stimulation (Fig. 12; Eugène and Taxi, 1994). In mammals, several authors (see section V,A) were able to substitute motor or even sensory fibers for preganglionic fibers and to obtain "heterogeneous" ganglionic synapses with a reasonable rate of success. However, it is clear that not all abnormal connections are possible in mammals and that certain nerves are more efficient than others

(Östberg and Vrbovà, 1982). Moreover, the postganglionic fibers of frog abdominal ganglia are not able to reestablish synapses with sympathetic neurons of the other chain, contrary to results obtained by Ceccarelli *et al.* (1974) in cat SCG. This probably means that the frog sympathetic ganglia are devoid of cholinergic neurons, which would exist in small numbers in mammals.

All these data suggest a higher degree of connection specificity between preganglionic fibers and sympathetic neurons in amphibians than in mammals which reinforces the conclusion drawn from the experiments of deafferentation and reinnervation by preganglionic fibers (see Section IV,D). However, other properties of nerve cells may also be involved. For instance, Purves and Lichtman (1978) noticed that different sorts of foreign cholinergic axons make a different number of synapses with target sympathetic neurons. Thus, the poor performance of motor axons in reinnervating SCG or frog ganglia might be due to an intrinsic inability to produce a large number of synaptic boutons.

Several difficulties, especially acute in the frog, can be imagined in the process of synapse reestablishment and could explain why the number of synapses after regeneration is generally much lower than the normal value, with neurons remaining nonreinnervated. First, in the frog, a foreign nerve fiber must find its way to the specific neuron it must innervate. This situation is completely different from that in mammals, in which each fiber makes contacts with several nerve cells through the endings of its terminal arborization. In the frog, the pathway to a neuron is formed by the Schwann cell sheath of the degenerating preganglionic fibers, which ends in the glial apparatus surrounding the unique process of the neuron. Perhaps only a limited number of regenerating foreign fibers enter a preexisting Schwann sheath, whereas others are not directed to a neuron because they regenerate their own Schwann cell sheath, as preganglionic fibers are able to do, when they join the sympathetic trunk after ablation of a fairly long segment.

Moreover, even if a foreign fiber finds its way to a neuron, it may find itself having no "affinity" for the neuronal surface because its surface molecules are not able to form bonds in sufficient number with those of the target neuron, the set of molecules responsible for the special adhesive properties of the synaptic areas being perhaps different in preganglionic and motor neurons.

Another fact may seem puzzling at first sight: whereas the reinnervation with somatic fibers seems relatively easy in mammals (see above), we obtained it only in 7 of 42 frogs, among which 3 received both somatic and preganglionic axons. However, intracellular postsynaptic responses were higher in frog reinnervated neurons than in the guinea pig, indicating better synaptic transmission in the frog. An explanation of this apparent paradox could lie in the polyneuronal type of reinnervation detected in the frog (see Fig. 12) and reduced in the guinea pig (Purves, 1976b), and also in the type of synaptic organization: all the synaptic boutons of one reinnervating fiber are concentrated on the same neuron in the frog, the reinnervation being of an "all or nothing" type. On the other hand, in the guinea pig, each sympathetic neuron receives one or a few boutons from reinnervating fibers and the synaptic deficit in reinnervation is distributed on the dendrites of all neurons, each remaining far from its normal rate of innervation.

## VI. General Conclusion

This comparative study of the effects of different experimental situations on synapses in sympathetic ganglia was limited in fact to frog versus some laboratory mammals, primarily because of the scarcity of studies concerning other vertebrates, mainly for practical reasons. For instance, it would be very interesting to compare the results of the same type of experiments in urodelan and anuran amphibians, which represent different levels of evolution, because the anurans are a highly specialized group more adapted to terrestrian life. But sympathetic ganglia in urodelans have not been studied, probably because they are small and not easy to operate on.

From all the data reported and discussed here, the following may be derived.

The consequences of axotomy on synaptic transmission and structure are similar in anurans and mammals. In both groups, the bulk of synapses are disrupted, with loss of the postsynaptic differentiations of the synaptic complexes. This is the result of a dramatic reorientation of protein metabolism in neurons, the priority being given to structural molecules of the regenerating axon over proteins involved in transmission and neuronal shape, for instance. However, usually a certain proportion of synapses remains intact. It is one of the aspects of the variability of the responses of neurons to axotomy, which can affect all the parameters of the response, for reasons that are not yet understood. Moreover, in the frog, there is a good amount of evidence that transmission can occur between preganglionic fibers and the neuronal surface at "simple contacts" devoid of the morphological differentiations of synaptic complexes. Such a possibility has not yet been considered in mammalian ganglia.

After deafferentation of the sympathetic ganglia in frogs, as in mammals, the postsynaptic differentiations of the synaptic complexes persist in about one-third of the normal synapses. Although divergent opinions have been expressed on the interpretation of this observation, it seems likely that two kinds of postsynaptic differentiations can be distinguished, according to their stability after deafferentation: the labile and the stable differentiations. This might eventually be related to cyclic or random changes intervening in the distribution of synapses, as has been postulated from *in vivo* observations in parasympathetic ganglia (Purves and Lichtman, 1987). In the frog, it seems that differentiations provided with a subsynaptic bar are especially stable.

The regeneration of preganglionic fibers after sectioning leads to a recovery of ganglionic transmission, but the number of reestablished synapses is generally less than the normal number. All the vacated postsynaptic differentiations were reused for new synapses in mammals, whereas a number of them were not reoccupied in the frog, probably as a consequence of the more rigid organization of the relationships between preganglionic fibers and neurons, because each neuron generally receives its synapses from only one preganglionic fiber in the frog, whereas mammalian neurons are always polyinnervated. Furthermore, from experiments involving both deafferentation and axotomy, it appears that adult ganglionic neurons are able to differentiate new synaptic specializations during the regeneration processes in both mammals and the frog.

In the case of chronic deafferentation of ganglia, the evolution is different in mammals and the frog. The formation of intrinsic (adrenergic) synapses, presumably between axon collaterals and sympathetic neurons, was observed in mammals, but never in the frog. This suggests a higher plasticity of adult neurons in mammalian ganglia, although it is not known at present whether these synapses are functional or not, and whether adrenergic receptors are present at these synapses.

The possibility to replace preganglionic fibers by nerve fibers of another origin, for instance somatic motor nerve fibers, is clearly better in mammals than in the frog, in which this operation is rarely successful. In all cases, whatever the origin of the fibers, it seems that the heterogeneous synapses formed are cholinergic. Once more, these data show that the plasticity of synaptic connections is higher in mammals than in the frog. This lack of plasticity of neuronal connections in the frog is also underlined by the limitations observed in the substitution of preganglionic C fibers for B fibers after selective sectioning.

What could be the evolutionary significance of this relative lack of plasticity in the organization of frog sympathetic ganglia, compared to that of mammals? It is well known that the structure of frog sympathetic ganglia has "primitive" features, as they are built in the same way as many invertebrate ganglia (perikarya in the periphery, nerve fibers in the center), but synapses are axosomatic and not in a neuropil; in mammals, the perikarya and neuropil are mixed. Another primitive characteristic may be the specificity of different neuronal types with regard to the preganglionic fibers, whereas clear-cut functional types of neurons were never established in the paravertebral ganglia of mammals. However, it cannot be discarded that the rigid organization of frog ganglia may be related to the high specialization of anurans. Data concerning the organization of ganglia in urodelans could permit a better understanding of the situation in anurans. In contrast to the frog, the relative plasticity in mammals, as well as the increase in cell number in their ganglia and the polyinnervation of each of their neurons, appear to be evolutionary gains of potentials for adaptation.

## References

Acheson, G. H., and Remolina, J. (1955). The temporal course of the effects of post-ganglionic axotomy on the inferior mesenteric ganglion of the cat. *J. Physiol. (London)* **127**, 603–616.

Acheson, G. H., and Schwarzacher, H. G. (1956). Correlations between the physiological changes and the morphological changes resulting from axotomy in the inferior mesenteric ganglion of the cat. *J. Comp. Neurol.* **106**, 247–265.

Adams, P. R., Jones, S. W., Pennefather, P., Brown, D. A., Koch, C., and Lancaster, B. (1986). Slow synaptic transmission in frog sympathetic ganglia. *J. Exp. Biol.* **124**, 259–285.

Alkadhi, K. A., and McIsaac, R. J. (1973). Non-nicotinic transmission during ganglionic block with chlorisondamine and nicotine. *Eur. J. Pharmacol.* **24**, 78–85.

Ashe, J. H., and Libet, B. (1981). Orthodromic production of non-cholinergic slow depolarizing response in the superior cervical ganglion of the rabbit. *J. Physiol. (London)* **320**, 333–346.

Barr, M. L. (1940). Axon reaction in motoneurons and its effect on the end bulbs of Held-Auerbach. *Anat. Rec.* **77**, 367–374.

Barron, K. D. (1983). Comparative observations on the cytologic reactions of central and peripheral nerve cells to axotomy. *In* "Spinal Cord Reconstruction" (C. C. Kao, R. P. Bunge, and P. J. Reier, eds.), pp. 7–40. Raven Press, New York.

Bianchine, J. R., Levy, L., and Farah, A. E. (1964). Chromatolysis in the superior cervical ganglion of the cat. *Exp. Mol. Pathol.* **3**, 128–140.

Black, I. B., Hendry, I. A., and Iversen, L. L. (1971). Transynaptic regulation of growth and development of adrenergic neurons in mouse sympathetic ganglion. *Brain Res.* **34**, 229–240.

Black, I. B., Hendry, I. A., and Iversen, L. L. (1972). The role of post-synaptic neurons in the biochemical maturation of presynaptic cholinergic nerve terminals in a mouse sympathetic ganglion. *J. Physiol. (London)* **221**, 149–159.

Black, I. B., Chikaraishi, D. M., and Lewis, E. J. (1985). Trans-synaptic increase in RNA coding for tyrosine hydroxylase in a rat sympathetic ganglion. *Brain Res.* **339**, 151–153.

Blinzinger, K., and Kreutzberg, G. (1968). Displacement of synaptic terminals from regenerating motoneurons by microglial cell. *Z. Zellforsch. Mikrosk. Anat.* **85**, 145–157.

Bodian, D., and Mellors, R. C. (1945). The regenerative cycle of motoneurons with special reference to phosphatase activity. *J. Exp. Med.* **81**, 469–488.

Boyle, F., and Gillespie, J. S. (1970). Accumulation and loss of noradrenaline central to a constriction on adrenergic nerves. *Eur. J. Pharmacol.* **12**, 77–84.

Branton, W. D., Phillips, H. S., and Jan, Y. N. (1986). The LHRH family of peptide messengers in the frog nervous system. *Prog. Brain Res.* **68**, 205–215.

Brown, D. A. (1969). Responses of normal and denervated cat superior cervical ganglion to some stimulant compounds. *J. Physiol. (London)* **201**, 225–236.

Brown, G. L., and Pascoe, J. E. (1954). The effect of degenerative section of ganglionic axons on transmission through the ganglion. *J. Physiol. (London)* **123**, 565–573.

Brzin, M., Tennyson, V. M., and Duffy, P. E. (1966). Acetylcholinesterase in frog sympathetic and dorsal root ganglia: A study by electron microscope cytochemistry and microgasometric analysis with the magnetic diver. *J. Cell Biol.* **31**, 215–242.

Cannon, W. B., and Rosenblueth, A. (1936). The sensitization of a sympathetic ganglion by preganglionic denervation. *Am. J. Physiol.* **116**, 408–413.

Cannon, W. B., and Rosenblueth, A. (1949). "The Supersensitivity of Denervated Structures: A Law of Denervation." Macmillan, New York.

Cassel, J. F., Clark, A. L., and McLachlan, E. M. (1986). Characteristics of phasic and tonic sympathetic ganglion cells of the guinea pig. *J. Physiol. (London)* **372**, 457–483.

Causey, G., and Barton, A. A. (1958). Synapses in the superior cervical ganglion and their changes under experimental conditions. *Exp. Cell Res., Suppl.* **5**, 338–346.

Causey, G., and Hoffman, H. (1956). The ultrastructure of the synaptic area in the superior cervical ganglion. *J. Anat.* **90**, 502–507.

Ceccarelli, B. (1968). La transmissione sinaptica nel ganglio cervicale superiore di gatto dopo reinnervazione vagale. *Atti Accad. Med. Lomb.* **23**, 1–18.

Ceccarelli, B., Clementi, F., and Mantegazza, P. (1971). Synaptic transmission in the superior cervical ganglion of the cat after reinnervation by vagus fibres. *J. Physiol. (London)* **216**, 87–98.

Ceccarelli, B., Clementi, F., Mantegazza, P., Perri, V., and Sacchi, O. (1974). Reinnervation of superior cervical ganglion of the cat by postganglionic sympathetic fibres. *Int. Congr. Electron Microsc. 8th, Canberra, 1974*, Vol. II, pp. 322–323.

Cheah, T. B., and Geffen, L. B. (1973). Effects of axonal injury on norepinephrine, tyrosine hydroxylase and monoamine oxidase levels in sympathetic ganglia. *J. Neurobiol.* **4**, 443–452.

Chen, S. S. (1971). Transmission in superior cervical ganglion of the dog after cholinergic suppression. *Am. J. Physiol.* **221**, 209–213.

Cole, A. E., and Shinnick-Gallager, P. (1980). Alpha-adrenoceptor and dopamine receptor antagonists do not block the slow inhibitory postsynaptic potential in sympathetic ganglia. *Brian Res.* **187**, 226–230.

Cole, A. E., and Shinnick-Gallager, P. (1984). Muscarinic inhibitory transmission in mammalian sympathetic ganglia mediated by increased potassium conductance. *Nature (London)* **307**, 270–271.

Couteaux, R. (1961). Principaux critères morphologiques et cytologiques utilisables aujourd'hui pour définir les divers types de synapses. *In* "Actualités Neurophysiologiques" (A. Monnier, ed.), Ser. 3, pp. 145–173. Masson, Paris.

Dail, N. G., and Evans, A. P. (1978). Ultrastructure of adrenergic terminals and SIF cells in the superior cervical ganglion of the rabbit. *Brain Res.* **148**, 469–477.

Dames, W., Joó, F., Fehér, O., Toldi, J., and Wolff, J. R. (1985). γ-Aminobutyric acid enables synaptogenesis in the intact superior cervical ganglion of the adult rat. *Neurosci. Lett.* **54**, 159–164.

de Castro, F. (1930). Recherches sur la dégénération et la régénération du système nerveux sympathique. Quelques observations sur la constitution des synapses dans les ganglions. *Trav. Lab. Rech. Biol. Univ. Madrid* **26**, 357–456.

de Castro, F. (1934). Note sur la régénération fonctionnelle hétérogénique dans les anstomoses des nerfs pneumogastrique et hypoglosse avec le sympathique cervical. *Trav. Lab. Rech. Biol. Univ. Madrid* **29**, 397–416.

de Castro, F. (1937). Sur la regénération fonctionnelle dans le sympathique (anastomoses croisées avec des nerfs de type iso- et hétéromorphes). Une référence spéciale sur la constitution des synapses. *Trav. Lab. Rech. Biol. Univ. Madrid* **31**, 271–345.

de Castro, F. (1942). Modelacion de un arco reflejo en el simpatico, uniendolo con la raiz aferente central del vago. Nuevas ideas sobra la sinapsis. *Trab. Inst. Cajal Invest. Biol.* **34**, 217–301.

de Castro, F. (1951). Aspects anatomiques de la transmission synaptique ganglionnaire chez les Mammifères. *Arch. Int. Physiol.* **59**, 479–513.

Diaz, J., and Pécot-Dechavassine, M. (1990). Nerve sprouting induced by a piece of peripheral nerve placed over a normally innervated frog muscle. *J. Physiol.* (*London*) **421**, 123–133.

Dodd, J., and Horn, J. P. (1983a). A reclassification of B and C neurones in the ninth and tenth paravertebral sympathetic ganglia of the bullfrog. *J. Physiol.* (*London*) **334**, 255–269.

Dodd, J., and Horn, J. P. (1983b). Muscarinic inhibition of sympathetic C neurones in the bullfrog. *J. Physiol.* (*London*) **334**, 271–291.

Dun, N. J., and Karczmar, A. G. (1980). A comparative study of the pharmacological properties of the positive potential recorded from the superior cervical ganglia of several species. *J. Pharmacol. Exp. Ther.* **215**, 455–460.

Dun, N., Nishi, S., and Karczmar, A. G. (1976a). Alteration in nicotinic and muscarinic responses of rabbit superior cervical ganglion cells after chronic preganglionic denervation. *Neuropharmacology* **15**, 211–218.

Dun, N., Nishi, S., and Karczmar, A. G. (1976b). Electrical properties of the membrane of denervated mammalian sympathetic ganglion cells. *Neuropharmacology* **15**, 219–223.

Dunant, Y., and Dolivo, M. (1967). Relations entre les potentiels synaptiques lents et l'excitabilité du ganglion sympathique chez le Rat. *J. Physiol.* (*Paris*) **59**, 281–294.

Dunn, P. M., and Marshall, L. M. (1985). Lack of nicotinic supersensitivity in frog sympathetic neurones following denervation. *J. Physiol.* (*London*) **363**, 211–225.

Eccles, J. C., Libet, B., and Young, R. R. (1958). The behaviour of chromatolysed motoneurons studied by intracellular recording. *J. Physiol.* (*London*) **143**, 11–40.

Eccles, R. M. (1952). Responses of isolated curarized sympathetic ganglia. *J. Physiol.* (*London*) **117**, 196–217.

Eccles, R. M., and Libet, B. (1961). Origin and blockade of the synaptic responses of curarized sympathetic ganglia. *J. Physiol.* (*London*) **157**, 484–503.

Elfvin, L. G. (1963). The ultrastructure of the superior cervical sympathetic ganglion of the cat. II. The structure of the preganglionic end fibers and the synapses as studied by serial sections. *J. Ultrastruct. Res.* **8**, 441–476.

Eränkö, O., and Härkönen, M. (1965). Monoamine-containing small cells in the superior cervical ganglion of the rat and an organ composed of them. *Acta Physiol. Scand.* **63**, 511–512.

Eugène, D. (1987). Fast non-cholinergic depolarizing postsynaptic potentials in neurons of rat superior cervical ganglia. *Neurosci. Lett.* **78**, 51–56.

Eugène, D., and Taxi, J. (1988). GABA-ergic innervation in rat isolated superior cervical ganglia (SCG) studied by combined electrophysiological and immunocytochemical techniques. *J. Physiol. (London)* **406**, 175P.

Eugène, D., and Taxi, J. (1990). Synapses ganglionnaires et transmission synaptique après axotomie des neurones sympathiques chez la Grenouille. *C. R. Séances Acad. Sci.* **310**, 599–606.

Eugène, D., and Taxi, J. (1991). Effects of axotomy on synaptic transmission and structure in frog sympathetic ganglia. *J. Neurocytol.* **20**, 404–419.

Eugène, D., and Taxi, J. (1993). Synaptic transmission and morphology in frog sympathetic ganglia anastomosed to the sixth spinal nerve. *J. Physiol. (London)* **473**, 135P.

Eugène, D., and Taxi, J. (1994). Reinnervation of frog sympathetic ganglia with somatic nerve fibers. *J. Auton. Nerv. Syst.* In press.

Fawcett, J. W. (1992). Intrinsic neuronal determinants of regeneration. *Trends Neurosci.* **15**, 5–8.

Feldman, D. H. (1988). Synaptic specificity in frog sympathetic ganglia during reinnervation, sprouting and embryonic development. *J. Neurosci.* **8**, 4367–4378.

Flett, D. L., and Bell, C. (1991). Topography of functional subpopulations of neurons in the superior cervical ganglion of the rat. *J. Anat.* **177**, 55–66.

Francini, F., and Urbani, F. (1973). Organization of efferent extraspinal sympathetic pathways in *Rana esculenta. Arch. Fisiol.* **70**, 149–165.

Fujiwara, M., and Kurahashi, K. (1976). Cholinergic nature of the primary afferent vagus synapsed in cross anastomosed superior cervical ganglia. *Life Sci.* **19**, 1175–1180.

Fujiwara, M., Kurahashi, K., Mizuno, N., and Nakamura, Y. (1978). Involvement of nicotinic and muscarinic receptors in synaptic transmission in cat superior cervical ganglion reinnervated by vagal primary afferent axons. *J. Pharmacol. Exp. Ther.* **205**, 77–90.

Gaupp, E. (1899). "Ecker's and Wiedersheim's Anatomie des Frosches." Vieweg, Braunschweig.

Gautron, J., and Gisiger, V. (1976). Effet de la section du nerf préganglionnaire sur la localisation de l'acétylcholinestérase du ganglion cervical supérieur du Rat. *C. R. Hebd. Séances Acad. Sci.* **283**, 1505–1507.

Geist, F. D. (1993). Chromatolysis of efferent neurons. *Arch. Neurol. Psychiatry* **29**, 88–103.

Gisiger, V., Vigny, M., Gautron, J., and Rieger, F. (1978). Acetylcholinesterase of rat sympathetic ganglion: Molecular forms, localization and effects of denervation. *J. Neurochem.* **30**, 501–516.

Gordon, T., Kelly, M. E. M., Sanders, E. J., Shapiro, J., and Smith, P. A. (1987). The effects of axotomy on bullfrog sympathetic neurones. *J. Physiol. (London)* **392**, 213–229.

Gray, E. G. (1963). Electron microscopy of presynaptic organelles of the spinal cord. *J. Anat.* **97**, 101–106.

Greif, K. F. (1986). Plasticity of expression of a synaptic vesicle antigen in adult rat superior cervical ganglion. *J. Neurosci.* **6**, 3628–3633.

Griffith, W. H., Hills, J. M., and Brown, D. A. (1988). Substance P-mediated membrane currents in voltage-clamped guinea pig inferior mesenteric ganglion cells. *Synapse* **2**, 432–441.

Gustafsson, B. (1979). Changes in motoneurone electrical properties following axotomy. *J. Physiol. (London)* **293**, 197–215.

Guth, L., and Bailey, C. J. (1961). Pupillary function after alteration of the preganglionic sympathetic innervation. *Exp. Neurol.* **3**, 325–332.

Guth, L., and Bernstein, J. J. (1961). Selectivity in the re-establishment of synapses in the superior cervical sympathetic ganglion of the cat. *Exp. Neurol.* **4**, 59–69.

Hamberger, B., Norberg, K.-A., and Sjöqvist, F. (1965). Correlated studies of monoamines and acetylcholinesterase in sympathetic ganglia, illustrating the distribution of adrenergic and cholinergic neurons. *In* "Pharmacology of Cholinergic and Adrenergic Transmission" (G. B. Koelle, M. V. Douglas, and A. Carlsson, eds.), pp. 41–54. Pergamon, Oxford.

Hamori, J., Lang, E., and Simon, L. (1968). Experimental degeneration of the preganglionic fibres in the superior cervical ganglion of the cat. *Z. Zellforsch. Mikrosk. Anat.* **90**, 37–52.

Härkönen, M. H. A. (1964). Carboxylic esterases, oxidative enzymes and catecholamines in the superior cervical ganglion of the rat and the effect of pre- and postganglionic nerve division. *Acta Physiol. Scand., Suppl.* **237**, 1–94.

Härkönen, M. H. A., and Kauffman, F. C. (1973). Metabolic alterations in the axotomized superior cervical ganglion of the rat. I. Energy metabolism. *Brain Res.* **65**, 127–139.

Härkönen, M. H. A., and Kauffman, F. C. (1974). Metabolic alterations in the axotomized superior cervical ganglion of the rat. II. The pentose phosphate pathway. *Brain Res.* **65**, 141–157.

Hayashi, H., Ohsumi, K., Fujiwara, M., and Mizuno, N. (1983). Do the substance P-ergic vagus afferents in cats reinnervate the deafferented superior cervical ganglion? *Brain Res.* **270**, 178–180.

Hendry, I. A. (1975a). The retrograde trans-synaptic control of the development of cholinergic terminals in sympathetic ganglia. *Brain Res.* **86**, 483–487.

Hendry, I. A. (1975b). The effects of axotomy on the development of the rat superior cervical ganglion. *Brain Res.* **90**, 235–244.

Henningsen, I., Liestøl, K., Mæhlen, J., and Njå, A. (1985). The selective innervation of guinea-pig superior cervical ganglion cells by sprouts from intact preganglionic axons. *J. Physiol. (London)* **358**, 239–253.

Hillarp, N. Å. (1946). Structure of the synapse and the peripheral innervation apparatus of the autonomic nervous system. *Acta Anat., Suppl.* **IV**, 1–153.

Hökfelt, T. (1991). Neuropeptides in perspective: The last ten years. *Neuron* **7**, 867–879.

Hökfelt, T., Elfvin, L. G., Schultzberg, M., Fuxe, K., Said, S. I., Mutt, V., and Goldstein, M. (1977a). Immunohistochemical evidence of vasoactive intestinal polypeptide-containing neurons and nerve fibers in sympathetic ganglia. *Neuroscience* **2**, 885–896.

Hökfelt, T., Elfvin, L. G., Schultzberg, M., Goldstein, M., and Nilsson, G. (1977b). On the occurrence of substance P-containing fibers in sympathetic ganglia: Immunohistochemical evidence. *Brain Res.* **132**, 29–41.

Holtzmann, E., Novikoff, A. B., and Villaverde, H. (1967). Lysosomes and GERL in normal and chromatolytic neurons of the rat ganglion nodosum. *J. Cell Biol.* **33**, 419–436.

Honma, S. (1970). Functional differentiation in sB and sC neurons of toad sympathetic ganglia. *Jpn. J. Physiol.* **20**, 281–295.

Horn, J. P. (1992). The integrative role of synaptic cotransmission in the bullfrog vasomotor C system: Evidence for a synaptic gain hypothesis. *Can. J. Physiol. Pharmacol.* **70**, S19–S26.

Horn, J. P., and Stofer, W. D. (1988). Spinal origins of preganglionic B and C neurons that innervate paravertebral sympathetic ganglia nine and ten of the bullfrog. *J. Comp. Neurol.* **268**, 71–83.

Horn, J. P., and Stofer, W. D. (1989). Preganglionic and sensory origins of calcitonin gene-related peptide-like and substance P-like immunoreactivities in bullfrog sympathetic ganglia. *J. Neurosci.* **9**, 2543–2561.

Horn, J. P., Stofer, W. D., and Fatherazi, S. (1987). Neuropeptide Y-like immunoreactivity in bullfrog sympathetic ganglia is restricted to C cells. *J. Neurosci.* **7**, 1717–1727.

Horn, J. P., Fatherazi, S., and Stofer, W. D. (1988). Differential projections of B and C sympathetic axons in peripheral nerves of the bullfrog. *J. Comp. Neurol.* **278**, 570–580.

Huber, G. C. (1900). A contribution on the minute anatomy of the sympathetic ganglia of different classes of Vertebrates. *J. Morphol.* **16**, 27–90.

Hunt, C. C., and Nelson, P. G. (1965). Structural and functional changes in the frog sympathetic ganglion following cutting of the presynaptic nerve fibres. *J. Physiol.* (*London*) **177**, 1–20.

Hunt, C. C., and Riker, W. K. (1966). Properties of frog sympathetic neurons in normal ganglia and after axon section. *J. Neurophysiol.* **29**, 1096–1114.

Hyatt-Sachs, H., Schreiber, R. C., Bennett, T. A., and Zigmond, R. E. (1993). Phenotypic plasticity in adult sympathetic ganglia *in vivo:* Effects of deafferentation and axotomy on the expression of vasoactive intestinal peptide. *J. Neurosci.* **13**, 1642–1653.

Jacob, M. H. and Berg, D. K. (1987). Effects of preganglionic denervation and postganglionic axotomy on acetylcholine receptors in the chick ciliary ganglion. *J. Cell Biol.* **105**, 1847–1854.

Jacob, M. H., and Berg, D. K. (1988). The distribution of acetylcholine receptors in chick ciliary ganglion neurons following disruption of ganglionic connections. *J. Neurosci.* **8**, 3838–3849.

Jan, L. Y., and Jan, Y. N. (1982). Peptidergic transmission in sympathetic ganglia of the frog. *J. Physiol.* (*London*) **327**, 219–246.

Jan, L. Y., Jan, Y. N., and Brownfield, M. S. (1980). Peptidergic transmitters in synaptic boutons of sympathetic ganglia. *Nature* (*London*) **288**, 380–382.

Jan, Y. N., Jan, L. Y., and Kuffler, S. W. (1979). A peptide as a possible transmitter in sympathetic ganglia of the frog. *Proc. Natl. Acad. Sci. U.S.A.* **76**, 1501–1506.

Jan, Y. N., Jan, L. Y., and Kuffler, S. W. (1980). Further evidence for peptidergic transmission in sympathetic ganglia. *Proc. Natl. Acad Sci. U.S.A.* **77**, 5008–5012.

Jänig, W., and McLachlan, E. M. (1992). Characteristics of function-specific pathways in the sympathetic nervous system. *Trends Neurosci.* **15**, 475–481.

Jassar, B. S., Pennefather, P. S., and Smith, P. A. (1993). Changes in sodium and calcium channel activity following axotomy of B-cells in bullfrog sympathetic ganglion. *J. Physiol.* (*London*) **472**, 203–231.

Johnson, M. I., Ross, C. D., and Bunge, R. P. (1980). Morphological and biochemical studies on the development of cholinergic properties in cultured sympathetic neurons. II. Dependence on post-natal age. *J. Cell Biol.* **84**, 692–704.

Joó, F., Dames, W., and Wolff, L. R. (1980). Effects of prolonged sodium bromide administration on the fine structure of dendrites in the superior cervical ganglion of adult rat. *Prog. Brain Res.* **51**, 109–115.

Karczmar, A. G., Koketsu, K., and Nishi, S. (1986). "Autonomic and Enteric Ganglia. Transmission and Its Pharmacology." Plenum, New York and London.

Kása, P., Toldi, J., Farkas, Z., Joó, F., and Wolff, J. R. (1987). Inhibition by sodium bromide of acetylcholine release and synaptic transmission in the superior cervical ganglion of the rat. *Neurochem. Int.* **11**, 443–449.

Kása, P., Joó, F., Dobó, E., Wenthold, R. J., Ottersen, O. P., Storm-Mathisen, J., and Wolff, J. R. (1988). Heterogeneous distribution of GABA-immunoreactive nerve fibers and axon terminals in the superior cervical ganglion of adult rat. *Neuroscience* **26**, 635–644.

Kelly, M. E. M., Gordon, T., Shapiro, J., Smith, P. A. (1986). Axotomy affects calcium-sensitive potassium conductance in sympathetic neurones. *Neurosci. Lett.* **67**, 163–168.

Kelly, M. E. M., Bisby, M. A., and Lukoviak, K. (1988). Regeneration restores some of the altered electrical properties of axotomized bullfrog B-cells. *J. Neurobiol.* **19**, 357–372.

Kelly, M. E. M., Bulloch, A. G. M., Lukowiak, K., and Bisby, M. A. (1989a). Regeneration of frog sympathetic neurons is accompanied by sprouting and retraction of intraganglionic neurites. *Brain Res.* **477**, 363–368.

Kelly, M. E. M., Traynor, P., and Smith, P. A. (1989b). Amphibian sympathetic ganglia as a model system for investigating regeneration in the vertebrate peripheral nervous system. *Comp. Biochem. Physiol. A* **93**, 133–140.

Kessler, J. A., and Black, I. B. (1979). The role of axonal transport in the regulation of enzyme activity in sympathetic ganglia of adult rats. *Brain Res.* **171**, 415–424.

Kessler, J. A., and Black, I. B. (1982). Regulation of substance P in adult rat sympathetic ganglia. *Brain Res.* **234**, 182–187.

Kiraly, M., Favrod, P., and Matthews, M. R. (1989). Neuroneuronal interconnections in the rat superior cervical ganglion: Possible anatomical bases for modulatory interactions revealed by intracellular horseradish peroxidase labelling. *Neuroscience* **33**, 617–642.

Kirkpatrick, J. B. (1968). Chromatolysis in the hypoglossal nucleus in the rat: An electron microscopic analysis. *J. Comp. Neurol.* **132**, 189–212.

Koelle, G. B. (1951). The elimination of enzymatic diffusion artifacts in the histochemical localization of cholinesterases and a survey of their cellular distributions. *J. Pharmacol. Exp. Ther.* **103**, 153–171.

Koelle, G. B. (1955). The histochemical identification of acetylcholinesterase in cholinergic, adrenergic and sensory neurons. *J. Pharmacol. Exp. Ther.* **114**, 167–184.

Koelle, G. B. (1988). Enhancement of acetylcholinesterase synthesis by glycyl-L-glutamine: An example of a small peptide that regulates differential transcription? *Trends Pharmacol. Sci.* **9**, 318–321.

Koelle, G. B., Massoulié, J., Eugène, D., and Melone, M. A. B. (1988). Effects of glycyl-L-glutamine in vitro on the molecular forms of acetylcholinesterase in the preganglionically denervated superior cervical ganglion of the cat. *Proc. Natl. Acad. Sci. U. S. A.* **85**, 1686–1690.

Koenig, H., and Droz, B. (1971). Effect of nerve section on protein metabolism of ganglion cells and preganglionic nerve endings. *Acta Neuropathol., Suppl.* **5**, 119–125.

Koketsu, K. (1986). Inhibitory transmission: Slow inhibitory postsynaptic potential. *In* "Autonomic and Enteric Ganglia. Transmission and Its Pharmacology" (A. G. Karczmar, K. Koketsu, and S. Nishi, eds.), pp. 210–223. Plenum, New York and London.

Kondo, H., Kuramoto, H., Wainer, B. H., and Yanaihara, N. (1985). Evidence for the coexistence of acetylcholine and enkephalin in the sympathetic preganglionic neurons of rats. *Brain Res.* **335**, 309–314.

Kuba, K., Tsuji, S., and Minota, S. (1989). Transduction mechanisms for muscarinic regulation of ion channels. *In* "Biosignal Transduction Mechanisms" (M. Kasai *et al.*, eds.), pp. 85–112. Jpn. Sci. Soc. Press, Tokyo/Springer-Verlag, Berlin.

Kuno, M., and Llinas, R. (1970a). Enhancement of synaptic transmission by dendritic potentials in chromatolysed motoneurons of the cat. *J. Physiol. (London)* **210**, 807–821.

Kuno, M., and Llinas, R. (1970b). Alterations of synaptic action in chromatolysed motoneurones of the cat. *J. Physiol. (London)* **210**, 823–838.

Kuramoto, H., and Fujita, T. (1986). An immunohistochemical study of calcitonin gene-related peptide (CGRP)-containing nerve fibers in the sympathetic ganglia of bullfrogs. *Biomed. Res.* **7**, 349–357.

Laiwand, R., Werman, R., and Yarom, Y. (1988). Electrophysiology of degenerating neurones in the vagal motor nucleus of the guinea-pig following axotomy. *J. Physiol. (London)* **404**, 749–766.

Lakos, I. (1970). Ultrastructure of chronically denervated superior cervical ganglion in the cat and rat. *Acta Biol. Acad. Sci. Hung.* **21**, 425–427.

Langley, J. N. (1897). On the regeneration of pre-ganglionic and of post-ganglionic visceral nerve fibres. *J. Physiol. (London)* **22**, 215–230.

Langley, J. N. (1898). On the union of cranial autonomic visceral fibres with the nerve cells of the superior cervical ganglion. *J. Physiol. (London)* **23**, 240–270.

Langley, J. N., and Anderson, H. K. (1904). The union of different kinds of nerve fibres. *J. Physiol. (London)* **31**, 365–391.

Langley, J. N., and Orbeli, L. A. (1910). Observations on the sympathetic and sacral autonomic system of the frog. *J. Physiol. (London)* **41**, 450–482.

Lascar, G., and Taxi, J. (1992). Complementary histochemical observations on sympathetic ganglia, especially the coeliac plexus of the frog. *Acta Histochem. Cytochem.* **25**, 71–76.

Lascar, G., Taxi, J., and Kerdelhué, B. (1982). Localisation immunocytochimique d'une substance du type gonadolibérine (LH-RH) dans les ganglions sympathiques de Grenouille. *C. R. Séances Acad. Sci.* **294**, 175–179.

Lawrentjew, B. J. (1925). Ueber die Erscheinungen des Degeneration und Regeneration im sympathischen Nervensystem. *Z. Mikrosk.-Anat. Forsch.* **2**, 201–223.

Levinsohn, G. (1903). Ueber des Verhalten des Ganglion cervicale supremum nach Durchschneidung seiner prae-bezw. postcellulären Fasern. *Arch. Anat. Physiol. Anat. Abt.* 438–459.

Libet, B., Chichibu, S., and Tosaka, T. (1968). Slow synaptic responses and excitability in sympathetic ganglia of the bullfrog. *J. Neurophysiol.* **31**, 383–395.

Lieberman, A. R. (1971). The axon reaction. A review of the principal features of perikaryal response to axon injury. *Int. Rev. Neurobiol.* **14**, 49–124.

Liestøl, K., Mæhlen, J., and Njå, A. (1986). Selective synaptic connections: Significance of recognition and competition in mature sympathetic ganglia. *Trends Neurosci.* **9**, 1–4.

Liestøl, K., Mæhlen, J., and Njå, A. (1987). Two types of synaptic selectivity and their interrelation during sprouting in the guinea-pig superior cervical ganglion. *J. Physiol. (London)* **384**, 233–245.

Liestøl, K., Mæhlen, J., and Njå, A. (1992). A model of selective synapse formation in sympathetic ganglia. *J. Neurobiol.* **24**, 263–279.

Lipscombe, D., and Rang, H. P. (1988). Nicotinic receptors of frog ganglia resemble pharmacologically those of skeletal muscle. *J. Neurosci.* **8**, 3258–3265.

Love, J. A., and Szurszewski, J. H. (1987). The electrophysiological effects of vasoactive intestinal polypeptide in the guinea-pig inferior mesenteric ganglion. *J. Physiol. (London)* **394**, 67–84.

Mæhlen, J., and Njå, A. (1981). Selective synapse formation during sprouting after partial denervation of the guinea-pig superior cervical ganglion. *J. Physiol. (London)* **319**, 555–567.

Mæhlen, J., and Njå, A. (1982). The effects of electrical stimulation on sprouting after partial denervation of guinea-pig sympathetic ganglion cells. *J. Physiol. (London)* **322**, 151–166.

Mæhlen, J., and Njå, A. (1984). Rearrangement of synapses on guinea-pig sympathetic ganglion cells after partial interruption of the preganglionic nerve. *J. Physiol. (London)* **348**, 43–56.

Marshall, L. M. (1985). Presynaptic control of synaptic channel kinetics in sympathetic neurones. *Nature (London)* **317**, 621–623.

Marshall, L. M. (1986). Different synaptic channel kinetics in sympathetic B and C neurons of the bullfrog. *J. Neurosci.* **6**, 590–593.

Masuko, S., and Chiba, T. (1988). Projection pathways, co-existence of peptides and synaptic organization of nerve fibers in the inferior mesenteric ganglion of the guinea-pig. *Cell Tissue Res.* **253**, 507–516.

Matsumura, M., and Koelle, G. B. (1961). The nature of synaptic transmission in the superior cervical ganglion following reinnervation by the afferent vagus. *J. Pharmacol. Exp. Ther.* **131**, 28–46.

Matthews, M. R. (1974). Ultrastructure of ganglionic junctions. *In* "The Peripheral Nervous System" (J. I. Hubbard, ed.), pp. 111–150. Plenum, New York and London.

Matthews, M. R. (1976). Synaptic and other relationships of small granule-containing cells (SIF cells) in sympathetic ganglia. *In* "Chromaffin, Enterochromaffin and Related Cells" (R. E. Coupland and T. Fujita, eds.), pp. 131–146. Elsevier, New York.

Matthews, M. R. (1983). The ultrastructure of junctions in sympathetic ganglia of Mammals. *In* "Autonomic Ganglia" (L. G. Elfvin, ed.), pp. 111–150. Wiley, Chichester and New York.

Matthews, M. R. and Clowry, G. J. (1987). Experiments on the connectivity of axotomized rat sympathetic neurones. *Exp. Brain Res.* **16**, 297–304.

Matthews, M. R., and Nash, J. R. G. (1970). An efferent synapse from a small granule-containing cell to a principal neurone in the superior cervical ganglion. *J. Physiol. (London)* **210**, 11P–14P.

Matthews, M. R., and Nelson, V. H. (1975). Detachment of structurally intact nerve endings from chromatolytic neurones of rat superior cervical ganglion during the depression of synaptic transmission induced by post-ganglionic axotomy. *J. Physiol. (London)* **245**, 91–135.

Matthews, M. R., and Östberg, A. (1973). Effects of preganglionic nerve section upon the afferent innervation of the small granule-containing cells in the rat superior cervical ganglion. *Acta Physiol. Pol.* **24**, 215–223.

Matthews, M. R., and Raisman, G. (1969). The ultrastructure and somatic efferent synapses of small granule-containing cells in the superior cervical ganglion. *J. Anat.* **105**, 255–282.

Matthews, M. R., and Raisman, G. (1972). A light and electron microscopic study of the cellular response to axonal injury in the superior cervical ganglion of the rat. *Proc. R. Soc. London Ser. B* **181**, 43–79.

Matthews, M. R., Connaughton, M., and Cuello, A. C. (1987). Ultrastructure and distribution of substance P-immunoreactive sensory collaterals in the guinea pig prevertebral sympathetic ganglia. *J. Comp. Neurol.* **258**, 28–51.

McLachlan, E. M. (1974). The formation of synapses in mammalian sympathetic ganglia reinnervated with preganglionic or somatic nerves. *J. Physiol. (London)* **237**, 217–242.

McLachlan, E. M., and Meckler, R. L. (1989). Characteristics of synaptic input to three classes of sympathetic neurone in the coeliac ganglion of the guinea-pig. *J. Physiol. (London)* **415**, 109–129.

McLennan, H. (1954). Acetylcholine metabolism of normal and axotomized ganglia. *J. Physiol. (London)* **124**, 113–116.

Minota, S., Eguchi, T., and Kuba, K. (1989). Nicotinic acetylcholine receptor-ion channels involved in synaptic currents in bullfrog sympathetic ganglion cells and effects of atropine. *Pflügers Arch.* **414**, 249–256.

Murata, Y., Chiba, T., Kumamoto, E., and Kuba, K. (1989). Synaptic structure and axon collaterals of type B neurons in bullfrog sympathetic ganglia: Intracellular horseradish peroxidase (HRP)-labeling study. *Neurosci. Res.* **7**, 33–42.

Murray, J. G., and Thompson, J. W. (1957). The occurrence and function of collateral sprouting in the sympathetic nervous system of the cat. *J. Physiol. (London)* **135**, 133–162.

Nacimiento, W., and Kreutzberg, G. W. (1990). Cytochemistry of 5'-nucleotidase in the superior cervical ganglion of the rat: Effects of pre- and postganglionic axotomy. *Exp. Neurol.* **109**, 362–373.

Nagata, Y., Mikoshiba, K., and Tsukada, Y. (1973). Effect of potassium ions on glucose and phospholipid metabolism in the rat's cervical sympathetic ganglia with and without axotomy. *Brain Res.* **56**, 259–269.

Nagata, Y., Ebisu, H., Tamaru, M., Fujita, K., and Koide, T. (1989). Decrease of atrial natriuretic peptide content in rat superior cervical sympathetic ganglion after denervation and axotomy. *J. Neurochem.* **52**, 1570–1575.

Newberry, N. R., and Connolly, G. P. (1989). Selective antagonism of muscarinic potentials on the superior cervical ganglion of the rat. *Neuropharmacology* **28**, 487–493.

Nicholson, F. M. (1924). Morphologic changes in nerve cells following injury to their axons. *Arch. Neurol. Psychiatry* **11**, 680–697.

Nishi, S., Soeda, H., and Koketsu, K. (1965). Studies on sympathetic B and C neurons and patterns of preganglionic innervation. *J. Cell. Comp. Physiol.* **66**, 19–32.

Nishi, S., Soeda, H., and Koketsu, K. (1967). Release of acetylcholine from sympathetic preganglionic nerve terminals. *J. Neurophysiol.* **30**, 114–134.

Nissl, F. (1892a). Ueber die Veränderungen der Ganglienzellen am Facialiskern des Kaninchens nach Ausreissung der Nerven. *Allg. Z. Psychiatr. Ihre Grenzgeb.* **48**, 197–198.

Nissl, F. (1892b). Ueber experimentell erzeugte Veränderungen am den Vorderhornzellen des Rückenmarkes beim Kaninchen. *Allg. Z. Psychiatr. Ihre Grenzgeb.* **48**, 675–681.

Njå, A., and Purves, D. (1977a). Specific innervation of guinea-pig superior cervical ganglion cells by preganglionic fibres arising from different levels of the spinal cord. *J. Physiol. (London)* **264**, 565–583.

Njå, A., and Purves, D. (1977b). Re-innervation of guinea-pig superior cervical ganglion cells by preganglionic fibres arising from different levels of the spinal cord. *J. Physiol. (London)* **272**, 633–651.

Njå, A., and Purves, D. (1978a). The effects of nerve growth factor and its antiserum on synapses in the superior cervical ganglion of the guinea-pig. *J. Physiol. (London)* **277**, 53–75.

Njå, A., and Purves, D. (1978b). Specificity of initial synaptic contacts made on guinea-pig superior cervical ganglion cells during regeneration of the cervical sympathetic trunk. *J. Physiol. (London)* **281**, 45–62.

Okamoto, T., Kurahashi, K., and Fujiwara, M. (1988). Cholinergic transmission in the superior cervical ganglion reinnervated by peripheral vagal stump cut below the nodose ganglion in cats. *J. Pharmacol. Exp. Ther.* **245**, 990–994.

Östberg, A., and Vrbová, G. (1982). Competitive reinnervation of the rat superior cervical ganglion by foreign nerves. *Neuroscience* **7**, 3177–3189.

Östberg, A. J. C., Raisman, G., Field, P. M., Iversen, L. L., and Zigmond, R. E. (1976). A quantitative comparison of the formation of synapses in the rat superior cervical sympathetic ganglion by its own and by foreign nerve fibres. *Brain Res.* **107**, 445–470.

Palay, S. L. (1956). Synapses in the central nervous system. *J. Biophys. Biochem. Cytol.* **2**, 193–202.

Palay, S. L. (1958). The morphology of synapses in the central nervous system. *Exp. Cell Res. Suppl.* **5**, 275–293.

Peng, Y.-Y., and Horn, J. P. (1991). Continuous repetitive stimuli are more effective than bursts for evoking LHRH release in bullfrog sympathetic ganglia. *J. Neurosci.* **11**, 85–95.

Pick, J. (1970). "The Autonomic Nervous System." Lippincott, Philadelphia.

Pilar, G., and Landmesser, L. (1972). Axotomy mimicked by localized colchicine application. *Science* **177**, 1116–1118.

Proctor, W., and Roper, S. (1982). Competitive elimination of foreign motor innervation on autonomic neurones in the frog heart. *J. Physiol. (London)* **326**, 189–200.

Proctor, W., Roper, S., and Taylor, B. (1982). Somatic motor axons can innervate autonomic neurones in the frog heart. *J. Physiol. (London)* **326**, 173–188.

Purves, D. (1975). Functional and structural changes in mammalian sympathetic neurones following interruption of their axons. *J. Physiol. (London)* **252**, 429–463.

Purves, D. (1976a). Functional and structural changes in mammalian sympathetic neurones following colchicine application to post-ganglionic nerves. *J. Physiol. (London)* **259**, 159–175.

Purves, D. (1976b). Competitive and non-competitive re-innervation of mammalian sympathetic neurones by native and foreign fibres. *J. Physiol. (London)* **261**, 453–475.

Purves, D., and Lichtman, J. W. (1978). Formation and maintenance of synaptic connections in autonomic ganglia. *Physiol. Rev.* **58**, 821–862.

Purves, D., and Lichtman, J. W. (1985). "Principles of Neural Development." Sinauer Assoc., Sunderland, MA.

Purves, D., and Lichtman, J. W. (1987). Synaptic sites on reinnervated nerve cells visualized at two different times in living mice. J. Neurosci. 7, 1492–1497.

Purves, D., and Njå, A. (1976). Effect of nerve growth factor on synaptic depression after axotomy. Nature (London) 260, 535–536.

Purves, D., and Thompson, W. (1979). The effects of post-ganglionic axotomy on selective synaptic connexions in the superior cervical ganglion of the guinea-pig. J. Physiol. (London) 297, 95–110.

Quilliam, J. P., and Tamarind, D. L. (1972). Electron microscopy of degenerative changes in decentralised rat superior cervical ganglion. Micron 3, 454–472.

Raisman, G., and Matthews, M. R. (1972). Degeneration and regeneration of synapses. In "The Structure and Function of Nervous Tissue" (G. H. Bourne, ed.), Vol. 4, pp. 61–104. Academic Press, New York and London.

Raisman, G., Field, P. M., Östberg, A. J. C., Iversen, L. L., and Zigmond, R. E. (1974). A quantitative ultrastructural and biochemical analysis of the process of reinnervation of the superior cervical ganglion in the adult rat. Brain Res. 71, 1–16.

Ramcharan, E. J., and Matthews, M. R. (1991). Distribution of functional muscarinic binding on sympathetic neurones and its modification after axotomy. J. Auton. Nerv. Syst. 33, 139–140.

Ramón y Cajál, S. (1911). "Histologie du Système Nerveux de l'Homme et des Vertébrés" (Trad. Azoulay). Maloine, Paris.

Ramsay, D. A., and Matthews, M. R. (1985). Denervation-induced formation of adrenergic synapses in the superior cervical sympathetic ganglion of the rat and the enhancement of this effect by postganglionic axotomy. Neuroscience 16, 997–1026.

Rao, M. S., Sun, Y., Vaidyanathan, U., Landis, S. C., and Zigmond, R. E. (1993). Regulation of substance P is similar to that of vasoactive intestinal peptide after axotomy or explantation of the rat superior cervical ganglion. J. Neurobiol. 24, 571–580.

Repérant, J., Rio, J. P., Ward, R., Miceli, D., Vesselkin, N. P., Hergueta, S., and Lemire, M. (1991). Sequential events of degeneration and synaptic remodelling in the viper optic tectum following retinal ablation. A degeneration, radioautographic and immunocytochemical study. J. Chem. Neuroanat. 4, 397–419.

Roivainen, R. (1991). Increase in protein kinase-C-β-like immunoreactivity (PKC-β-LI) in the rat superior cervical ganglion after decentralization. Neurosci. Res. 11, 292–296.

Russell, N. J., and Riker, W. K. (1977). Early functional and ultrastructural changes following preganglionic denervation of sympathetic ganglia. Proc. West. Pharmacol. Soc. 20, 9–13.

Sanchez-Vives, M., Valdeolmillos, M., Martinez, S., and Gallego, R. (1993). Axotomy-induced changes in Ca$^{2+}$ homeostasis in rat sympathetic ganglion cells. Eur. J. Neurosci. 6, 9–17.

Sargent, P. B. (1983). The number of synaptic boutons terminating on Xenopus cardiac ganglion cells is directly correlated with cell size. J. Physiol. (London) 343, 85–104.

Sargent, P. B., and Dennis, M. J. (1981). The influence of normal innervation upon abnormal synaptic connections between frog parasympathetic neurons. Dev. Biol. 81, 65–73.

Sargent, P. B., and Pang, D. Z. (1988). Denervation alters the size, number, and distribution of clusters of acetylcholine receptor-like molecules on frog cardiac ganglion neurons. Neuron 1, 877–886.

Sargent, P. B., Bryan, G. K., Streichert, L. C., and Garrett, E. N. (1991). Denervation does not alter the number of neuronal bungarotoxin binding sites on autonomic neurons in the frog cardiac ganglion. J. Neurosci. 11, 3610–3623.

Schmidt, H., and Stefani, E. (1976). Reinnervation of twitch and slow muscle fibres of the frog after crushing the motor nerves. J. Physiol. (London) 258, 99–123.

Schultzberg, M., Hökfelt, T., Terenius, L., Elfvin, L. G., Lundberg, J. M., Brandt, J., Elde, R. P., and Goldstein, M. (1979). Enkephalin immunoreactive nerve fibres and cell bodies in sympathetic ganglia of the guinea-pig and rat. *Neuroscience* **4,** 249–270.

Schultzberg, M., Hökfelt, T., Lundberg, J. M., Dalsgaard, C. J., and Elfvin, L. G. (1983). Transmitter histochemistry of autonomic ganglia. *In* "Autonomic Ganglia" (L. G. Elfvin, ed.), pp. 205–233. Wiley, Chichester and New York.

Seto-Oshima, A., Sano, M., Kitajima, S., Kawamura, N., Yamazaki, Y., and Nagata, Y. (1987). The effect of axotomy and denervation on calmodulin content in the superior cervical sympathetic ganglion of the rat. *Brain Res.* **410,** 292–298.

Shapiro, J., Gurtu, S., Gordon, T., and Smith, P. A. (1987). Axotomy increases the conduction velocity of C-cells in bullfrog sympathetic ganglia. *Brain Res.* **410,** 186–188.

Shaw, G., Winialski, D., and Reier, P. (1988). The effect of axotomy and deafferentation on phosphorylation dependent antigenicity of neurofilaments in rat superior cervical ganglion neurons. *Brain Res.* **460,** 227–234.

Siegrist, G., Dolivo, M. Dunant, Y., Foroglou-Kerameus, C., De Ribeaupierre, F., and Rouiller, C. (1968). Ultrastructure and function of the chromaffin cells in the superior cervical ganglion of the rat. *J. Ultrastruct. Res.* **25,** 381–407.

Siklós, L., Párducz, A., Halász, N., Rickmann, M., Joó, F., and Wolff, J. R. (1990). An unbiased estimation of the total number of synapses in the superior cervical ganglion of adult rats established by the disector method. Lack of change after long-lasting sodium bromide administration. *J. Neurocytol.* **19,** 443–454.

Simmons, M. A. (1985). The complexity and diversity of synaptic transmission in the prevertebral sympathetic ganglia. *Prog. Neurobiol.* **24,** 43–93.

Sinicropi, D. V., Kauffman, F. C., and Burt, D. R. (1979). Axotomy in rat sympathetic ganglia: reciprocal effects on muscarinic receptor binding and 6-phosphogluconate dehydrogenase activity. *Brain Res.* **161,** 560–565.

Skok, V. I. (1964). Conduction in tenth ganglion of the frog sympathetic trunk. *Fed. Proc., Fed. Am. Soc. Exp. Biol.* **24,** Transl. Suppl. T363–T367.

Smirnow, A. (1890). Die Struktur der Nervenzellen im Sympathicus der Amphibien. *Arch. Mikrosk. Anat.* **35,** 407–424.

Smith, P. A., and Weight, F. F. (1986). The pathway for the slow inhibitory postsynaptic potential in bullfrog sympathetic ganglia. *J. Neurophysiol.* **56,** 823–834.

Smith, P. A., Shapiro, J., Gurtu, S., Kelly, M. E. M., and Gordon, T. (1988). The response of ganglionic neurones to axotomy. *In* "The Current Status of Peripheral Nerve Regeneration" (T. Gordon, R. B. Stein, and P. A. Smith, eds.), pp. 15–23. Alan R. Liss, New York.

Smolen, A. J. (1988). Morphology of synapses in the autonomic nervous system. *J. Electron Microsc. Tech.* **10,** 187–204.

Smolen, A., and Raisman, G. (1980). Synapse formation in the rat superior cervical ganglion during normal development and after neonatal deafferentation. *Brain Res.* **181,** 315–323.

Sotelo, C. (1968). Permanence of postsynaptic specializations in the frog sympathetic ganglion cells after denervation. *Exp. Brain Res.* **6,** 294–305.

Streichert, L. C., and Sargent, P. B. (1990). Differential effects of denervation on acetylcholinesterase activity in parasympathetic and sympathetic ganglia of the frog. *Rana pipiens. J. Neurobiol.* **21,** 938–949.

Sumner, B. E. H., and Sutherland, F. I. (1973). Quantitative electron microscopy on the injured hypoglossal nucleus in the rat. *J. Neurocytol.* **2,** 315–328.

Tamarind, D. L., and Quilliam, J. P. (1971). Synaptic organisation and other ultrastructural features of the superior cervical ganglion of the rat, kitten and rabbit. *Micron* **2,** 204–234.

Taxi, J. (1961a). La distribution des cholinestérases dans divers ganglions du système nerveux autonome des Vertébrés. *Bibl. Anat.* **2,** 73–89.

Taxi, J. (1961b). Etude de l'ultrastructure des zones synaptiques dans les ganglions sympathiques de Grenouille. *C. R. Hebd. Séances Acad. Sci.* **252,** 174–176.

Taxi, J. (1962). Etude au microscope électronique de synapses ganglionnaires chez quelques vertébrés. *Proc., Int. Cong. Neuropathol., 4th, 1961,* Vol. II, pp. 197–203.

Taxi, J. (1965). Contribution à l'étude des connexions des neurones moteurs du système nerveux autonome. *Ann. Sci. Nat., Zool. Biol. Anim.* [12] **7,** 413–674.

Taxi, J. (1976). Morphology of the autonomic nervous system. *In* "Frog Neurobiology" (R. Llinas and W. Precht, eds.), pp. 93–150. Springer-Verlag, Berlin and New York.

Taxi, J. (1979a). Degeneration and regeneration of frog ganglionic synapses. *Neuroscience* **4,** 817–823.

Taxi, J. (1979b). Chromaffin and chromaffin-like cells in the autonomic nervous system. *Int. Rev. Cytol.* **57,** 283–343.

Taxi, J., and Babmindra, V. P. (1972). Light and electron microscopic studies of normal and heterogeneously regenerated ganglionic synapses of the dog. *J. Neural Transm.* **33,** 257–274.

Taxi, J., and Eugène, D. (1991). Ultrastructural changes in the shape of neurons and related structures in frog sympathetic ganglia after axotomy. *Biol. Cell.* **72,** 75–82.

Taxi, J., Gautron, J., and L'Hermite, P. (1969). Données ultrastructurales sur une éventuelle modulation adrénergique de l'activité du ganglion cervical supérieur du Rat. *C. R. Hebd. Séances Acad. Sci.* **269,** 1281–1284.

Ternaux, J. P., Falempin, M., Palouzier, B., Chamoin, M. C., and Portalier, P. (1989). Presence of cholinergic neurons in the vagal afferent system: Biochemical and immunohistochemical approaches. *J. Auton. Nerv. Syst.* **28,** 233–242.

Titmus, M. J., and Faber, D. S. (1990). Axotomy-induced alterations in the electrophysiological characteristics of neurons. *Prog. Neurobiol.* **35,** 1–51.

Tokimasa, T. (1984). Calcium-dependent hyperpolarizations in bullfrog sympathetic neurons. *Neuroscience* **12,** 929–937.

Toldi, J., Farkas, Z., Fehér, O., Dames, W., Kása, P., Gyurkovits, K., Joó, F., and Wolff, J. R. (1986). Promotion by sodium bromide of functional synapse formation from foreign nerves in the superior cervical ganglion of adult rat with intact preganglionic nerve supply. *Neurosci. Lett.* **69,** 19–24.

Tosaka, T., Chichibu, S., and Libet, B. (1968). Intracellular analysis of slow inhibitory and excitatory postsynaptic potentials in sympathetic ganglia of the frog. *J. Neurophysiol.* **31,** 396–409.

Tranzer, J. P., and Thoenen, H. (1967). Electron microscope localization of 5-hydroxydopamine (3,4,5-trihydroxy-phenyl-ethylamine), a new "false" sympathetic transmitter. *Experientia* **23,** 743–745.

Traynor, P., Dryden, W. F., and Smith, P. A. (1992). Trophic regulation of action potential in bullfrog sympathetic neurones. *Can. J. Physiol. Pharmacol.* **70,** 826–834.

Tsubomura, T., Kurahashi, K., Oikawa, H., and Fujiwara, M. (1988). Acetylcholine content in superior cervical ganglion following reinnervation of vagal afferent fibers in cat. *Life Sci.* **42,** 1049–1058.

Watanabe, H. (1983). The organization and fine structure of autonomic ganglia of amphibia. *In* "Autonomic Ganglia" (L. G. Elfvin, ed.), pp. 183–201. Wiley, Chichester and New York.

Watanabe, H., and Burnstock, G. (1976). Junctional subsurface organs in frog sympathetic ganglion cells. *J. Neurocytol.* **5,** 125–136.

Watanabe, H., and Burnstock, G. (1978). Postsynaptic specializations at excitatory and inhibitory cholinergic synapses. *J. Neurocytol.* **7,** 119–133.

Watson, W. E. (1974). Cellular responses to axotomy and to related procedures. *Br. Med. Bull.* **30,** 112–115.

Weight, F. F., and Weitsen, H. A. (1977). Identification of small intensely fluorescent (SIF) cells as chromaffin cells in bullfrog sympathetic ganglia. *Brain Res.* **128,** 213–226.

Weitsen, H. A., and Weight, F. F. (1977). Synaptic innervation of sympathetic ganglion cells in the bullfrog. *Brain Res.* **128,** 197–211.

Williams, T. H. W. (1967). Electron microscopic evidence for an autonomic interneuron. *Nature (London)* **214,** 309–310.

Williams, T. H. W., Chiba, T., Black, A. C., Bhalla, R. C., and Jew, J. (1976). Species variation in SIF cells of superior cervical ganglia: Are they two functional types? *In* "SIF Cells. Structure and Function of the Small, Intensely Fluorescent Sympathetic Cells" (O. Eränkö, ed.), Fogarty Int. Cent. Proc., Vol. 30, pp. 143–162. DHEW, N. I. H., Washington, D. C.

Wolfe, D. E., Potter, L. T., Richardson, K. C., and Axelrod, J. (1962). Localizing tritiated norepinephrine in sympathetic axons by electron microscope autoradiography. *Science* **138,** 440–442.

Wolff, J. R., Kása, P., Dobó, E., Párducz, A., and Joó, F. (1992). The GABAergic innervation of paravertebral sympathetic ganglia. *In* " GABA Outside the CNS" (S. L. Erdö, ed.), pp. 45–63. Springer-Verlag, Berlin.

Wolff, J. R., Joó, F., and Dames, W. (1978). Plasticity of dendrites shown by continuous GABA administration in superior cervical ganglion of adult rat. *Nature (London)* **274,** 72–74.

Wu, K., and Black, I. B. (1987). Regulation of molecular components of the synapse in the developing and adult rat superior cervical ganglion. *Proc. Natl. Acad. Sci. U. S. A.* **84,** 8687–8691.

Wu, K., and Black, I. B. (1988). Transsynaptic impulse activity regulates postsynaptic density molecules in developing and adult rat superior cervical ganglion. *Proc. Natl. Acad. Sci. U. S. A.* **85,** 6207–6210.

Yarosh, C. A., Agosta, C. G., and Ashe, J. H. (1988). Modification of nicotinic ganglionic transmission by muscarinic slow postsynaptic potentials in the in vitro rabbit superior cervical ganglion. *Synapse* **2,** 174–182.

Yavari, P., and Weight, F. F. (1988). Pharmacological studies in frog sympathetic ganglion: Support for the cholinergic monosynaptic hypothesis for slow IPSP mediation. *Brain Res.* **452,** 175–183.

Yawo, H. (1987). Changes in the dendritic geometry of mouse superior cervical ganglion cells following postganglionic axotomy. *J. Neurosci.* **7,** 3703–3711.

Yokota, R., and Burnstock, G. (1983). Decentralisation of neurones in the pelvic ganglion of the guinea-pig: Reinnervation by adrenergic nerves. *Cell Tissue Res.* **232,** 399–411.

Yokota, R., and Yamauchi, A. (1974). Ultrastructure of the mouse superior cervical ganglion, with particular reference to the pre- and postganglionic elements covering the soma of its principal neurons. *Am. J. Anat.* **140,** 281–298.

# Chondrocyte Differentiation

Ranieri Cancedda,*,† Fiorella Descalzi Cancedda,*,‡ and Patrizio Castagnola*

* Centro di Biotecnologie Avanzate, Istituto Nazionale per la Ricerca sul Cancro, 16132 Genoa, Italy, † Istituto di Oncologia Clinica e Sperimentale, Universitá di Genova, 16132 Genoa, Italy, and ‡ Istituto Internazionale di Genetica e Biofisica, Consiglio Nazionale delle Ricerche, 80125 Naples, Italy

Data obtained while investigating growth plate chondrocyte differentiation during endochondral bone formation both *in vivo* and *in vitro* indicate that initial chondrogenesis depends on positional signaling mediated by selected homeobox-containing genes and soluble mediators. Continuation of the process strongly relies on interactions of the differentiating cells with the microenvironment, that is, other cells and extracellular matrix. Production of and response to different hormones and growth factors are observed at all times and autocrine and paracrine cell stimulations are key elements of the process. Particularly relevant is the role of the TGF-$\beta$ superfamily, and more specifically of the BMP subfamily. Other factors include retinoids, FGFs, GH, and IGFs, and perhaps transferrin. The influence of local microenvironment might also offer an acceptable settlement to the debate about whether hypertrophic chondrocytes convert to bone cells and live, or remain chondrocytes and die. We suggest that the ultimate fate of hypertrophic chondrocytes may be different at different microanatomical sites.

**KEY WORDS:** Chondrogenesis, Osteogenesis, Cartilage growth plate, Type X collagen, TGF-$\beta$, Bone morphogenetic protein, Transferrin, Homeogenes, Limb bud development.

## I. Introduction

A chondrocyte can be defined as a cell embedded in a cartilage matrix. Although cartilage has long been the subject of study, little information is available on mechanisms controlling chondrocyte differentiation. Traditionally cartilages are classified as hyaline, elastic, and fibrous on the basis

265

of their morphological and histological appearance and their developmental history. In this article we consider only differentiation of chondrocytes from hyaline cartilages. In the body, hyaline cartilage is found on joint surfaces (articular cartilage), in a small number of bones remaining cartilaginous throughout the life of the animal (permanent cartilage), and in the initial cartilaginous models of vertebral column, pelvis, and extremity bones, which are formed during embryogenesis and subsequently replaced by bone. This last type of cartilage persists in the growth plate of long bones after birth, until sexual maturity is reached. Cartilage formation in the embryo involves the progeny of a small number of mesenchymal stem cells. Their proliferation, commitment, acquiring of the chondrocyte phenotype, and formation of a unique tissue as cartilage depend on both genomic potential (master and controlling genes) and local microenvironment (paracrine regulation, extracellular matrix, and other extrinsic factors) and on signals released by the cell itself, for which the specific receptor is present on the cell surface (autocrine regulation). When a cartilaginous model of a bone is at first organized and subsequently replaced by bone tissue, this occurs by endochondral ossification, through an orderly sequence of chondrocyte differentiation, beginning with cell proliferation and ending with chondrocyte hypertrophy and death or further maturation to osteoblast-like cells. Differentiation stages are characterized by changes in cell proliferation, extracellular matrix production, and cell morphology. Endochondral ossification begins (Fig. 1) during formation of long bone in the embryo. After birth and before adulthood, it continues in the long bone cartilaginous growth plates, allowing bone growth in length. In the adult animal a similar process occurs in the callus during bone fracture repair and in degenerating joints, whereas the cells do not differentiate beyond the stage of resting chondrocytes in normal articular cartilage. Chondrocyte differentiation is regulated by a number of humoral hormones and factors and by locally produced cytokines. Cartilage and bone organization are characterized by profound changes in the nature and amount of extracellular matrix macromolecules. Collagen is the major component of the extracellular matrix. Cartilage and bone consist of genetically distinct collagens, which are specific products of phenotypic expression by differentiated cells. The collagen gene family

FIG. 1   Diagram of endochondral bone formation. In the early stages of embryogenesis mesenchymal cells in the limb bud condense to form a cartilaginous model of the bone (A–D). Chondrocyte hypertrophy begins in the diaphysis of the cartilaginous bone and initial bone is formed at the boundary between hypertrophic chondrocytes and surrounding undifferentiated stacked cells (E). Subsequently, blood vessels enter the region and cartilaginous matrix is substituted by new bone matrix (F). While marrow cavity is formed, formation of epiphyseal growth plates occurs (G). Secondary ossification centers are formed within epiphyseal cartilages (G and H).

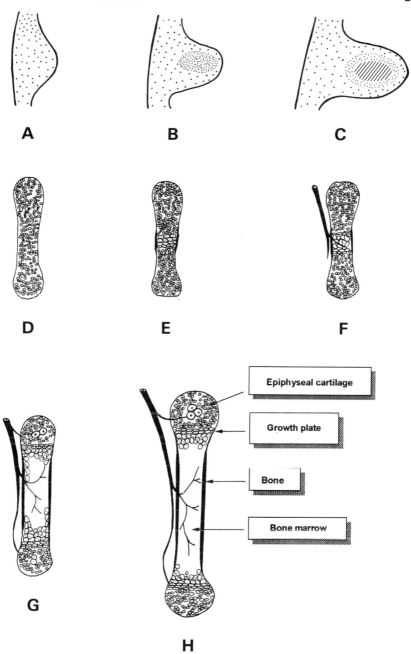

A  B  C

D  E  F

G  H

Epiphyseal cartilage

Growth plate

Bone

Bone marrow

consists of at least 30 genes making up a minimum of 18 different collagen types. Four of these collagen types (II, IX, X, and XI) have traditionally been considered specific for cartilage.

Despite the great number of similarities existing in their metabolism and the number of differentiation markers (including collagens) that they share, chondrocytes from articular cartilage, from permanent cartilage, and from growth plate must be regarded as cells with a different developmental history. In this article we focus on growth plate chondrocyte differentiation and only marginal information is given on other chondrocyte types. Data available both on avian and mammalian cells are reported. It should be kept in mind that differences exist in avian and mammalian growth plates with respect to cell organization and mechanisms by which the cartilaginous model is replaced by bone tissue. Therefore information derived from avian systems cannot always be extrapolated to mammalian systems and vice versa.

## II. The Mesenchymal Stem Cell

Chondrocytes are derived from multipotent mesenchymal cells. Precursor cells have been cloned and subsequently shown to be able to differentiate *in vitro* into a number of different mesenchymal phenotypes (Caplan, 1991). The differentiation processes involve four defined steps: (1) proliferation, (2) predifferentiation growth arrest, (3) reversible nonterminal differentiation, and (4) terminal differentiation associated with the irreversible loss of proliferative capacity. In a large number of cases, primary cultures of mesenchymal cells with chondrogenic potential were obtained from stage 24–25 (Hamburger and Hamilton, 1951) chick embryo limb buds (Hascall *et al.*, 1976; Ahrens *et al.*, 1977). Primary cultures of limb bud mesenchymal cells are induced to differentiate to a cartilage phenotype by modifications in the culture conditions or in the culture medium composition (see Section IV,A).

In the early stages of differentiation, the chondrocyte phenotype is unstable. Several research groups have shown dedifferentiation of chondrocytes, that is, the cells dramatically change their synthesis of extracellular matrix macromolecules by switching from the production of cartilage-specific proteoglycans and type II (and other cartilage-specific) collagens to the production of type I collagen characteristic of prechondrogenic cells (Coon, 1966; Schiltz *et al.*, 1973; Mayne *et al.*, 1976; Muller *et al.*, 1977). Dedifferentiation is accelerated by bromodeoxyuridine (Levitt and Dorfman, 1972; Schiltz *et al.*, 1973; Mayne *et al.*, 1975), the carcinogen phorbol myristate acetate (Pacifici and Holtzer, 1977; Lowe *et al.*, 1978), fibronectin (West *et al.*, 1979), and viral transformation (Yoshimura *et al.*, 1981; Adams *et al.*, 1982; Gionti *et al.*, 1983). Culturing of early differentiation stage chondrocytes on an adherence-permissive substratum per se induces dedifferentiation. When plated in standard tissue culture dishes, chondrocytes from early-stage chick embryo tibias adhere to the substratum, assume a fibroblastic morphology, and dedifferentiate (Castagnola *et al.*, 1986). Interestingly, these dedifferentiated cells can be expanded as such in culture and maintain a chondrogenic potential. When transferred into suspension culture, dedifferentiated cells reacquire the chondrocyte phenotype and resume differentiation to hypertrophic cells.

Although bone marrow represents a preferential tissue from which to derive cultures of mesenchymal stem cells with osteogenic potential, little information is available on cultures of chondroprogenitor cells established from bone marrow. Marrow stromal cells from embryonic and neonatal chicks produced clonally derived chondrocytic colonies (Berry *et al.*, 1992). The ability of chick bone marrow cells to form chondrocytic colonies decreased during development. Rabbit chondrocyte-enriched cultures were obtained from mesenchymal stem cells of adult rabbit bone marrow for the purpose of obtaining implants to be used for joint surface reconstruction (Robinson *et al.*, 1993). Rabbit marrow cells (Bab *et al.*, 1986) and neonatal pig bone marrow cells (Thomson *et al.*, 1993) inoculated into diffusion chambers were implanted intraperitoneally into athymic mice for, respectively, 3 and 6–8 weeks. Bone, cartilage, and fibrous tissues were formed within the chamber.

Pluripotent mesenchymal stem cells have also been identified within the connective tissues of skeletal muscle, fat, cartilage, and bone. When cultured in the presence of dexamethasone cells from clones, indifferently derived from all tissues, differentiated in a time- and concentration-dependent manner into muscle, fat, cartilage, and bone phenotypes (Young *et al.*, 1993). When bone marrow- or muscle connective tissue-derived cells were exposed to bone morphogenetic protein (BMP) and associated noncollagenous matrix proteins, transferred to diffusion chambers, and transplanted into the abdominal wall of syngeneic rats, cells differentiated into cartilage and chondroosteoid on the inside of diffusion chamber, whereas new bone developed on the membrane surfaces adjacent to the vascularized outside (Kataoka and Urist, 1993).

Chondrogenic differentiation may also occur starting from uncommitted periosteum-derived mesenchymal stem cells. Cultured periosteal cells from chick bone presented an undifferentiated phenotype when maintained as a monolayer, but revealed chondroosteogenic potential when inoculated in athymic, nude mice (Nakahara *et al.*, 1990). Chick periosteum-derived cells, not entering the chondrogenic cell lineage during normal bone development and growth, presented chondrogenic potential in high-density cultures (Nakata *et al.*, 1992). The developmental potential of the same cells was clonally assessed with an agar gel culture system (Nakase *et al.*, 1993). Both chondrogenic and osteogenic colonies were observed; transforming growth factor $\beta$1 (TGF-$\beta$1) shortened the time course of chondrogenesis and increased colony-forming efficiency of chondrogenic colonies. Transforming growth factor $\beta$1 also induced cellular heterogeneity in the periosteum when injected onto the outer periostea of rat parietal bones (Taniguchi *et al.*, 1993). Transforming growth factor $\beta$1 induced intramembranous ossification of the parietal bone in neonatal rats, and it induced endochondral ossification in adult animals. Therefore different responses of mesenchymal cells in the periosteum may depend on the age of the animal. In addition, the findings described above suggest the existence of a progenitor stem cell capable of both chondrogenic and osteogenic differentiation. The existence of a similar precursor has also been postulated in cells outgrown from chips of chick embryonic bones (Manduca *et al.*, 1992).

Over the years a few mesenchymal cell lines derived from different rodent embryos and able to differentiate *in vitro* have been obtained. The mouse cell line C3H-10T1/2 (10T1/2) has the capacity to differentiate into myoblasts, adipocytes, chondrocytes, and osteoblasts. 5-Azacytidine converts 10T1/2 cells into three stably determinate, but undifferentiated, stem cell lineages that can differentiate into myofibers, chondrocytes, and adipocytes (Konieczny and Emerson, 1984). It was proposed that 5-azacytidine converts 10T1/2 cells by hypomethylation of "determination" regulatory loci that establish lineages of stem cells with a restricted differentiation potential. Bone morphogenetic protein 2 (BMP-2) caused a dose-dependent differentiation of 10T1/2 cells into fat, cartilage, and bone cells; low concentrations favor adypocytes and high concentrations favor chondrocytes and osteoblasts (E. A. Wang *et al.*, 1993). Permanent transfection of genes encoding BMP-2 and BMP-4 in the same cells induced differentiation into osteoblasts, chondrocytes, and adipocytes (Ahrens *et al.*, 1993).

The multipotential cell line RCJ 3.1, clonally derived from fetal rat calvaria, differentiated in a time-dependent manner when it was cultured in the presence of ascorbic acid, $\beta$-glycerophosphate, and dexamethasone (Grigoriadis *et al.*, 1988, 1989). The formation of adipocytes and chondrocytes was dependent on the addition of dexamethasone. Muscle cells and osteoblasts also developed in the absence of dexamethasone, although at a low frequency. A monopotential chondrogenic cell line, RCJ 3.1 C5.18 (C5.18), was isolated from the multipotential cell line RCJ 3.1. Continuous or pulse exposure of the C5.18 line to retinoic acid (0.01–100 n$M$) inhibited chondrocyte phenotype expression at all periods tested and resulted in the disappearance of preexisting cartilage nodules (Lau *et al.*, 1993).

The CFK2 cell line derived from fetal rat calvaria during extended culture of the cells, without subculture, expresses a differentiated chondrocyte phenotype including production of type II collagen, link protein, and cartilage-specific proteoglycan core protein (Bernier *et al.*, 1990). The expression of the chondrocyte phenotype is regulated by several soluble modulators, including retinoic acid and parathyroid hormone (PTH) (Bernier and Goltzman, 1993).

In summary, undifferentiated mesenchymal cells with chondrogenic potential exist in different tissues and regions of the body. Some of these precursors also give rise to other lineages and in particular to muscle, fat, and bone lineages. Apparently, during development of the differentiation process, precursor cells progressively lose their multipotency. The existence of precursors from which only adipocytes, chondrocytes, and osteoblasts are derived, and the contemporary existence of precursors capable of giving rise only to chondrocytes and osteoblasts, suggest that during differentiation of mesenchymal cells a branching for muscle cells, and a branching for adipocytes, exist.

Last, it should be recalled that studies on mesenchymal stem cells are important not only to clarify the developmental pathway of differentiated cells of mesodermal origin as chondrocytes, but also have a great biotechnological relevance. The isolation, *in vitro* expansion, and site directed delivery of autologous stem cells can in fact provide the background for a new approach to cell self-repair of damaged tissues.

## III. Cartilage Formation during Embryo Development

### A. Onset of Chondrogenesis

### 1. Positional Signaling

The development of the vertebrate limbs and of skeletal elements within may be regarded as the cells interpreting their position in a three-dimensional system. The limb bud starts as a small bulge and grows out as a mesenchyme mass encased in ectoderm. This ectoderm has a thickening along the distal rim, called the apical ectodermal ridge (AER). Limb growth along the proximodistal axis is due to the proliferation of cells in the mesenchyme immediately under the ectoderm (progress zone) in response to signals from the AER (Summerbell *et al.*, 1973). The response to the ectoderm-derived signals is mediated by the extracellular matrix (ECM) (Solursh *et al.*, 1984). Information on the nature of signals from the AER is given in Section VI. Skeletal elements differentiate within the mesenchyme in a proximodistal sequence. Positional information along the anteroposterior axis is determined by a zone at the posterior margin of the bud called the zone of polarizing activity (ZPA) (Tickle *et al.*, 1975). It is suggested that cells are assigned positional information when they are in the progress zone. In the case of the anteroposterior axis, position is determined by distance from the ZPA (Fig. 2). Grafts of ZPA from several species, including chick, mouse, human, and hamster, to the anterior region of chick buds stimulated a mirror image duplication of digits (Tickle, 1980).

## 2. Condensation of Mesenchymal Cells

Elements of the vertebrate skeleton are initiated as cell condensations. In the proximal and central zone of the chick embryo limb bud, at stage 22 (Hamburger and Hamilton, 1951), prechondrogenic mesenchymal cells

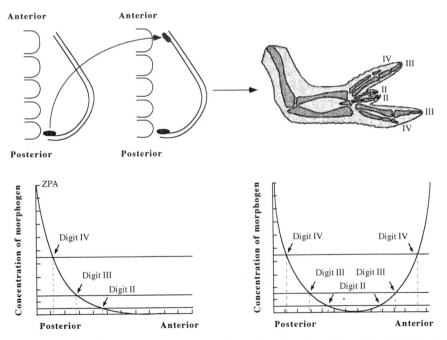

FIG. 2   *Top:* Specification of the anteroposterior axis in the chick limb. When the ZPA is grafted to anterior region of limb bud, duplicated digits, as mirror image of normal digits, are formed during development. *Bottom:* Models for a diffusible morphogen induced by the ZPA in normal limb (left) and in grafted limb (right). As the diffusible morphogen reaches certain concentrations it instructs cells to make specific digits.

become closely packed and, after establishing cell–cell contacts and gap junctions, start to differentiate into chondrocytes (Thorogood and Hinchliffe, 1975; Ede, 1983; Kelley and Fallon, 1978, 1983). This process, called precartilage condensation, is characterized by the neoproduction of sulfated proteoglycans and the switch from type I to type II collagen synthesis (Ede, 1983). Thereafter chondrocytes proliferate and secrete increasing amounts of ECM macromolecules until each single cell is completely surrounded by a matrix.

Kulyk *et al.* (1989a) have reported a level of fibronectin mRNA and protein in the proximal central core regions of chick limb, where condensation and cartilage matrix deposition are being initiated, about fourfold higher than in the distal subridge region, where undifferentiated nonchondrogenic mesenchymal cells are located. They have suggested a role for fibronectin in the initial cell condensation preceding the onset of chondrogenesis.

Affinity-purified antibodies raised against rat liver gap junctional proteins were used to block communications between chick limb mesenchymal cells and therefore to investigate the role of gap junctions in patterning of the chick limb (Allen *et al.*, 1990). A mixture of cells from the posterior polarizing region and from anterior mesenchyme (1 : 9) was grafted into chick wing buds (see Section III,A,1). When tissue from either the posterior polarizing region or anterior mesenchyme was loaded separately with antibodies before grafting, there was little effect on respecification of the digit pattern. On the other hand, loading both tissues caused a significant decrease in duplication. This suggests that gap junctions may enable polarizing region cells to communicate directly with adjacent anterior mesenchyme.

Oberlender and Tuan (1994) have shown that the cell adhesion protein N-cadherin is expressed in the developing chick limb bud during chondrogenesis, the maximal level of expression being at the time of active cellular condensation. Injection of a monoclonal antibody directed against the binding region of N-cadherin into embryonic limb buds resulted in a significant perturbation of chondrogenesis. In chick embryo limb buds, also the neural cell adhesion molecule (N-CAM) was transiently expressed by mesenchymal cells in the precartilaginous condensation, but was lost on differentiation into cartilage (Widelitz *et al.*, 1993).

It should be noted that intrinsic differences in different precartilage mesenchymes can contribute to variability in size and shape of different skeletal elements. For example, a correlation between morphogenetic differences in chick fore- and hindlimb mesenchyme and mechanisms of skeletal pattern formation has been reported (Downie and Newman, 1994).

## 3. Retinoids as Morphogens

In addition to the ZPA, a number of tissues acting as organizers during embryo development, such as the Hensen's node, the derived notochord, and the floor plate, induce the formation of extradigits when they are grafted in the chick limb bud (Hoffmann and Eichele, 1994). It was initially proposed that the ZPA and other organizers produce a morphogen that diffuses and forms a gradient in the limb bud (Tickle *et al.*, 1975). The body pattern observed in the limb should be the result of differences in morphogen concentration, intrinsic in the gradient concept, in different limb regions. The discovery that the local application of a carrier soaked in all-*trans*-retinoic acid (RA) induced the same digit duplication as ZPA grafts cast some light on the nature of the morphogen (Tickle *et al.*, 1982, 1985). Indeed, Thaller and Eichele (1987) have determined the concentration of all-*trans*-RA in developing chick limb bud and found that it formed a concentration gradient from 50 to 20 n$M$ along the posteroanterior axis. Following additional grafting experiments, it has been proposed that all-*trans*-RA is probably not equivalent to the morphogen released by the ZPA, but instead it was responsible for converting tissues near the grafted area to a ZPA, which would in turn generate secondary morphogens (Noji *et al.*, 1991; Wanek *et al.*, 1991). Several other natural retinoids were tested for their ability to induce digit duplication. 3,4-Didehydro-all-*trans*-RA was as efficient as all-*trans*-RA (Thaller and Eichele, 1990) whereas 9-*cis*-RA was 20 to 30 times more active than all-*trans*-RA (Thaller *et al.*, 1993). Interestingly, cells derived from mesoderm of chick limb buds presented in culture a different proliferative response to RA depending on their distal or proximal position and on the embryo age (Ide and Aono, 1988). This may reflect differences in their content of RA receptors. To understand the role of retinoids during limb development one should in fact consider the pattern of expression of retinoid receptors and cellular binding proteins. Most of the information available was obtained by several laboratories by *in situ* hybridization. In the limb bud of early mouse embryo, retinoic acid receptor $\alpha$ (RAR-$\alpha$) and RAR-$\gamma$ are uniformly expressed in the mesenchyme, whereas RAR-$\beta$ are restricted to the most proximal region of the limb bud (Dollé *et al.*, 1989). A proximal localization of RAR-$\beta$ in the limb was also observed by several laboratories in chick embryos (Noji *et al.*, 1991; Smith and Eichele, 1991; Schofield *et al.*, 1992). In addition in chick limbs RAR-$\beta$ was localized in the apical ectodermal ridge. The absence of RAR-$\beta$ in the progress zone suggests that this receptor is not required for anteroposterior patterning. Expression of the cell retinoic acid-binding proteins CRABP I and CRABP II in the limb buds has also been reported by several groups (Maden *et al.*, 1988, 1989; Dollé *et al.*, 1989; Ruberte *et al.*, 1992). Apparently the expression of these proteins is differently modulated in different regions during limb development. It has been proposed that the existence of opposite gradients of ligand and binding proteins could determine a steeper gradient of free retinoic acid available for receptor binding. Additional studies are necessary to better understand the role of RARs, CRABPs, and possibly other retinoic acid receptors, such as RXR, in controlling mesenchymal cell determination and chondrogenesis induction.

## 4. Homeobox-Containing Genes

The discovery of the vertebrate homologs of the *Drosophila* segmentation and homeotic genes has been one of the most relevant in development biology. Homeobox-containing genes play a key role in determining the whole vertebrate body plan, including skeletal segments, during embryogenesis. In particular, it has been shown that during limb bud development, before differentiation of mesenchymal cells into different cell lineages occurs, several homeogenes are already expressed in distinct spatial patterns (Izpisua-Belmonte and Duboule, 1992; Morgan and Tabin, 1993). The area of expression for each gene often partially overlaps with areas of expression of other homeogenes. Clusters or families of homeogenes are indeed activated during limb development in a sequential time-ordered pattern. In each gene cluster, the 3' genes are generally expressed earlier than 5' genes. As a result, in the limb buds at all stages, groups of mesenchymal cells, otherwise histologically identical, can be distinguished on the basis of the particular combination and levels of expressed homeogenes. The subsequent fate of these groups of cells, that is, the development or nondevelopment into each skeletal segment, is therefore already determined at a very early limb bud stage. Alterations of homeogene pattern of expression lead to skeletal abnormalities. A new nomenclature for homeobox genes has been proposed (Fig. 3; Scott, 1993). Throughout this section and the whole review we have adopted the new nomenclature.

*a. Hox Clusters*  *Hox* genes are expressed in very early chick limb bud mesenchyme, but expression ceases with condensation (Yokouchi *et al.,* 1991b). The different families of *Hox* genes play a role in establishing the patterns that are subsequently expressed by condensations. It is relevant that *Hoxa* genes are expressed by mesenchyme in the proximodistal limb territories where the proximodistal sequence of limb cartilage is established and that, on the other hand, *Hoxd* genes, which are coordinately expressed along the anteroposterior axis in partially overlapping regions during wing development with maximal expression in the posterior region (Izpisua-Belmonte *et al.,* 1991; Mackem and Mahon, 1991; Rogina *et al.,* 1992), play a role in the branching and bifurcating patterning within limb

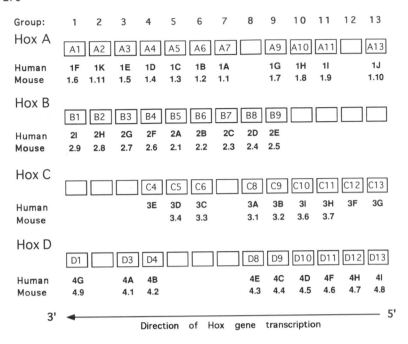

FIG. 3 Mammalian Hox complexes and new and old names of genes. The new names (boxed) are shown in the order in which they are located along the chromosome. Most commonly used old names are indicated below each box. For additional synonyms see Table 1 in Scott (1993).

regions. Similar results were also obtained when expression of *Hoxa* and *Hoxd* genes were investigated in developing mouse limbs (Haack and Gruss, 1993). It is interesting to note that in the limb of the chick talpid (*ta3/ta3*) mutation, characterized by polydactyly and defects in developing wing anteroposterior polarity, the *Hoxd* genes are expressed right across the anteroposterior axis instead of being expressed just posteriorly (Izpisua-Belmonte *et al.*, 1992).

Distribution of Hoxd-9 and Hoxc-6 proteins was investigated in mouse, chicken, and *Xenopus* limbs with specific antibodies (Oliver *et al.*, 1989). The Hoxd-9 forms a gradient, maximal in the distal and posterior regions, in developing fore- and hindlimb buds. A gradient of Hoxc-6 with an opposite polarity, that is, maximal in proximal and anterior regions, is detectable in the forelimb but not in the hindlimb (Oliver *et al.*, 1988).

Application of RA anteriorly in the chick wing bud, a treatment producing duplicated digits, substantially increases Hoxc-6 expression (Oliver *et al.*, 1990). The effect of RA application shows that only anterior cells are competent to express Hoxc-6. It has been demonstrated that RA is a potent stimulator of *Hox* gene expression (Simeone *et al.*, 1990). Retinoic acid induces 3′-located genes more rapidly and at lower concentrations than 5′ genes. Interestingly, in the limb buds, the anterior local application of RA induces *de novo* transcription of the *Hoxd* genes (Nohno *et al.*, 1991; Izpisua-Belmonte *et al.*, 1991). The obtained mirror image patterns of *Hoxd* gene expression correlate with the mirror image patterns of subsequently developed digits. Morgan *et al.* (1992) have shown that using retroviral vectors to expand the domain of expression of the *Hoxd-11* gene in the developing chick limb leads to reproducible pattern alterations consistent with a posterior homeotic transformation.

In mutant mice in which the *Hoxd-13* gene was disrupted via homologous recombination in embryonic stem cells, in addition to several skeletal alterations along all body axes, several abnormalities were observed in limbs, such as bone length reduction, phalange loss, bone fusions, and the presence of extra elements (Dollé *et al.*, 1993). Mice homozygous for a targeted mutation of the *Hoxa-2* gene contained multiple cranial skeletal defects, including cleft palate (Gendron-Maguire *et al.*, 1993).

A different localization has been shown for the expression of the *Evx-1* gene, a member of the *Hoxa* family located at the 5′ extremity and transcribed in the opposite direction from the other members of the family. Evx-1 mRNA was detected in the progress zone in the distal limb immediately after formation of the apical ectodermal ridge (Niswander and Martin, 1993). The ridge, or exogenously added fibroblast growth factor (FGF), is required for induction and maintenance of *Evx-1* expression. At the time the ridge regresses *Evx 1* becomes undetectable in the distal mesenchyme.

***b. Other Homeobox-Containing Genes***  A chicken homeobox-containing gene homologous to the *Drosophila* muscle segment homeobox (*msh*) gene was isolated and named *Msx-1* (the same gene in the past was called *Hox7*). By *in situ* hybridization Yokouchi *et al.* (1991a) have shown Msx-1 expression in the somatopleure in early-stage embryos. *Msx-1* expression by the mesenchymal cells derived from the somatopleure progressively diminished in the posterior half of the limb bud and then in the presumptive cartilage-forming mesenchyme. At the same time remarkable *Msx-1* expression was observed in the apical ectodermal ridge (AER) controlling growth and development of the progress zone of the limb and in mesenchyme in the proximal anterior nonchondrogenic periphery (see also Coelho *et al.*, 1993a). Therefore *Msx-1* expression was found in non-cartilage-forming mesenchymal cells. Local treatment with RA generated duplication of digits in the region of presumptive nonforming mesenchyme and at the same time suppressed the expression of *Msx-1* in the same areas.

The polydactylous mutant chick embryos, talpid2 and diplopodia-5, are characterized by expanded subridge mesoderm surmounted by thickened ectodermal ridge promoting formation of digits from both the anterior and the posterior mesoderm. In these mutants *Msx-1* is expressed throughout the entire expanded subectodermal mesoderm (Coehlo *et al.*, 1993b).

*Msx-2*, formerly *Hox8*, is expressed in the AER and in the anterior mesenchyme. At later developmental stages both *Msx-1* and *Msx-2* are expressed in interdigital spaces (W. B. Upholt, personal communication).

A homeobox-containing gene called *Prx-1*, and related to the *Drosophila paired* gene and to mouse *Pax* genes, was isolated from a chick limb bud cDNA library and used for *in situ* hybridization analysis. At early stages of limb development, distal mesenchymal cells express *Prx-1* with a gradient along the proximal distal axis (Nohno *et al.*, 1993). Later, *Prx-1* is expressed only in the interdigital and perichondral regions. On the other hand, during mouse limb bud development, expression of the homeobox gene *goosecoid*, first detected in undifferentiated tissue, persists also when the proximal limb bud undergoes morphogenesis to form limb structures (Gaunt *et al.*, 1993).

A homeogene that is expressed both in prechondrocytic mesenchymal cells and in cartilage primordia is *Cart-1*, a gene containing a paired-type homeodomain initially isolated from rat cDNA (Zhao *et al.*, 1993). The same gene is expressed in a well-differentiated chondrosarcoma tumor.

## B.  Chondrocyte Maturation and Cartilage Hypertrophy

Soon after the cartilaginous model of the bone is organized, chondrocytes in the center region of the model start to become larger in size (hypertrophic chondrocytes) and to secrete and organize a different extracellular matrix. Perhaps the most dramatic changes observed in the extracellular matrix of hypertrophic cartilage is the appearance of type X collagen. Type X collagen was first identified in the medium of cultured chick embryo chondrocytes and is considered a specific marker of hypertrophic chondrocytes (Schmid and Conrad, 1982; Capasso *et al.*, 1982; Gibson *et al.*, 1982). In the developing bone, type X collagen is synthesized by chondrocytes after they have become hypertrophic and before mineralization of the extracellular matrix occurs.

Hypertrophic cartilage is surrounded by a layer of undifferentiated stacked cells, which during development will give origin to osteoblasts.

Initial bone trabeculae are formed at the boundary between hypertrophic chondrocytes and these undifferentiated cells (see also Section V,C). Hypertrophic cartilage is more susceptible to blood vessel invasion from the periphery. As a result of the interplay between hypertrophic chondrocytes, cells in the surrounding stacked cell layer, and cells that have migrated into the area with the blood vessels, new bone is formed and hypertrophic cartilage is either replaced by bone or reabsorbed, thus giving room to bone marrow. As a consequence, a front between the newly formed bone and the remaining cartilage is determined. As replacement of cartilage by bone spreads in both directions from the center to the extremities, chondrocytes near the ossification begin to proliferate and new cartilage, which in turn will be replaced by bone, is formed. In this way a gradient of differentiating chondrocytes from resting to hypertrophic is established on both sides of the bone. These regions are called epiphyseal growth plates. In each growth plate, from the periphery to the center of the bone, the following histological regions are observed (Fig. 4): (1) the reserve zone, containing spherical cells with little or no cell division, (2) the proliferative zone, where cells divide to give rise to columns of flattened cells secreting hyaline extracellular matrix rich in type II collagen, (3) the zone of maturation, where the cells round up and begin to enlarge into hypertrophic chondrocytes, (4) the upper hypertrophic zone, characterized by cells that have enlarged 5- to 10-fold, by synthesis of type X collagen, and by reduction in matrix volume per total tissue volume, and (5) the lower hypertrophic zone, where calcification of the extracellular matrix occurs mainly in the longitudinal septa. (Holtrop, 1972; Gilbert, 1991).

The time required by one cell to pass from the upper proliferative zone to the lowermost hypertrophic zone has been estimated to be up to 3 days (Hunziker et al., 1987). As long as chondrocytes in the growth plate continue to proliferate, the bone keeps growing. At the time sexual maturity is reached, the high levels of estrogen and testosterone lead all chondrocytes to hypertrophy. All cartilage is replaced by bone and bone growth ceases.

In avian embryos the characteristic histological picture described above is not observed (Fig. 5). The cell arrangement is not in regular, vertical columns and cartilage calcification and reabsorption occur late after initial bone is already formed along walls of marrow tunnels and within cartilage (Stocum et al., 1979; Leach and Gay, 1987; Roach and Shearer, 1989).

Quantitative and qualitative changes occur in the composition and structure of extracellular matrix in the growth plate during chondrocyte differentiation. The collagen produced in larger amounts by chondrocytes is type II collagen. It may represent up to 25% of the dry weight in cartilage. Type II collagen has been considered and should continue to be considered

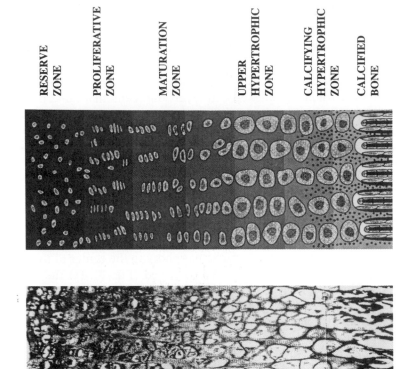

RESERVE ZONE

PROLIFERATIVE ZONE

MATURATION ZONE

UPPER HYPERTROPHIC ZONE

CALCIFYING HYPERTROPHIC ZONE

CALCIFIED BONE

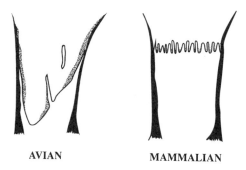

AVIAN                    MAMMALIAN

FIG. 5  Endochondral bone formation in avian and mammalian embryos. A diagram of the metaphysis is shown. A well-defined growth plate is observed only in the mammalian embryo. In the avian embryo marrow tunnels are formed directly in the (uncalcified) hypertrophic cartilage.

the most characteristic marker of cartilage. Nevertheless it should be mentioned that, during early embryogenesis, type II collagen is much more widely distributed than previously thought and probably has some as yet undefined function unrelated to chondrogenesis (Kosher and Solursh, 1989). During long bone development type II collagen was found by immunofluorescence in all cartilaginous structures (von der Mark et al., 1976). By in situ hybridization of human growth plate it was shown that type II collagen is mainly present in the proliferative, maturing, and upper hypertrophic zones (Sandberg and Vuorio, 1987). In chick the same collagen is mostly concentrated in the proliferative zone (Leboy et al., 1988). There is little evidence of type II collagen synthesis in the lower hypertrophic zone both in avian and mammalian growth plate. Nevertheless, in the growth plate of the bovine fetus, a progressive increase in the C-propeptide of the type II collagen (chondrocalcin) in the maturing and hypertrophic zones was observed (Alini et al., 1992). It has been proposed that chondrocalcin, being a calcium-binding protein, may accelerate mineral growth (Poole, 1991). A location similar to that of type II collagen was shown for type IX collagen (Reese et al., 1982; von der Mark et al., 1982; Irwin et al., 1985; Muller-Glauser et al., 1986; Linsenmayer et al., 1991) and for type XI collagen (Burgeson and Holliser, 1979; Eyre and Wu, 1987). Type X collagen has a different location because it concentrates in the lower maturing and upper hypertrophic zone (Capasso et al., 1984; Schmid and Lynsenmayer, 1985a,b; Gibson et al., 1986; Leboy et al., 1988; Iyama et al., 1991; Linsenmayer et al., 1991; Kirsch and von der

FIG. 4  Growth plate of mammalian long bone. A picture of a histological section and a diagram showing the different zones of the growth plate are presented.

Mark, 1991). At this time, type X collagen represents about half of the collagen produced by chondrocytes (Reginato *et al.*, 1986).

Growth plate chondrocytes synthesize large amounts of chondroitin 4- and 6-sulfate and keratan sulfate (Hascall, 1981; Hardingham and Fosang, 1992). These sulfate chains are not unique to cartilage, but are present in particularly elevated amounts in this tissue. Their abundance confers the characteristic Alcian Blue and toluidine stainability to the tissue. In cartilage most of the chondroitin- and keratan sulfate chains are linked to serine residues of the cartilage specific about 250,000-kDa core protein (Dorfman *et al.*, 1979; Hascall, 1981). The core protein contains a hyaluronic acid-binding domain, thus forming large aggregates with this molecule. In the proximal tibial growth plate of the bovine fetus it has been observed that aggrecan, the large cartilage proteoglycan, and hyaluronic acid concentrations progressively increase in the extracellular matrix of maturing and hypertrophic zones, being maximal at the time of mineralization (Alini *et al.*, 1992). The small, ubiquitous proteoglycan decorin, containing a single chondroitin sulfate or dermatan sulfate chain, is produced in the epiphyseal cartilage and in the resting zone, but not in the proliferating, maturing, and hypertrophic zones (Poole and Pidoux, 1986).

The binding of chondroitin and keratan sulphate proteoglycans to hyaluronic acid is stabilized by a 45,000-kDa link protein that is specific for cartilage and should be regarded as a chondrocyte-specific differentiation marker (Fife *et al.*, 1985). Using RNA *in situ* hybridization, the temporal and spatial expression of link protein in developing chick buds was investigated, together with the expression of the cartilage matrix protein (CMP; another cartilage-specific protein), type II collagen, and proteoglycan core protein (Stirpe and Goetinck, 1989). Transcripts of type II collagen were first detected at the 23-Hamburger and Hamilton stage, transcripts of link and proteoglycan core proteins at stage 25, and transcripts of cartilage matrix protein at stage 26. This study indicates that the temporal expression of each of these genes during limb bud development is independent of each other.

With development of chondrocyte hypertrophy the expression by chondrocytes of proteins generally considered as markers of the osteoblastic phenotype has been shown by several groups (see also Section V).

During maturation of hypertrophic chondrocytes profound rearrangements of the extracellular matrix occur. Metalloproteinases play a role in the degradation and remodeling of growth plate extracellular matrix. Metalloproteinase expression by growth plate chondrocytes has been shown by immunolocalization in newborn to 6-week-old rabbits (Brown *et al.*, 1989) and in the growth plate of rachitic rats (Dean *et al.*, 1985). In one study it was shown that production of metalloproteinases was linked to calcification of the chondrocyte culture (Brown *et al.*, 1993).

Increased amounts of active collagenase and insufficient levels of the tissue inhibitor of metalloproteinases (TIMP) may account for the reduced collagen content seen in the lower hypertrophic chondrocyte zone of growth plates of rats treated with different drugs (Dean *et al.*, 1989). *In vitro* rabbit articular chondrocytes constitutively express the 72-kDa gelatinase (MMP-2) and in a regulated manner the 92-kDa gelatinase (MMP-9) (Ogata *et al.*, 1992). Stromelysin has been purified from human articular cartilage (Gunja-Smith *et al.*, 1989).

## C.  Hypertrophic Cartilage Mineralization

The lowermost part of hypertrophic cartilage undergoes calcification. Deposition of calcium mineral is easily detectable in mammal growth plates. It starts in discrete focal sites of the territorial matrix and then spreads through the longitudinal septa. Transverse septa and pericellular matrix around chondrocytes usually remain uncalcified. In avian embryos the growth plate is poorly organized and the cartilage does not calcify prior to resorption (Roach and Shearer, 1989). Mineralization of hypertrophic cartilage extracellular matrix most probably favors its degradation by cells, such as osteoclasts, whose degradative activities are activated by interaction with a calcified matrix (R. Cancedda, F. Descalzi Cancedda, and P. Castagnola, unpublished results).

In chick embryos the source of calcium is the eggshell calcium carbonate. In embryos developed in shell-less cultures, the cartilaginous skeleton failed to calcify and to mature into bone (Tuan and Lynch, 1983). In mammals the maternofetal blood circulation can easily provide the calcium required for cartilage calcification and bone development. There is a general consensus that calcium deposited in the hypertrophic cartilage extracellular matrix derives from the inside of the chondrocytes. The large proteoglycans of cartilage offer a high concentration of negative charges that may bind calcium. A subsequent local increase in phosphate concentration, due to alkaline phosphatase activity, could displace calcium from proteoglycans and, once a critical concentration of calcium phosphate is reached, this mineral could precipitate as hydroxyapatite (Hunter, 1987; Poole, 1991).

Deposition of mineral requires the participation of alkaline phosphatases and an initial nucleation agent. Alkaline phosphatase is synthesized in the growth plate cartilage before mineral deposition (Osdoby and Caplan, 1981). Cartilage calcification involves matrix vesicles released by selected areas of chondrocyte membrane at defined differentiation stages. The membrane vesicles, which are rich in alkaline phosphatase, adhere to the molecules of the extracellular matrix. The presence of alkaline phosphatase within vesicles in high concentrations is in itself a strong indication of a role for matrix vesicles in the mineralization process, but it does not necessarily imply that matrix vesicles are sites of initial mineral nucleation. A marked decrease in alkaline phosphatase activity within matrix vesicles is observed during $Ca^{2+}$ uptake, suggesting that the enzyme becomes profoundly impaired by the mineralization process (Genge et al., 1988). Exploration of the constituents of vesicles reveals several protein components responsible for vesicle adherence to extracellular matrix. In particular, major fragments of the link protein and the hyaluronic acid-binding region of matrix proteoglycans may be responsible for the association to hyaluronic acid (Wu et al., 1991a), and annexin II and annexin V (previously identified also as anchorin II) may be responsible for the binding to type II and type X collagen (Genge et al., 1991, 1992; Wu et al., 1991b). The concentration of calcium and phosphate in these membrane vesicles is particularly high: not only do they contain calcium from the chondrocytes, but, once in place, they can also concentrate calcium and phosphate from the surrounding fluids. Evidence has been given that annexin V is a new type of ion-selective $Ca^{2+}$ channel protein (Genge et al., 1992).

Despite the above evidence there is no general consensus that membrane vesicles are the sites where the earliest mineral is detected and that they represent the primary sites of calcification. Although matrix vesicles are intimately associated with focal mineralizing sites, they do not consistently occupy a central position in mineralized foci, as would be expected if mineral deposition initiated within them and subsequently spread in the extracellular matrix (Poole and Pidoux, 1989).

Mineral–matrix interactions regulate hydroxyapatite in bones. It has been shown that mineralization in vitro is also affected by collagen, providing a template for hydroxyapatite deposition, and by noncollagenous matrix protein with high affinity for hydroxyapatite (Boskey, 1992).

A suggestion that type X collagen plays a role during cartilage calcification was derived by studies of Habuchi et al. (1985), who reported a close association of type X collagen with matrix vesicles. Type X collagen has been also associated with mineralization by other workers (Berry and Shuttleworth, 1989; Kwan et al., 1989; Oshima et al., 1989; Thomas et al., 1990; Coe et al., 1992), but there is no evidence yet that this collagen is directly involved in nucleation or mineral growth. The molecule is also present in the uncalcified membranes of the avian eggshell (Schmid et al., 1991) and can be expressed independently of cell hypertrophy (Pacifici et al., 1991a,b). Although in the growth plates the number and size of bony trabeculae composed of calcified hypertrophic cartilage were reduced, no mineralization abnormalities were detected in transgenic mice carrying a dominant negative type X collagen mutation (Jacenko et al., 1993a).

## IV. Chondrocyte Differentiation *in Vitro:* From Prechondrogenic Cells to Hypertrophic Chondrocytes

A number of *in vitro* models have been developed that recapitulate the ordered sequence of events observed during differentiation and maturation of cartilage *in vivo*. Despite differences in the source of starting cell population and the variety of culture conditions used, apparently in all models a decreased cell–substratum interaction associated with a change in the cell shape (rounding up) is required for induction and/or maintenance of the chondrocytic phenotype. In most cases cells were maintained in the presence of fetal calf serum, but serum-free medium has also been developed. For the induction of chondrogenesis, cell–cell interactions similar to the ones observed *in vivo* when mesenchymal cells initially condense in the limb bud are also required. In several cases, when culture conditions allow extracellular matrix assembly, a deposition of calcium mineral in the extracellular matrix is observed at late culture time, highly reminiscent of the cartilage mineralization occurring *in vivo*.

Besides helping to clarify molecular and cellular mechanisms that underlie chondrocyte differentiation and cartilage maturation, these models permit the identification of new markers that could be used in the future to extend our knowledge on these processes in physiological and pathological situations.

### A. Induction and Maintenance of Chondrocyte Phenotype

### 1. Decreased Cell–Substratum Interaction and Change in Cell Shape

Differences in the cell substratum interactions dramatically change the behavior of cultured chondroprogenitors and chondrocytes. *In vitro* spontaneous chondrogenesis in cultures of prechondrogenic cells derived from chick limb was observed only when the cells were plated at high cell density or maintained as micromass culture (Ahrens *et al.*, 1977; Solursh *et al.*, 1978; Gay and Kosher, 1984). Similarly, when rabbit or rat growth plate chondrocytes were cultivated as pelleted mass, not only persistence of the chondrocytic phenotype but also progression of differentiation to hypertrophic chondrocytes was observed (Kato *et al.*, 1988; Ballock *et al.*, 1993).

Chondrocytic phenotype can also be maintained or reexpressed by culturing cells in three-dimensional matrices. At variance with the large majority of other cells, except tumor and transformed cells, chondrocytes also proliferate and are metabolically active when grown in agarose or soft agar.

Dedifferentiated rabbit articular chondrocytes reexpress the differentiated phenotype during suspension culture in firm gels of 0.5% agarose (Benya and Schaffer, 1982). Chick embryo sternal chondrocytes fully express chondrocyte markers when grown within agar (Bruckner et al., 1989). Embryonic rat mesenchymal cells isolated from muscle, embedded in agarose, and treated with an extract from bovine demineralized bone powder proliferated and synthesized a cartilage-type matrix (Thompson et al., 1985). Chick embryo sternal chondrocytes have also been cultured within three-dimensional collagen gels (Gibson et al., 1982). Chondrocytes growing within the gel tend to form colonies reminiscent of the cellular organization in growth plate cartilage. Articular and growth plate chondrocytes from both fetal and adult rabbit, bovine, and pig have been cultured within three-dimensional matrices using calcium alginate beads (Ramdi et al., 1993; Olney et al., 1993; Grandolfo et al., 1993). Under these conditions chondrocytes are able to grow and divide for several days inside the beads. Entrapped cells maintain their differentiated phenotype over time. It should be mentioned that alginate gels can be easily dissolved to obtain cell populations for further studies.

Similarly to culture in three-dimensional gels, a culture of chondrocytes in suspension maintains the chondrocytic phenotype (Horwitz and Dorfman, 1970).

Induction of chondrogenesis in prechondrogenic cells can be easily obtained by transferring the cells in suspension culture. When dedifferentiated chondrocytes from early-stage chick embryo tibias are transferred to suspension culture on agarose-coated dishes, the cells not only revert to the chondrocyte phenotype, but they also resume their maturation and progress to hypertrophic chondrocytes producing type X collagen (Fig. 6) (Castagnola et al., 1986). To determine whether the phenomena observed during the in vitro differentiation were really due to changes in the differentiation state of the cells and to rule out a selection of a cell subpopulation, development of hypertrophic chondrocytes was investigated in cultures

FIG. 6 In vitro chondrogenesis from dedifferentiated chick embryo cells. Phase-contrast pictures of dedifferentiated chondrocytes derived from 6.5-day-old embryo tibias and grown as adherent cells for 3 weeks (a) and of the same cells transferred to suspension culture on an agarose gel (b) and cultured for 7 (c) and 17 (d and e) days. Culture was performed either in the absence (a–d) or in the presence (e) of ascorbic acid. Immediately after transfer to suspension cells aggregate; only later do cell aggregates loosen and release single isolated hypertrophic chondrocytes. In the presence of ascorbic acid hypertrophic chondrocytes are not released by the cell aggregates and small fragments of in vitro-made cartilage are organized (see Fig. 8).

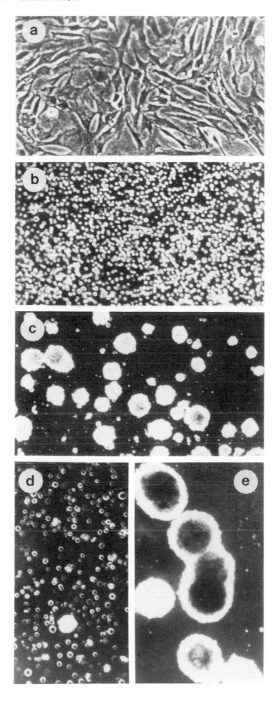

obtained starting from selected clones of dedifferentiated cells (Quarto et al., 1990).

To study the properties and the developmental pathway of chondrocytes at each differentiation stage, a different approach was followed by Carey et al. (1993). They have in fact fractionated bovine growth plate chondrocytes by a density gradient separation and cultured each fraction independently of each other.

The microenvironment in which chondrocytes are maintained during culture certainly plays a major role in determining their differentiation pathway. The relevance of culture conditions in determining chondrocyte differentiation is demonstrated by work showing that chondrocytes from chick sternal caudal zone (permanent cartilage) (Solursh et al., 1986; Castagnola et al., 1987) or articular cartilage (Pacifici et al., 1991a) express type X collagen and other hypertrophic chondrocyte markers, when exposed to a permissive microenvironment (i.e., suspension culture).

As limb mesenchymal cells differentiate into chondrocytes they initiate the synthesis of type II collagen and stop synthesizing type I collagen (Kosher et al., 1986). A striking increase in cytoplasmic type II collagen mRNA is coincident with in vitro cell condensation.

A qualitative and quantitative comparison of collagens expressed by cultured chick chondrocytes obtained from cartilage undergoing endochondral bone formation and from cartilage remaining as such was carried out by Gerstenfeld et al. (1989). The two chondrocyte cultures differed quantitatively and qualitatively in total collagen synthesis, procollagen processing, and distribution of collagen types.

Transcription activity and mRNA level of collagens during in vitro differentiation of endochondral chondrocytes were determined by run-off experiments and by Northern and slot blots (Castagnola et al., 1988). Type I collagen mRNA was highly expressed in the dedifferentiated cells and rapidly decreased during culture. The levels of the cartilage specific type II and type IX collagen mRNAs were negligible in the dedifferentiated cells, rapidly increased during the first week, reached the maximum level in the second week, and thereafter began to decrease. The expression of the hypertrophic cartilage-specific type X collagen mRNA progressively increased during culture and reached its maximum value after 3–4 weeks. Addition of dimethylsulfoxide to the culture medium interfered with the type X collagen expression (Manduca et al., 1988). All changes in the steady state levels of collagen mRNAs, except type II collagen mRNA, could be attributed to changes in the rate of transcription (Dozin et al., 1990). In the case of type II collagen mRNA, an additional stabilization of the transcript was in fact observed. This stabilization did not occur at an early stage of differentiation when type II collagen starts to accumulate. In addition, a very early (within 48 hr of cell transfer to suspension) and transient upregulated expression of type VI collagen was observed (Quarto *et al.*, 1993). It has been proposed that differentiation of growth plate chondrocytes from prechondrogenic cells to hypertrophic chondrocytes proceeds through different steps: a stage I (proliferating and differentiating chondrocytes) characterized by high levels of type II and IX collagens and a stage II (hypertrophic chondrocytes) characterized by the highest level of type X collagen mRNA (Castagnola *et al.*, 1988). Stage I may be further subdivided into an early phase (stage Ia) characterized by a high and transient type VI collagen synthesis and a later phase (stage Ib) in which type II and IX collagens are predominant (Fig. 7) (Quarto *et al.*, 1993). A rapid induction of type X collagen contemporary with the maturation of chondrocyte into hypertrophic cells was also described by Adams *et al.* (1991) in cultures of chick vertebral chondrocytes.

Data on collagen expression during the cartilage development mentioned above represent only a selected sample of the large number of contributions on this topic published by several laboratories. When considering the literature on this subject one should always keep in mind that identification of mRNA sequences complementary to collagen genes in one cell does not necessarily mean that the particular collagen is synthesized in the cell. For example, an untranslated form of $\alpha 1(1)$ collagen mRNA, which is 120 bases shorter and with the first 94 bases different from the calvaria mRNA, was found in chick chondrocytes grown in suspension (Bennett *et al.*, 1989).

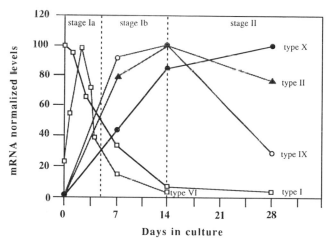

FIG. 7  Expression of collagens during *in vitro* differentiation of chick embryo chondrocytes. Dedifferentiated cells passaged for 3 weeks in adherent culture conditions were detached and transferred in suspension for 4 weeks. Definition of chondrocyte differentiation stages is given in text. (Redrawn from Quarto *et al.*, 1993.)

In addition one should consider the possibility of an alternative splicing of the primary gene transcript resulting in mRNA encoding different forms of the same collagen. For example, in the case of type II collagen, the existence of an additional exon, encoding 69 amino acids and located at the $NH_2$ portion of the molecule (exon 2), has been reported only in the mRNA population of chondrocytes from human fetal skeleton, juvenile costal cartilage, and bovine articular cartilage (Ryan and Sandell, 1990). The removal of exon 2 from the pre-mRNA, and consequently of the $NH_2$ peptide from the protein, may be an important step in chondrogenesis. By *in situ* hybridization it was observed that the mRNA without exon 2 (called type IIB) was associated with frank chondrocytes expressing the cartilage proteoglycan aggrecan, whereas the exon $2^+$ mRNA (type IIA) was expressed by prechondrocytes and immature chondrocytes (Sandell *et al.*, 1991). Therefore type II collagen exon 2 may be a marker for a distinct chondrocyte population.

The existence of two transcription sites generating alternative forms of mRNA has been shown in the case of chick, mouse, and human type IX collagen (Nishimura *et al.*, 1989; Muragaki *et al.*, 1990). It should be noted that only one of these mRNA forms is expressed in cartilage cells. An unusual pattern of alternatively spliced exons during chondrogenesis has also been found in the case of fibronectin mRNA (Bennett *et al.*, 1991).

The differentiating chondrocyte cultures have also been extremely useful in investigating the expression of other known cartilage and bone proteins such as anchorin CII (annexin V) (Castagnola and Cancedda, 1991), cartilage link protein and cartilage matrix protein (Szabo *et al.*, 1995), and osteopontin (Castagnola *et al.*, 1991), and in identifying new differentiation markers such as Ch21, a new protein highly expressed by hypertrophic chondrocytes and diretly secreted into the culture medium (Descalzi Cancedda *et al.*, 1988). By combined cDNA and peptide sequencing, the primary structure of the secreted form of the Ch21 protein has been established (Descalzi Cancedda *et al.*, 1990a,b). Computer-assisted analysis of that sequence revealed that this protein belongs to the superfamily of lipocalins, low molecular weight proteins sharing two basic frameworks for the binding and transport of small hydrophobic molecules. The complete sequence of the specific mRNA was determined, and, by blot and *in situ* hybridizations, it was observed that during chondrocyte differentiation, parallel and proportional increases in the levels of mRNA and gene transcription account for the progressive increase in the protein (Dozin *et al.*, 1992a). The expression of the protein is not limited to chondrocytes because bone fragments or isolated osteoblasts cultured *in vitro* produce and secrete the Ch21 protein and basal levels of the mRNA are found in noncollagenous embryonal tissues, the main source being the granulocytes. Although preliminary data reveal that the

Ch21 protein is capable of binding retinoids (R. Cancedda, F. Descalzi Cancedda, and M. Malpeli, unpublished results), both the function of this protein during skeletal tissue development and the nature of its ligand remain to be elucidated.

Tenascin is a large extracellular matrix glycoprotein localized in a few embryonic tissues (Bourdon and Ruoslahti, 1989). Although both antiadhesion and proadhesion signals exist on the molecule, tenascin interferes with attachment and spreading of several cultured cell types. Production of tenascin by forming cartilage nodules in cultures of chick embryo limb bud cells was shown by antibody staining (Mackie *et al.*, 1987). The same cells grown on a substrate of tenascin produced more cartilage nodules than cultures grown on tissue culture plastic. Interestingly tenascin is observed *in vivo* in condensing mesenchyme of cartilage anlagen, but not in the surrounding mesenchyme (Mackie *et al.*, 1987). With continuous development the protein was undetectable in the growth plate and its expression remained associated with articular cartilage development (Pacifici *et al.*, 1993).

An increasing body of evidence has shown that in all tissues and cell systems cell–substratum interactions are mostly mediated by specific membrane receptors for extracellular matrix proteins called integrins. Durr *et al.* (1993) have reported on the analysis of integrins in human fetal cartilage and cultured fetal chondrocytes. Collagen-binding $\alpha2\beta1$ integrin was located on chondrocyte surface. Anti-$\alpha1$ antibodies reduced binding of the cells to collagen, although $\alpha1$ could not be unequivocally identified on chondrocytes. Chondrocyte adhesion to fibronectin could be inhibited by anti-$\alpha5$. $\alpha2$ was not detected in fetal cartilage, in contrast to $\alpha5$, $\alpha v$-, and $\beta1$. This suggests that $\alpha2\beta1$ integrin may mediate chondrocyte–pericellular collagen matrix interactions *in vitro,* but may not be involved in chondrocyte–collagen interactions in the intact cartilage.

The spreading loss/shape change-dependent induction of chondrogenesis observed in all culture systems is probably mediated by modifications in the actin cytoskeleton architecture. When chick limb bud cells cultured at subconfluent densities (conditions under which they do not undergo spontaneous chondrogenesis) are treated with the microfilament-disruptive agent cytochalasin D, the cells round up, lose their actin cables, and undergo chondrogenesis (Zanetti and Solursh, 1984). Similarly, dihydrocytochalasin B (DHCB) induced reexpression of the cartilage phenotype in dedifferentiated rabbit articular chondrocytes, without a requirement for cell rounding (Benya *et al.*, 1988; Brown and Benya, 1988). Dihydrocytochalasin B also enhanced TGF $\beta$-induced reexpression of the differentiated chondrocyte phenotype in the same cells (Benya and Padilla, 1993).

It has been shown that, in association with other proteins, tensin mediates the microfilament–integrin link. In cultured chick embryo chondrocytes, tensin expression is dependent on substrate adhesion and is turned off after 1 week in suspension culture (van de Werken et al., 1993). Similarly, tensin is also regulated during chondrocyte differentiation in vivo. Interestingly, under the same culture conditions, vimentin, a major component of intermediate filaments, presented the opposite behavior.

## 2. Supplements to Culture Medium

The addition of high concentrations, usually 10%, of fetal calf serum (FCS) to the medium used for culturing chondrocytes is a common practice. Fetal calf serum by itself acts as a strong mitogen and induces the entire cascade of terminal differentiation (Bruckner et al., 1989). In the absence of fetal calf serum the chondrocytes remain viable only when seeded at high density. Under this condition chondrocytes deposit cartilage matrix, but do not proliferate and do not express markers of hypertrophic chondrocytes such as type X collagen (Bruckner et al., 1989).

Cultured chondrocytes are sensitive to toxic compounds derived from molecular oxygen, such as hydroxyl radicals and $H_2O_2$. The fetal calf serum added to the culture medium protects chondrocytes against this toxicity. In dense cultures in agarose, primary chick embryo sternal chondrocytes survive in serum-free medium because they release low molecular mass molecules supporting their own viability (Tschan et al., 1990). In low-density cultures this activity can be replaced by pyruvate and sulfydryl compounds such as cysteine or diothioerythritol or by catalase, an enzyme decomposing $H_2O_2$.

Different laboratories have used defined media to investigate chondrocyte differentiation from mesenchymal cells or to culture already differentiated chondrocytes. In most cases the expression of the cartilage phenotype has proved difficult to induce or maintain (Kujawa et al., 1989). Serum-free media that support expression of differentiated function by chondrocytes have been developed (Bohme et al., 1992; Quarto et al., 1992b). Insulin-like growth factor I (IGF-I) (Bohme et al., 1992) or a high insulin concentration (Quarto et al., 1992b) was an absolute requirement for cell differentiation. Thyroid hormones did not allow cell proliferation but induced chondrocyte maturation to hypertrophic chondrocytes (Bohme et al., 1992; Quarto et al., 1992a). Dexamethasone supported cell viability and modulated some differentiated functions (Quarto et al., 1992a). Additional information on soluble modulators required for chondrocyte proliferation and differentiation is given in Section VI.

Culture media utilized by several groups to culture chondrocytes include ascorbic acid. Ascorbic acid is an absolute requirement for the in vitro extracellular matrix assembly, but it is often not required for the development of the chondrocyte differentiation program. Nevertheless, a supplement of ascorbic acid is necessary to obtain chondrogenic differentiation in suspension cultures of mouse dedifferentiated cells (Dozin *et al.*, 1992b). In addition the presence of an organized extracellular matrix surrounding the cells, that is, a supplement of ascorbic acid added to the medium, is required by differentiated chondrocytes to express genes such as those encoding transferrin and retinoic acid-induced heparin-binding factor (RIHB) (Gentili *et al.*, 1994; Castagnola *et al.*, 1995).

## B. Cell—Cell Interactions during Initial Cell Aggregation

Precartilage condensation is an absolute requirement for chondrogenesis to occur. A correlation between aggregation ability and formation of cartilage nodules by cultured cells has been established (von der Mark and von der Mark, 1977). However, the molecular mechanisms that activate chondrogenesis, following cell aggregation, are still unknown. Different researchers have focused on different aspects, including cell–cell contact (Solursh and Reiter, 1980; Solursh, 1983), cell–matrix interactions (Dessau *et al.*, 1980; Tomasek *et al.*, 1982; Mackie *et al.*, 1987), and change in cell shape (Zanetti and Solursh, 1984). For all mechanisms experimental evidence has been given; nevertheless, the question still remains to be answered whether each of these mechanisms is capable of triggering the chondrogenesis process or whether chondrogenesis is the result of a synergistic effect of all mechanisms. In addition it must be considered that some of these phenomena may not induce chondrogenesis but instead may be a consequence of the process.

In high-density micromass cultures of chick limb bud mesenchymal cells, fibronectin expression increases about fivefold during cell condensation, remains at a high level during initial cartilage matrix deposition, and subsequently declines when a uniform mass of cartilage is formed (Kulyk *et al.*, 1989a). A monoclonal antibody against the $NH_2$ terminal of fibronectin reduced the number of condensations by more than 50%, as did the oligopeptide Gly-Arg-Gly, which is a repeated motif in that fibronectin domain (Frenz *et al.*, 1989).

During the condensation process, intimate cell–cell interactions occur. In micromass cultures of chick limb bud mesenchymal cells, it was observed that extensive gap junctional communications are formed between differentiating cartilage cells, whereas the nonchondrogenic cells differentiating into connective tissue do not exhibit intercellular communication via gap junction (Coelho and Kosher, 1991). An aggregation-inhibiting antiserum preventing cell condensation in cultures of chick limb bud cells and recognizing several cell surface components was raised by immunization of a rabbit with limb bud cells (Bee and von der Mark, 1990). In experiments in which cells were grown in carboxymethylcellulose to slow down condensation, a significant slowing down in the differentiation pathway was observed (Tacchetti et al., 1992).

Therefore such cell–cell interactions may involve direct contact of cell membrane and be mediated via interaction with peri- or extracellular matrix. This problem has been investigated by taking advantage of the culture obtained starting from dedifferentiated chick embryo cells (Castagnola et al., 1986). Immediately after their transfer into suspension culture dedifferentiated cells aggregate and only later do the cell aggregates loosen and release single hypertrophic chondrocytes. The initial cell aggregation is mediated by extracellular matrix proteins such as fibronectin (S. Tavella, P. Castagnola, C. Tacchetti, and R. Cancedda, unpublished results). Thereafter, in the aggregates, cell–cell interactions are established through cell adhesion proteins such as N-CAM and N-cadherin.

N-CAM and $Ca^{2+}$-dependent N-cadherin are molecules that play a central role in the establishment of cell–cell contacts coincident with permissive events during the development of many organs and systems (Detrick et al., 1990; Fujimori et al., 1990; Chuong and Edelman, 1985a,b; Jiang and Chuong, 1992). Using cultures of chick limb bud cells, it has been shown that $Ca^{2+}$ influences chondrocyte aggregation (Bee and von der Mark, 1990) and that adhesion molecules such as N-CAM and tenascin are induced by growth factors promoting chondrogenesis (Jiang et al., 1993). By Northern blot, Western blot, and immunocytochemistry, we have investigated the expression of N-CAM and N-cadherin during the in vitro differentiation of chondrocytes (Tavella, 1994). The timing of appearance of N-cadherin and N-CAM suggests that these molecules are developmentally regulated in differentiating chondrocytes. N-CAM and N-cadherin appear first on the surface of cells. Later, N-CAM becomes restricted to cells at the aggregate periphery. In the same area a number of cell–cell contacts are observed by electron microscopy. The expression of N-CAM and the expression of type II collagen by chondrocytes were mutually exclusive, suggesting that in the chondrocyte aggregates a sorting out between differentiating and nondifferentiating cells occurs.

That N-cadherin is functionally required for mediating cell–cell interactions among mesenchymal cells that are important for chondrogenesis was confirmed by experiments in which a rat monoclonal antibody directed against N-cadherin inhibited overt chondrogenesis, in micromass cultures of chick limb bud mesenchymal cells (Oberlender and Tuan, 1994). Other experiments also suggest that N-CAM is involved in the formation of precartilaginous condensations. Aggregation of dissociated mesenchymal chick limb bud cells was inhibited by incubating the cells with Fab' fragments of anti-N-CAM antibodies (Widelitz *et al.,* 1993). In addition, in micromass cultures, which recapitulated *in vitro* the *in vivo* events, N-CAM was enriched in condensations of 2-day cultures but it was diminished and distributed around cartilage nodules after 4 days of culture. Electroporation of cells with N-CAM expression vector before culture resulted in larger cartilage nodules and greater chondrogenic differentiation.

Therefore cell aggregation plays a major role in the activation of the differentiation program in chondrogenic cells.

## C. Assembly and Turnover of Extracellular Matrix

Only in the presence of ascorbic acid, a cofactor of the prolyl and lysyl hydroxylases, are collagen chains correctly hydroxylated and form stable triple helices that, in turn, assemble to form stable collagen fibrils (Prockop *et al.,* 1979). When the culture medium of differentiating chondrocytes was supplemented with ascorbic acid, cell clusters failed to open and chondrocytes remained aggregated (Tacchetti *et al.,* 1987). After 1 week, these cell aggregates have the histological and ultrastructural appearance of mature cartilage (Fig. 8). Cartilage-specific collagens can be identified in the extracellular matrix by specific antibodies. A tissue maintaining the organization of mammal growth plate cartilage was obtained *in vitro* when rabbit or rat chondrocytes were cultured as a pelleted mass in the presence of ascorbic acid (Kato *et al.,* 1988; Ballock *et al.,* 1993).

Little information is available on the production of metalloproteinases capable of matrix degradation by cultured chondrocytes. We have unpublished evidence that a 62-kDa metalloproteinase, identified as type IV collagenase, is constitutively expressed at a low level by proliferating and hypertrophic chondrocytes in suspension culture. Differentiating chondrocytes also secrete a 56-kDa metalloproteinase, tentatively identified as stromelysin.

## D. Mineral Deposition in Cartilage Extracellular Matrix
## *in Vitro*

Chondrocyte cultures have been extensively used to investigate chondrocyte differentiation. However, only in a few circumstances has mineraliza-

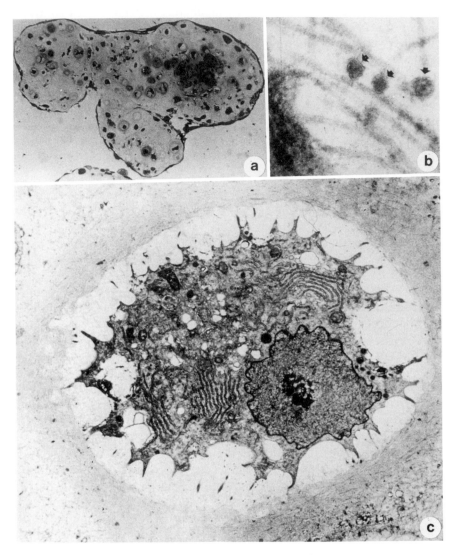

FIG. 8  Morphology of *in vitro*-made cartilage. Section of cell aggregates obtained by culturing dedifferentiated chick embryo chondrocytes in suspension in the presence of ascorbic acid for 17 days (a). Cells in the aggregates are contained in lacunae separated by an abundant extracellular matrix. A layer of elongated cells, reminiscent of a perichondrium, surrounds the aggregates. Electron micrographs reveal details of a chondrocyte within a lacuna (c) and of matrix vesicles present in the territorial matrix surrounding chondrocytes (b). Magnification: (a) ×81, (b) ×45,000, and (c) ×4500.

tion of the extracellular matrix been observed. The calcification of the matrix in some cases required the addition of high concentrations of inorganic phosphate (threefold over the concentration in standard culture medium) (Binderman et al., 1979; Suzuki et al., 1981). The mineral deposition in the presence of high phosphate should be considered a physicochemical phenomenon (Suzuki et al., 1981).

Mineralization of cultures occurred only when ascorbic acid was added to the culture medium, that is, when an extracellular matrix was correctly assembled, but the presence of capillaries and/or bone marrow cells was not required. Mineralization was observed in the matrix of the tissue closely resembling hypertrophic cartilage, which is formed when chick embryo chondrocytes are cultured in suspension in the presence of ascorbic acid (Tacchetti et al., 1989). Ultrastructural analysis of the in vitro-made cartilage revealed that numerous matrix vesicles, often considered as nucleation sites for cartilage calcification, are associated with collagen fibers. Indeed, it was shown that calcium minerals are deposited in the matrix of the in vitro-made cartilage and that calcification is preceded by a significant rise in the alkaline phosphatase activity. Induction of alkaline phosphatase by ascorbic acid in cultured chick chondrocytes has also been observed by Leboy et al. (1989). Mineralization in cartilage nodules formed in micromass cultures of chick limb bud mesenchymal cells has similarly been reported (Boskey et al., 1992). Kato et al. have shown that, in cultures of rabbit growth plate chondrocytes, calcification of the in vitro-formed cartilage was induced by hypertrophic chondrocytes themselves and suppressed by TGF-$\beta$1, which blocks chondrocyte maturation (Kato et al., 1988), and parathyroid hormone (Jikko et al., 1993).

In cultures of chick vertebral chondrocytes, mineralization occurred late in culture when proteoglycan and type II collagen synthesis was decreased and type X collagen synthesis continued to increase (Gerstenfeld and Landis, 1991). Although numerous vesicular structures could be detected, mineralized crystals were located in extracellular matrices principally associated with collagen fibrils and there was no clear evidence of mineral association with vesicles. In some studies the properties of crystals precipitated in vitro were compared with those of crystals formed in vivo in the growth plate hypertrophic cartilage and found to be similar (Jikko et al., 1993).

Deposition of mineral probably requires an initial nucleation agent and the participation of alkaline phosphatases. In vivo, cartilage calcification involves matrix vesicles released by chondrocytes at defined differentiation stages. We have already discussed the possible role of matrix vesicles in this process (Section III,C).

## V.  The Fate of Hypertrophic Chondrocytes

## A.  Apoptotic Death

To maintain appropriate cell number and tissue organization, during embryonic and adult development unnecessary or deleterious cells are eliminated by apoptosis, or programmed cell death (Collins and Lopez-Rivas, 1993; Lee *et al.*, 1993). Apoptosis has been described both *in vivo* and *in vitro* in a wide variety of cells and tissues. This process is characterized by specific morphological modifications (i.e., cell pyknosis and chromatin condensation) and DNA cleavage to nucleosomal fragments, and it depends on the activation of a specific set of novel "death genes."

Despite the potential interest in apoptosis as the mechanism responsible for hypertrophic chondrocyte destruction, little has been published on this topic. An ultrastructural study has given evidence that hypertrophic chondrocytes in the condylar cartilage of young rat may be disintegrated by apoptosis (Lewinson and Silberman, 1992). The expression of transglutaminase, an enzyme associated with programmed cell death, by chondrocytes undergoing terminal differentiation has been shown (Aeschlimann *et al.*, 1993).

Apoptosis plays instead a clear role in sculpturing the digits during limb development. DNA fragmentation associated with programmed cell death was observed in the interdigital tissue of the embryonic chick limb bud (Garcia-Martinez *et al.*, 1993). A progressive loss of the chondrogenic potential of the interdigital mesenchyme was observed before the onset of the degenerative process.

Frisch and Francis (1994) have reported that disruption of epithelial cell–matrix interactions induces apoptosis. They have termed this phenomenon *anoikis*. Overexpression of Bcl-2-protected cells against anoikis. Given the dramatic changes observed in the cell–matrix interactions during chondrocyte differentiation, it will certainly be interesting to investigate whether anoikis also plays a role during differentiation of chondrocytes and endochondral bone formation.

## B.  Further Differentiation *in Vitro*

### 1.  From Hypertrophic Chondrocytes to Osteoblast-like Cells

Transcription of the bone-related osteopontin gene in hypertrophic chondrocytes cultured in suspension and still expressing frank chondrocyte markers has been reported (Castagnola *et al.*, 1991). The induction of bone gene expression with development of chondrocyte hypertrophy *in vitro* has also been found by other groups. In particular, in chondrocytes derived from chick embryo vertebrae cultured in the presence of ascorbic acid and under conditions that promoted differentiation toward hypertrophy and extracellular matrix mineralization, synthesis of both osteocalcin and osteopontin has been described (Lian *et al.*, 1993). Osteopontin was already detected in early cultures in the absence of a calcified matrix. The maximal level of osteocalcin was observed with the peak in alkaline phosphatase and type X collagen, but although the increase over the earliest level of detection was about 50-fold, it still remained 1/100 of the level observed in cultures of frank osteoblasts. It should be noted that by *in situ* hybridization, the mRNA for osteonectin, a protein produced by bone cells but widely distributed, was detected in chick proliferating cartilage; lower levels of osteonectin mRNA were seen in the midhypertrophic region (Oshima *et al.*, 1989). Interestingly, cultured chick chondrocytes respond to heat shock by secreting large amounts of osteonectin (Neri *et al.*, 1992).

To answer whether a differentiation of hypertrophic chondrocytes toward an "osteoblast-like" phenotype could actually be recognized, chick hypertrophic chondrocytes were replated after hyaluronidase digestion on an anchorage-permissive substratum (Descalzi Cancedda *et al.*, 1992). Under these culture conditions hypertrophic chondrocytes acquired an elongated or star-shaped morphology, resumed cell proliferation (although at a low rate), and progressively stopped to deposit Alcian-positive extracellular matrix. At the same time the appearance of some cells with alkaline phosphatase activity was observed. Alkaline phosphatase positivity progressively extended to other cells, especially the ones with stellate and fibroblastic morphology. Eventually, after the cells had reached confluency, calcium mineral was deposited in the extracellular matrix.

The analysis of collagen types secreted by the cells during culture, both by electrophoresis of the pepsin-resistent domains and by immunoprecipitation with specific antibodies, was consistent with the morphological observations. Between the first and second weeks of culture the cells stopped synthesizing cartilage-specific type II and X collagens and switched to the synthesis of type I collagen (Descalzi Cancedda *et al.*, 1992; Gentili *et al.*, 1993). This finding was also confirmed by the measurement of the levels of the mRNAs extracted from cultured cells. It must be noted that the high expression of osteopontin, a protein proposed to act as a bridge between cells and apatitic mineral, was evidenced in the same series of experiments later in culture, when extracellular matrix mineralization occurred. Completely unexpected was the transient expression of ovotransferrin by the differentiating hypertrophic chondrocytes (see Section VI,F).

The presence of "posthypertrophic" chondrocytes producing type I collagen was also shown by Kirsch et al. (1992) in cultures of human chondrocytes.

## 2. Bonelike Mineralization

The first detectable calcification of the bonelike extracellular matrix was observed 5–6 weeks after hypertrophic chondrocyte replating and subsequently the mineralization extended throughout the culture, reaching its maximum after 7–8 weeks. Deposition of calcium mineral was enhanced by high cell densities. By electron microscopy the early foci of mineralization consisting of clusters of thin mineral crystals were identified (Gentili et al., 1993). At the edge of clusters, spreading of mineralization occurred along fibrils of collagen presenting the 64- to 70-nm periodicity, characteristic of type I collagen. Fibrils were coaligned and assembled in bundles in older cultures. No proteoglycan "granules," found in cartilage matrix, were ever observed.

On the basis of the above results and on the observations made by immunocytochemistry on sections of embryonic bones, the osteoblast-like cells derived from hypertrophic chondrocytes were proposed as an additional differentiation stage (stage III) of growth plate chondrocytes (Descalzi Cancedda et al., 1992).

It is interesting to note that retinoic acid, a known differentiating agent, highly accelerates the whole in vitro differentiation process from hypertrophic chondrocytes to osteoblast-like cells, including mineralization of the extracellular matrix (see Section VI,A).

## 3. Organ Culture of Hypertrophic Cartilage

In vitro cultures of cartilage pieces of different origin have shown that hypertrophic chondrocytes may have osteogenic potential. When perichondrium-free pieces of epiphyseal cartilage from quail embryo were incubated on chick embryo chorioallantoic membranes, some quail chondrocytes synthesized a bonelike matrix and underwent "transformation" into an osteocytic and osteoblastic type of cell (Kahn and Simmons, 1977). When femurs from 14-day chick embryos were cut through the region of hypertrophic cartilage and cultured for 12–18 days, the deposition of a new bone matrix inside cartilage lacunae was observed, thus suggesting that bone-forming cells were derived from the hypertrophic chondrocytes (Roach, 1992).

The mandibular condyle of a newborn mouse is composed of defined zones containing various cell types of the chondrogenic lineage that undergo osteogenic differentiation when cultured in vitro (Silberman et al., 1983; Strauss et al., 1990). In the fresh explant, young chondroblasts and hypertrophic chondrocytes are localized in the core and the basal part of the condyle. Only the basal portion of the cartilage is surrounded by a narrow ring of osteoid matrix. After a 2-week culture, an advanced stage of osteogenic differentiation and osteoid formation encircles the whole area of hypertrophic chondrocytes. The activation of genes characteristic of the osteogenesis process and the formation of new bone have been described in detail both in the chondroprogenitor cells and in the differentiated chondrocytes (Strauss *et al.*, 1990). The chondrogenic cells are reprogrammed to osteogenesis and sequentially express type I collagen, osteonectin, alkaline phosphatase, and osteocalcin. The presence of two nuclei in hypertrophic chondrocytes located in each of the cartilage lacunae indicates that they have reentered the cell cycle (Closs *et al.*, 1990). The expression of the oncogene *fos* by mature hypertrophic chondrocytes is in agreement with these cells reentering the cell cycle.

Modulation of the chondrocyte phenotype was observed in an organ culture system using fetal rat Meckel's cartilage (Richman and Diewert, 1988). Perichondrium was mechanically removed, and the cartilage was split and grafted into the anterior chamber of an adult rat eye. The observed pattern of development suggests that fully differentiated chondrocytes have the capacity to become osteocytes.

The fate of hypertrophic chondrocytes was also studied in cultures of metatarsal bones of fetal mice (Thesingh *et al.*, 1991). Because the periosteum had been stripped off, osteoclasts could not invade the long bone and resorb the lacunar walls. When bones were cocultured with pieces of cerebrum, the hypertrophic chondrocytes converted to osteoblasts. The stimulating factor from brain tissue was not identified.

## C. Further Differentiation *in Vivo* and Initial Bone Mineralization

Endochondral bone formation in the epiphyseal growth plate is due to a sequence of events including hypertrophic cartilage formation, its invasion by blood vessels, the erosion of the calcified cartilage, and its replacement by bone. There is a general consensus that, in the region of cartilage to osteoid transition of the growth plate, hypertrophic chondrocytes degenerate and die. The alternate view that hypertrophic chondrocytes can transdifferentiate into osteoblasts has long been held. It has been observed that in some areas of young rat mandibular condylar cartilage (Yoshioka and Yagi, 1988) and chick long bones (Roach and Shearer, 1989) hypertrophic chondrocytes transform into bone-producing cells and osteoid matrix is deposited within lacunae. *In vitro* cultures of cartilage pieces of different origin and species have shown that hypertrophic chondrocytes may express proteins characteristic of the "osteoblast phenotype" (see Section V,B,3). Expression of these proteins by hypertrophic chondrocytes has also been observed *in vivo* in selected cartilage regions. *In situ* hybridization and/or immunohistochemistry showed synthesis of type I collagen (Yasui *et al.*, 1984), osteopontin (Mark *et al.*, 1988; Franzen *et al.*, 1989), osteonectin (Metsäranta *et al.*, 1989; Pacifici *et al.*, 1990a,b), and bone sialoprotein (BSP; Bianco *et al.*, 1991) by human, rat, and chick hypertrophic chondrocytes.

When transplanted into a muscle, isolated syngeneic murine epiphyseal chondrocytes formed a cartilage in which endochondral ossification began at the end of the second week after transplantation (Moskalewski and Malejczyk, 1989). By week 7 to 8 posttransplantation the whole cartilage was replaced by an ossicle. The bone was formed by chondrocytes and not by some other cell contaminating the chondrocyte suspension.

A characteristic form of ossification, suggesting a deregulated maturation of chondrocytes to osteoblast-like cells, has also been observed in the most common of the human lethal dysplasias, thanatophoric dysplasia (W. A. Horton *et al.*, 1988). The presence of atypical and bizarre cartilage that often undergoes a characteristic irregular ossification—bizarre paraosteal osteochondromatous proliferation of bone (BPOP)—has been reported in several patients (Meneses *et al.*, 1993). It can develop on long bone and it is more frequent in the small bones of the hands and feet.

To assess whether the hypertrophic chondrocyte differentiation toward osteoblast-like cells, observed *in vitro* (Descalzi Cancedda *et al.*, 1992; Gentili *et al.*, 1993), could also be recognized *in vivo*, synthesis of type I collagen by hypertrophic chondrocytes was investigated by immunolocalization of the C-propeptide. Hypertrophic chondrocytes synthesizing type I collagen were located in the cartilage adjacent to the chondro–bone junction and surrounding areas where neovessel formation occurred (Galotto *et al.*, 1994). Type I collagen was also synthesized by more elongated cells clearly derived from hypertrophic chondrocytes and contributing to the initial deposition of the osteoid matrix. By enzyme cytochemistry it was shown that the same cells express alkaline phosphatase. Evidence that *in vivo* hypertrophic chondrocytes resume proliferation was obtained by bromodeoxyuridine (BrdU) labeling and subsequent staining of cells with specific monoclonal antibodies. Additional evidence of maturation of hypertrophic chondrocytes to osteoblast-like cells was derived by experiments in which specific monoclonal antibodies, obtained by injecting mice with whole hypertrophic chondrocytes or osteoblasts, were used to stain sections of different-aged chick embryo tibias (Fig. 9). A monoclonal antibody (LA5) was isolated and characterized that recognizes a hypertrophic chondrocyte membrane protein. In addition to staining hypertrophic chondrocytes surrounded by a type II- and type X collagen-stainable matrix, the LA5 antibodies also stained elongated chondrocytes at the cartilage–bone collar interface and cells incorporated in the first layer of bone and osteoid matrix (Galotto *et al.*, 1994). By taking advantage of other monoclonal antibodies it was shown that antigens were shared both by osteocytes and a subset of peripherically located hypertrophic chondrocytes and by preosteoblasts and hypertrophic chondrocytes (Galotto *et al.*, 1995).

## VI. Production of and Response to Soluble Mediators

Peptide growth factors have been implicated both in the induction of mesoderm and differentiation of the first cartilaginous model of the bone in the early embryo and in the growth and differentiation of chondrocytes within the epiphyseal growth plate leading to endochondral bone formation. Three major classes of peptide growth factors have been associated with these processes: the transforming growth factor $\beta$ family (TGF-$\beta$ and related molecules, including bone morphogenetic proteins), the fibroblast growth factor family (FGF), and the insulin-like growth factors (IGFs), including insulin. Other peptide growth factors, hormones, and soluble modulators reported to play a role in cartilage differentiation and metabolism include retinoic acid (RA), thyroid hormones [triiodothyronine (T$_3$) and thyroxine (T$_4$)], steroid hormones, growth hormone (GH), parathyroid hormone (PTH), platelet-derived growth factor (PDGF), transferrin (Tf), chondromodulin (ChM), and retinoic acid-induced heparin-binding growth factor (RIHB).

## A. Retinoic Acid

The vitamin A derivative retinoic acid (RA) is involved in normal cartilage development. Retinoic acid-soaked beads applied locally to the developing limb bud mimic the action of a graft of the posterior margin region (ZPA; zone of polarizing activity) and induce digit pattern duplication (see Section III,A,3). Retinoic acid added exogenously to developing embryos induces facial and cranial malformations (Tamarin *et al.*, 1984). In the early phases of chondrocyte differentiation, retinoic acid *in vitro* is generally believed to inhibit chondrogenesis and to induce dedifferentiation of chondrocytes.

In differentiated chick and rabbit chondrocytes, RA caused a dramatic change of morphology in a fibroblastic phenotype, accompanied by arrest of type II and IX collagen synthesis and stimulation of type I and type

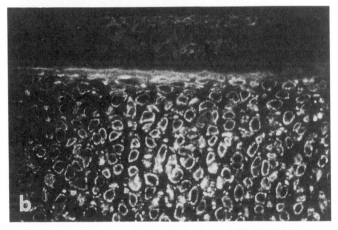

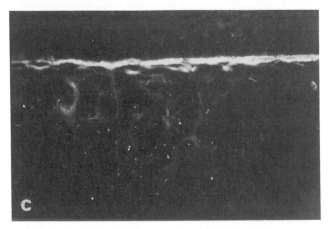

III collagens (Vasan and Lash, 1975; Shapiro and Poon, 1976; Pacifici *et al.*, 1980; Yasui *et al.*, 1986; Benya and Padilla, 1986; Horton *et al.*, 1987). Retinoic acid has a direct effect on gene expression and cell shape change is secondary. Retinoic acid-treated chick sternal chondrocytes suspended in methylcellulose remained rounded, but synthesized proteins characteristic of the dedifferentiated phenotype (Horton and Hassell, 1986).

In most of the experiments mentioned above a single, relatively high RA concentration was used. The effect of different RA concentrations on chondrogenesis was examined in serum-free cultures of chick limb bud mesenchymal cells (Paulsen *et al.*, 1988) and of chick craniofacial mesenchyme (Langille *et al.*, 1989). Concentrations of RA known to cause skeletal duplication *in vivo* dramatically enhanced *in vitro* chondrogenesis. On the other hand, high RA concentrations inhibited chondrogenesis. It is possible that variability in inhibition and induction of chondrogenesis observed by other researchers was due to differences in RA concentrations used.

At later stages of chondrocyte differentiation, retinoic acid promotes chondrocyte hypertrophy and type X collagen production (Oettinger and Pacifici, 1990). In cultures of immature chondrocytes from the permanent cartilage region of chick sternum, RA at "physiological concentrations" induced expression of maturation-associated markers (i.e., type X collagen and alkaline phosphatase) independent of cell hypertrophy (Pacifici *et al.*, 1991b). When immature chondrocytes from chick sterna were allowed to differentiate in culture and were treated with 10–100 n$M$ RA at different culture times, it was observed that more immature chondrocytes failed to activate type X collagen in response to RA. On the other hand, RA induced type X collagen in more developed chondrocytes (Iwamoto *et al.*, 1993a). A similar change in responsiveness to RA with chondrocyte development *in vivo* was determined by analyzing chondrocytes isolated from the cephalic region of different-aged chick sterna. Treatment of mature chondrocytes with RA caused an increase in alkaline phosphatase activity and promoted expression of other mineralization-related genes (Iwamoto *et al.*, 1993b). Under these conditions mineral deposition also occurred in the absence of ascorbic acid. Parathyroid hormone, a known modulator of maturation and mineralization of growth plate chondrocytes,

---

FIG. 9   Micrographs from double immunofluorescence of 13-day chick embryo tibias stained with LA5 monoclonal antibodies and with polyclonal antibodies directed against type I collagen. The series of pictures of the same section are as follows: (a) contrast phase; (b) LA5 antigen; (c) type I collagen. The hypertrophic chondrocyte characteristic of LA5 antigen is also detectable on the membranes of osteocytes entrapped in the type I collagen-rich osteoid. At the same time, synthesis of type I collagen is performed by several frank hypertrophic chondrocytes. Magnification: × 230.

can reverse the process. In conclusion it was shown that RA is a potent, developmentally regulated, ascorbate-independent promoter of chondrocyte maturation and matrix mineralization.

Addition of retinoic acid to the culture medium of *in vitro*, further differentiating hypertrophic chondrocytes, either at "high" concentration (0.5–1 m$M$) or at 50 n$M$ [a "physiological" concentration equivalent to concentrations measured in the chick limb bud (Eichele and Thaller, 1987; Thaller and Eichele, 1987)], accelerates differentiation of chondrocytes to osteoblast-like cells and organization of a bonelike matrix (Descalzi Cancedda *et al.*, 1992; Gentili *et al.*, 1993). In the presence of retinoic acid, cell proliferation was highly enhanced. Cells had already acquired a fibroblastic morphology after 2–3 days of retinoic acid treatment. Alkaline phosphatase-positive cells were detectable after about 6–7 days and their number continued to increase up to 8–10 days. Calcium mineral deposition started after 9–11 days. Cell growth arrested when mineralization of the extracellular matrix started. Large bundles of collagen fibrils with a periodicity characteristic of type I collagen and with evidence of discontinuous mineralization along their axis were identified by ultrastructural analysis of the extracellular matrix. The analysis of proteins secreted by the cells under these conditions were consistent with the morphological observations. A switch from type II to type I collagen was observed between day 2 and day 3 and type X collagen synthesis arrested between day 3 and day 4.

## B. Thyroid Hormones

Serum-free media that support expression of differentiated function by chick embryo sternal and growth plate chondrocytes have been developed. When the serum was replaced by thyroid hormones, chondrocytes did not proliferate, but matured into hypertrophic chondrocytes synthesizing type X collagen and alkaline phosphatase, provided that IGF-I or high-concentration insulin was present (Bohme *et al.*, 1992; Quarto *et al.*, 1992b). "Physiological" concentrations of $T_3$ ($10^{-11}$ $M$) were effective. It should be emphasized that the cocktail of hormones failed to support cell growth, and that chondrocytes replicated only in 10% fetal calf serum. Additional factors are therefore necessary during chondrogenesis to cell growth promotion.

## C. Transforming Growth Factor $\beta$ Superfamily

Transforming growth factor $\beta$ was first identified in tumor cells and subsequently found in a variety of normal tissues. Currently there are more than 20 members of the TGF-$\beta$ superfamily (Table I; Massagué, 1990; Hoffmann, 1991; Sporn and Roberts, 1992). All proteins share a conserved, approximately 100-amino acid carboxy-terminal domain containing 7 cysteines conserved in position. This carboxy-terminal domain is cleaved when two proteins, synthesized as larger preproteins, combine to form an active disulfide-bound homodimer or heterodimer. This article focuses both on the TGF-$\beta$s themselves and on the subfamily of the bone morphogenetic proteins (BMPs) that have been shown to play a major role in cartilage and bone formation during embryonic development.

## 1. Transforming Growth Factor $\beta$ Subfamily

High levels of TGF-$\beta$ mRNAs and proteins in the growth plate, epiphysis, and metaphysis of long bones have been shown by several groups by *in situ* hybridization (Sandberg *et al.*, 1988; Hoosein *et al.*, 1988; Gatherer *et al.*, 1990; Pelton *et al.*, 1990; Millan *et al.*, 1991) and by immunocytochemistry (Heine *et al.*, 1987; Jingushi *et al.*, 1990; Jakowlew *et al.*, 1991; Thorp *et al.*, 1992).

TABLE I

Transforming Growth Factor $\beta$ Superfamily

| Factor | Ref. |
|---|---|
| TGF-$\beta$1 | Massagué (1990); Sporn and Roberts (1990) |
| TGF-$\beta$2 | Massagué (1990); Sporn and Roberts (1990) |
| TGF-$\beta$3 | Massagué (1990); Sporn and Roberts (1990) |
| TGF-$\beta$4 | Massagué (1990); Sporn and Roberts (1990) |
| TGF-$\beta$5 | Massagué (1990); Sporn and Roberts (1990) |
| Inhibin $\alpha$ | Massagué (1990); Sporn and Roberts (1990) |
| Inhibin $\beta$A (activin A) | Thomsen *et al.* (1990) |
| Inhibin $\beta$B (activin B) | Thomsen *et al.* (1990) |
| Müllerian inhibiting substance | Massagué (1990); Sporn and Roberts, (1990) |
| Growth/differentiation factor 1 (GDF-1) | Lee (1990) |
| Growth/differentiation factor 5 (GDF-5) | Storm *et al.* (1994) |
| Growth/differentiation factor 6 (GDF-6) | Storm *et al.* (1994) |
| Growth/differentiation factor 7 (GDF-7) | Storm *et al.* (1994) |
| Vegetal specific 1 (Vgr-1) | Massagué (1990); Ruiz *et al.* (1990) |
| Bone morphogenetic protein 2 (BMP-2) | Wozney *et al.* (1990) |
| Bone morphogenetic protein 4 (BMP-4) | Wozney *et al.* (1990) |
| *Drosophila* decapentaplegic | Gelbart, (1989); Panganiban *et al.* (1990) |
| Bone morphogenetic protein 3 (BMP-3) | Wozney *et al.* (1990) |
| Bone morphogenetic protein 5 (BMP-5) | Wozney *et al.* (1990) |
| Bone morphogenetic protein 6 (BMP-6) | Wozney *et al.* (1990) |
| Bone morphogenetic protein 7 (BMP-7) | Wozney *et al.* (1990) |
| *Drosophila* 60A | Wharton *et al.* (1991) |

Transforming growth factor $\beta1$ and TGF-$\beta2$ elicit a striking increase in cartilage-specific proteoglycan accumulation in the matrix and type II collagen expression by high-density micromass cultures of cells derived from chick embryo stage 23/24 limb buds or from the prechondrogenic region comprising the distal subridge mesenchyme of stage 25 wing buds (Kulyk et al., 1989b). A brief exposure to TGF-$\beta$ at the beginning of the culture was sufficient to trigger chondrogenesis. In addition, TGF-$\beta$ also induced cartilage gene expression and cartilage matrix formation in low-density cultures of mesenchymal cells, a situation in which little or no chondrogenic differentiation occurs. Leonard et al. (1991) have repeated the experiment and proposed that activation of the chondrogenic pattern by TGF-$\beta$ is related to an enhancement of mesenchymal cell condensation possibly owing to stimulation of fibronectin expression. Experiments in which TGF-$\beta1$ was locally applied on developing chick wing buds indicated that exposure to the exogenous factor in vivo influences the development of limb skeletal elements in a stage- and position-dependent manner (Hayamizu et al., 1991). Schofield and Wolpert (1990) have confirmed the chondrogenesis induction by TGF-$\beta1$ and shown an additive effect of basic FGF (bFGF). A synergistic action of TGF-$\beta$ with bFGF was also observed for the induction of mesoderm in Xenopus embryos (Kimelman and Kirschner, 1987). Transforming growth factor $\beta1$ induced reexpression of the cartilage phenotype in dedifferentiated rabbit articular chondrocytes (Benya and Padilla, 1993). Dihydrocytochalasin B enhanced the TGF-$\beta$ effect.

In cultures of rabbit growth plate chondrocytes, TGF-$\beta1$ caused dose-dependent inhibition of DNA duplication and induced only low rates of colony formation. On the other hand, in the same cells, TGF-$\beta1$ potentiated DNA duplication and increased the efficiency of colony formation stimulated by purified bFGF (Hiraki et al., 1988; Iwamoto et al., 1989b). Therefore for TGF-$\beta$ a bifunctional effect on the proliferation of chondrocytes depending on the presence of bFGF has been proposed. In rabbit chondrocyte pellet cultures, TGF-$\beta$ also caused a dose-dependent stimulation of cartilage specific proteoglycan production (Hiraki et al., 1988) and, when added to the culture at late stage, suppressed almost completely the increase in alkaline phosphatase activity and blocked the conversion of maturing chondrocyte into calcifying chondrocytes (Kato et al., 1988). Using an in vitro model of rat epiphyseal three-dimensional pellet cultures, Ballock et al. (1993) have shown that exogenous TGF-$\beta1$ reversibly prevents terminal differentiation of chondrocytes to hypertrophic cells and stabilizes the phenotype of the prehypertrophic chondrocyte. On the other hand, in the same cells maintained as a subconfluent monolayer, TGF-$\beta1$ inhibited gene expression for cartilage matrix proteins, demonstrating the importance of microenvironment in determining the action of growth factors on chondrocytes. Transforming growth factor $\beta$ acting in synergy with basic FGF arrests spontaneous differentiation in chemically defined medium of chick sternal cephalic chondrocytes to hypertrophic chondrocytes (Bohme *et al.*, 1994). Transforming growth factor $\beta$ acting through autocrine mechanisms also prevents hypertrophy in high-density agarose suspension cultures of chick sternal chondrocytes (Tschan *et al.*, 1993). Under these culture conditions sternal chondrocytes did not progress to the hypertrophic stage. Media conditioned by such cells also prevent serum-induced chondrocyte hypertrophy in chondrocytes cultured at low densities and caused a phenotypic modulation closely resembling that observed in cartilage cells cultured in monolayer, which included synthesis of type I collagen rather than cartilage-specific collagens. Addition to the conditioned media of monoclonal antibodies directed against recombinant human TGF-$\beta$2 abolished all these effects.

Transforming growth factor $\beta$ autoregulates its expression in several cell types. Transforming growth factor $\beta$1, 2, 3, and 4 are coordinately expressed in chick chondrocytes both *in vitro* and *in vivo* (Jakowlew *et al.*, 1991). The addition of TGF-$\beta$1, 2, and 3 to cultured sternal chondrocytes resulted in an increase in the levels of the mRNAs for TGF-$\beta$1, 2, and 3, but did not change the level of TGF-$\beta$4 mRNA (Jakowlew *et al.*, 1992). Interestingly, the same article reports that, in chondrocytes, retinoic acid increases the expression of TGF-$\beta$2 and 3 mRNAs and TGF-$\beta$2 protein. A modulation of the different TGF-$\beta$ forms by retinoic acid has also been observed during early mouse morphogenesis (Mahmood *et al.*, 1992).

## 2. Bone Morphogenetic Protein Subfamily

Bone morphogenetic proteins (BMPs) are a group of factors of the TGF-$\beta$ superfamily playing a major role in the control of chondrogenesis and osteogenesis (Reddi, 1992; Wozney, 1992). The BMPs were originally identified as molecules derived from bone and with the ability to induce ectopic cartilage and bone (Luyten *et al.*, 1989; Celeste *et al.*, 1990; Carrington and Reddi, 1991). More information on the ectopic cartilage and bone formation induced by BMPs is given in Section VIII,B of this article. This inducer activity of BMPs and the finding that the Bmp5 mutation in *short ear* mice causes defects in skeletal structures (Kingsley *et al.*, 1992) strongly suggest that BMPs are physiological inducers of cartilage and bone formation during embryonic development. It has been shown that the alterations, including different numbers of bones, observed in the limb of *brachypodism* mice are due to a mutation of GDF5, a new BMP-related factor identified by degenerate polymerase chain reaction (Storm *et al.*, 1994). The *short ear* mutation alters the size and shape of ears, sternum, ribs, and vertebral processes but does not interfere with limb bone formation or digit number and morphology. On the other hand, the *brachypodism* mutation does not alter ear, sternum, rib, and vertebral process formation. This specificity of skeletal alteration suggests that different members of the BMP family control the formation of different bones in the vertebrate skeleton. It must be noted that these genes are involved in much more than bone formation. By *in situ* hybridization BMP-2 expression in the mouse was reported in the apical ectodermal ridge of the limb bud, the atrioventricular canal of the heart, the whisker follicles, and tooth buds (Lyons *et al.*, 1990).

In cultures of chick limb bud mesenchymal cells BMP-3 and BMP-4 stimulate cartilage formation at all concentrations and at all culture times tested (Carrington *et al.*, 1991; Chen *et al.*, 1991). In this culture system BMP-4 is not only capable of overcoming the observed inhibitory effect of TGF-$\beta$1 and 2, but also reveals a strong synergistic effect with these factors in promoting chondrogenesis. In cultures of already differentiated chondrocytes BMP-2, BMP-3, and BMP-4 promote and maintain the cartilage phenotype (Hiraki *et al.*, 1991b; Luyten *et al.*, 1992).

### 3. Activin

When the role of activin (inhibin $\beta$) in chick limb development was analyzed, it was found that activin enhanced chondrogenesis, up to fivefold, in mesenchymal cell micromass cultures (Jiang *et al.*, 1993). Activin treatment resulted in a size increase of the precartilaginous condensations, but not in the cell number. Activin promoted expression of N-CAM in precartilaginous condensations and of tenascin in cartilage nodules.

### D. Fibroblast Growth Factor Family

The fibroblast growth factor (FGF) family presently includes seven members (Table II). These factors present a varying degree of homology and

TABLE II

Fibroblast Growth Factor Family

| Factor | Ref. |
| --- | --- |
| aFGF (FGF-1) | Gimenez-Gallego *et al.* (1985); Burgess *et al.* (1986) |
| bFGF (FGF-2) | Bohlen *et al.* (1984); Esch *et al.* (1985) |
| INT-2 (FGF-3) | Dickson *et al.* (1984); Brookes *et al.* (1989) |
| K-FGF, HST (FGF-4) | Sakamoto *et al.* (1986); Delli Bovi *et al.* (1987) |
| FGF-5 | Zhan *et al.* (1988) |
| FGF-6 | Marics *et al.* (1989) |
| KGF (FGF-7) | Rubin *et al.* (1989) |

similarly promote the proliferation of several cells of mesodermal and neuroectodermal origin and are angiogenic (Basilico and Moscatelli, 1992). Characteristically, the factors of this family interact with extracellular matrix. Signal transduction occurs via the binding of the factors to membrane receptors, but interaction of factors with extracellular matrix molecules and in particular with the heparan sulfate moieties of heparan sulfate proteoglycans (HSPGs) present on the cell surface and in the extracellular matrix (Säkselä *et al.*, 1988) may be particularly important to potentiate the activity of the factors. Fibroblast growth factors have a major effect on both prechondrogenic mesenchymal cells and already differentiated chondrocytes.

The apical ectodermal ridge (AER), the specialized epithelium covering the embryo limb, stimulates proliferation of the underlying mesenchyme, causing a direct limb outgrowth (Tabin, 1991). FGF-4 (kFGF) and bone morphogenetic protein 2 (BMP-2) are expressed in the AER at the time the tissue mediates the proliferative response of the mesenchyme. Using cultures of mouse embryo limb, Niswander and Martin (1993) have shown that FGF-4 stimulates proliferation of the mesenchyme in the limb bud. On the other hand, BMP-2 has an inhibitory effect. The extent of limb growth can be modulated by mixing the two factors, suggesting that limb growth is controlled by a combination of stimulatory and inhibitory signals from the AER. The same group has applied beads soaked in recombinant FGF-4 to the limb mesenchyme of chick embryo *in ovo* after removal of the ridge (Niswander *et al.*, 1993). When applied simultaneously to apical and posterior mesenchyme, FGF-4 stimulates proliferation of cells in the distal mesenchyme and maintains the polarizing activity in the posterior mesenchyme required for elaboration of skeletal elements in the normal proximodistal sequence.

Basic FGF (FGF-2) is probably the most potent mitogen for chondrocytes (Kato *et al.*, 1983; Kato and Gospodarowicz, 1984). A large amount of a factor initially named cartilage-derived growth factor (CDGF) and subsequently identified as bFGF has been purified from neonatal bovine cartilage (Sullivan and Klagsburn, 1985). Therefore cartilage contains a large amount of bFGF. When rabbit or chick chondrocytes were grown in soft agar in the presence of bFGF, a high level of colony formation was induced (Kato *et al.*, 1987; Koike *et al.*, 1990). On the other hand, when the factor was added to cultures of pelleted chondrocytes at a late stage but before chondrocytes reached hypertrophy, bFGF prevented chondrocyte terminal differentiation, that is, synthesis of type X collagen, increase in alkaline phosphatase activity, and subsequent deposition of calcium minerals in the extracellular matrix (Kato and Iwamoto, 1990). With respect to this observation it must be noted that the number of bFGF receptors in chondrocytes decreases when the cells become hypertrophic (Iwamoto et al., 1991).


## E. Insulin-like Growth Factors and Growth Hormone

Insulin-like growth factors (IGFs) or somatomedins are growth hormone (GH)-dependent growth factors capable of promoting cartilage growth. Insulin-like growth factor type I is mainly synthesized in the liver and is the major circulating member of the family in rat. Insulin-like growth factor type II is particularly abundant in embryonic tissues and may have a more significant role in controlling prenatal growth. Limb development in chick embryo is characterized by lateral mesodermal condensation. The existence of a signal from the mesonephros capable of promoting limb outgrowth has been shown by laser ablation experiments (Geduspan and Solursh, 1992). The factor released by the mesonephros has been identified as IGF-I (Geduspan and Solursh, 1993).

Both IGF-I and IGF-II are synthesized by cartilage cells (Burch et al., 1986; Nilsson et al., 1986). Interestingly, changes in the numbers of IGF-I receptors were observed during in vitro maturation of rabbit growth plate chondrocytes (Suzuki, 1994). The maximum level was reached after 12 days of culture at the maturation stage. This suggests the existence of autocrine and paracrine loops involving IGFs during chondrocyte differentiation. Early cartilage is a nonvascularized tissue; in addition, serum concentration of IGFs does not correlate with disorders of growth. Therefore IGFs locally produced by chondrocytes are playing the major role in skeletal growth (Canalis et al., 1989).

The involvement of locally produced IGF-I in the growth of the mouse neonatal mandibular condyle has been shown (Maor et al., 1993). The distribution of IGF-I production was parallel to the distribution of the receptors for IGF-I, mainly in the younger zones of the condyle, that is, the chondroprogenitor and the chondroblast cell layers. This observation supports the hypothesis that cartilage development may be dependent on autocrine and paracrine activity of endogenous GH-independent IGF-I. In serum-free media the addition of insulin-like growth factor I (IGF-I) (Bohme et al., 1992) or high insulin concentration (Quarto et al., 1992b) was an absolute requirement for chick embryo chondrocyte differentiation. In the experiments by Quarto et al. the insulin could not be substituted by IGF-I; it remains to be investigated whether this was due to an intrinsic unresponsiveness of chondrocytes to IGF-I or to the production by the cells of IGF-binding proteins that compete with receptor binding. Indeed, the production of IGF-binding proteins (IGFBPs) by bovine fetal and adult articular and fetal growth plate chondrocytes grown in alginate beads has been observed (Olney *et al.*, 1993). In particular the synthesis of 39/43-kDa IGFBP-3, 33-kDa IGFBP-2, and 29- and 24-kDa IGFBP bands was demonstrated.

Pituitary growth hormone (GH) begins to regulate IGF-I production only after birth, i.e., 6 months of postnatal life in humans and 20 postnatal days in rats (Sara and Carlsson-Skwirut, 1988). Although fetal growth is considered to be independent of GH, it is possible that pituitary GH or nonpituitary GH-like molecules directly influence fetal growth through the GH receptor. While investigating differentiation of preadipose cells into adipocytes, Green *et al.* (1985) proposed a "dual effect theory" of GH action. According to this theory, tissue growth occurs in two steps: (1) differentiated cells are first formed from precursor stem cells, and (2) the number of young differentiated cells is then clonally expanded through a limited number of cell divisions. Growth hormone may stimulate the first step directly and the second step indirectly through its mediator IGF-I.

Local injection of GH into the tibial epiphyseal growth plate of hypophysectomized rats stimulates longitudinal bone growth (Isaksson *et al.*, 1982; Russell and Spencer, 1985). It has been suggested that the direct growth-promoting effect of GH on cartilage *in vivo* is mediated by local production of IGF-I (Schlechter *et al.*, 1986). Nevertheless, although additional experiments are necessary before definitive conclusions are reached, it is possible that GH also stimulates, both directly and indirectly, the chondrogenic lineage. Indeed, GH receptor immunoreactivity was detected in several rat embryo tissues, in particular skeletal tissues, including chondroprogenitor cells (Garcia-Aragon *et al.*, 1992). The presence of GH receptors was first demonstrated in developing rabbit tibial growth plate (Barnard *et al.*, 1988). In the neonatal period the receptors were limited to the mature hypertrophic zone and later they were more widespread and were also observed in the resting and proliferative zones. Additional investigation on localization of GH receptors in growth plate cells is probably necessary. Instead, IGF-I was immunologically detected in relatively high concentrations in the proliferative zone of rat growth plate cartilage but only weakly in the resting and hypertrophic zones (Nilsson *et al.*, 1986). After hypophysectomy the number of IGF-I-positive cells was strongly diminished but it was restored after GH treatment, suggesting an effect of GH on the induction of IGF-I synthesis by chondrocytes. More recently, Hunziker *et al.* (1994) have infused GH and IGF-I into hypophysectomized rats and concluded that both molecules are capable of stimulating growth plate chondrocytes at all stages of differentiation. In particular, at variance with previous reports, they have indicated that IGF-I is also capable of stimulating more immature cells. On the basis of *in vitro* studies, it has been proposed by Vetter *et al.* (1986) that the growth-stimulating activity of IGF-II is more effective on fetal chondrocytes whereas that of IGF-I is more effective on postnatal chondrocytes. Conflicting results have been obtained when the effect of GH on cultured chondrocytes was studied. It has been shown that both GH and IGF-I promote growth plate chondrocyte colony formation in agarose suspension culture (Lindahl *et al.*, 1986, 1987). Mainly on the basis of the analysis of formed colonies it was proposed that GH interacts with prechondrogenic or newly differentiated chondrocytes, whereas IGF-I interacts only with proliferative chondrocytes. Furthermore, Maor *et al.* (1989) have shown that human GH enhances proliferation, chondrogenesis, and osteogenesis of mouse chondroprogenitor cells cultured on collagen sponges. Results from another study suggest that an increased number of cell divisions during primary monolayer culture of tibial growth plate chondrocytes increases GH responsiveness in a subsequent suspension cultures (Ohlsson *et al.*, 1994). Furthermore, IGF-I primarily stimulated the growth of differentiated chondrocyte cell clones whereas GH mainly stimulated the growth of undifferentiated cell clones. Other workers failed to detect a direct effect of GH on chondrocytes at any differentiation stage (Isgaard, 1992). A GH-responsive chondrogenic cell line was established from fetal rat tibia (Ohlsson *et al.*, 1993). It is hoped that studies with this and similar cell lines will help to clarify the role of GH in chondrogenesis induction.

## F.  Transferrin

Hypertrophic chondrocytes undergoing differentiation to osteoblast-like cells *in vitro* synthesize and secrete large amounts of ovotransferrin (Gentili *et al.*, 1993). When the ovotransferrin mRNA level was determined in these cells, the mRNA was not detected in the starting hypertrophic chondrocytes, was present at high levels after 20 days of culture, and was barely detectable at the end of culture. A transient expression of the ovotransferrin was clearly observed, both at the protein and at the mRNA level, when chondrocytes were treated with retinoic acid. Maximal expression of ovotransferrin occurred after 2–3 days, when the switch from the synthesis of type II collagen to the synthesis of type I collagen was observed. It is noteworthy that, when culture conditions allowed extracellular matrix assembly (i.e., in the presence of ascorbic acid), ovotransferrin was also expressed by hypertrophic chondrocytes and derived osteoblast-like cells (Gentili *et al.*, 1994). In addition, ovotransferrin expression was also detected by *in vitro* differentiating chondrocytes within 2–3 days after transferring of the prechondrogenic dedifferentiated cells into suspension. By Western blot, ovotransferrin receptors were detected at a low level in extracts from differentiating and hypertrophic chondrocytes and at a higher level in extracts from hypertrophic chondrocytes undergoing differentiation to osteoblast-like cells and from mineralizing osteoblasts (Gentili *et al.*, 1994).

*In vivo,* the expression of ovotransferrin and its receptor during chick tibial development was investigated by immunocytochemistry and *in situ* hybridization (Gentili *et al.*, 1994). Ovotransferrin was first detectable in the 7-day cartilage rudiment. At later developmental stages, ovotransferrin localized in the articular cartilage, in the hypertrophic cartilage, and in the osteoid at the chondro–bone junction. The receptor was expressed at a low level by chondrocytes at all differentiation stages and at a high level by stacked osteoprogenitor cells in the diaphysis collar and by derived osteoblasts.

These findings suggest that autocrine and paracrine mechanisms involving ovotransferrin and its receptor may play a role in controlling chondrocytes differentiation, their maturation to osteoblast-like cells, and initial bone formation. Transferrin receptors are associated with virtually all actively proliferating cells. The possibility that ovotransferrin is responsible for the cell proliferation enhancement observed during hypertrophic chondrocyte differentiation should be considered. Transferrin and its receptor play a role in embryonic morphogenesis (Ekblom *et al.*, 1983). It has been shown that, in rat organ cultures of both developing kidneys and teeth, transferrin is required for cell proliferation and differentiation (Partanen *et al.*, 1984; Thesleff and Ekblom, 1984). As in the case of the maturation of hypertrophic chondrocytes to osteoblast-like cells, in both organs morphogenesis progression is characterized by changes in the proliferation rate of different cell populations and overt cell differentiation is preceded by active cell proliferation.

## G. Steroid Hormones

Among steroid hormones, dexamethasone (Dex) plays a major role in promoting proliferation of skeletal cell precursors. In cultures of mesenchymal cells derived from avian connective tissue matrices, Dex promotes differentiation of muscle, fat, cartilage, and bone cells in a time- and concentration-dependent manner (Young *et al.*, 1993). In RCJ 3.1, a clonally derived cell line obtained from fetal rat calvaria, the addition of Dex was an absolute requirement for the formation of adipocytes and chondrocytes; muscle cells and osteoblasts were also expressed at low frequency in the absence of Dex (Grigoriadis *et al.*, 1988).

In serum-free media, dexamethasone was not necessary for the progression of chondrogenesis but supported chick embryo chondrocyte viability and modulated some differentiated functions (Quarto *et al.*, 1992b). The effect of Dex on the expression of type X collagen was particularly interesting. The synthesis of the protein was in fact inversely proportional to the hormone concentration; however, Dex did not inhibit but rather delayed the time of appearance of the collagen.

It is widely accepted that *in vivo* the rise of sex hormones, testosterone and estrogen, coincident with puberty, promotes the hypertrophy of all chondrocytes and the resulting long bone growth arrest. Much less information is available on the effect of sex hormones on cultured growth plate chondrocytes. Chondrocytes from the male and female rat costochondral growth and resting cartilage zone were treated with testosterone; the hormone lowered cell number and DNA duplication in male cells and, when fetal calf serum was present, stimulated alkaline phosphatase activity in the cells from the growth zone (Schwartz *et al.*, 1994). These data are compatible with testosterone promoting differentiation of male chondrocytes *in vitro*.

Interestingly, interactions between steroid hormones and IGF-I in chondrocytes have been reported. In confluent cultures of rabbit costal chondrocytes, when one of the sex steroids (progesterone, testosterone, and 17$\beta$-estradiol) was added 24 hr before addition of IGF-I, the stimulation of DNA synthesis by the growth factor was enhanced 40–80% above that of the culture not pretreated with steroids (Itagane *et al.*, 1991). On the other hand, dexamethasone suppressed IGF-I induction of DNA synthesis by 60%. This suppression was higher when dexamethasone was added before IGF-I.

## H.  Vitamin $D_3$ Metabolites

The 1,25-$(OH)_2$ vitamin $D_3$, a hormonal metabolite of vitamin $D_3$, stimulates chondrogenesis in cultures of chick embryo limb bud mesenchymal cells (Tsonis, 1991). The effect is specific because vitamin $D_3$ or its derivative 24,25-$(OH)_2$ vitamin $D_3$ does not stimulate chondrogenesis. Cultured mesenchymal cells express 1,25-$(OH)_2$ vitamin $D_3$ receptors only following exposure to the hormone. *In vitro,* growth plate chondrocytes express 1,25-$(OH)_2$ vitamin $D_3$ receptors after they have stopped dividing and they have reached hypertrophy, whereas articular chondrocytes do not ( Jikko *et al.*, 1993). The affinity for the ligand and the number of receptors (11,000 per cell) were similar to those observed *in vivo* in the lower hypertrophic cartilage zone of growth plate (Iwamoto *et al.*, 1989a).

The 1,25-$(OH)_2$ vitamin $D_3$ DNA stimulated DNA synthesis and proliferation of cultured rabbit costal chondrocytes in the logarithmic growth phase in a dose-dependent manner, inhibiting at the same time their glycosaminoglycan synthesis (Takigawa *et al.*, 1988a). The 1,25-$(OH)_2$ vitamin $D_3$ also induced soft agar growth of chick embryo chondrocytes in the presence of fetal calf serum or bFGF (Sato *et al.*, 1990). Other vitamin $D_3$ metabolites such as 24,25-$(OH)_2$ vitamin $D_3$ and 25-(OH) vitamin $D_3$ did not stimulate chondrocyte growth (Sato *et al.*, 1990), although the former showed *in vitro* a stimulation of the expression of their differentiated phenotype (Takigawa *et al.*, 1988a; Hale *et al.*, 1986).

In cultures of pelleted rabbit growth plate chondrocytes, 1,25-$(OH)_2$ vitamin $D_3$ increased DNA content and proteoglycan synthesis while decreasing alkaline phosphatase activity (Kato *et al.*, 1990a). These reports suggest that 1,25-$(OH)_2$ vitamin $D_3$ also plays an important role in the chondrocyte terminal differentiation.

Sylvia *et al.* (1993) have shown that 1,25-$(OH)_2$ vitamin $D_3$ induced a rapid, dose-dependent stimulation of protein kinase C, beginning 3 min after addition and sustained until 90 min, in cultures of chondrocytes derived from the growth zone of rat chondrocostal cartilage, but not in cultures of chondrocytes derived from the resting zone.

## I. Additional Hormones and Growth Factors

Among growth factors, in addition to the ones already mentioned in previous sections, several others are certainly cooperating in controlling chondrocyte differentiation and metabolism. Retinoic acid-induced heparin-binding factor (RIHB) belongs to a family of heparin-binding and retinoic acid-induced proteins transiently expressed during embryogenesis (Vigny *et al.*, 1989; Urios *et al.*, 1991). RIHB expression by avian embryo prechondrogenic cells and chondrocytes was investigated. *In vivo* in chick wing limb the differentiation of the chondrogenic precursor was concomitant with the synthesis of RIHB mRNA (Duprez *et al.*, 1993). The factor was detected by immunofluorescence in the center of hindlimb buds at embryonic stage 27 and in embryo tibial chondrocytes at stage 36 (Castagnola *et al.*, 1994). *In vitro*, at physiological concentrations and within 5 days of culture, retinoic acid induces a burst of RIHB expression, both at the mRNA and protein level, in primary cultures of chick chondrocytes derived from 15-day-old embryo sterna (Cockshutt *et al.*, 1993). Differentiating chick embryo chondrocytes expressed RIHB mRNA and protein only when ascorbic acid was present in the culture medium (a condition promoting extracellular matrix assembly) (Castagnola *et al.*, 1994). Addition of the iron-kelating agent $\alpha,\alpha'$-dipyridyl, a condition overcoming the ascorbic acid action and impairing extracellular assembly, inhibited RIHB mRNA accumulation. Therefore, unless the culture is supplemented with retinoic acid, a correct extracellular matrix assembly is required for the expression of RIHB gene by chondrocytes.

Synthesis of proteoglycans, an indicator of chondrogenesis, was inhibited in cultures of chick limb bud mesenchymal cells by all three PDGF isoforms (AA, AB, and BB) (Chen *et al.*, 1992). Recombinant BMP-2B reversed the inhibitory effect of PDGF. All PDGF isoforms increased cell proliferation by 48 hr in both high- and low-density cultures. However, at later times cell proliferation was inhibited by PDGF AA and PDGF AB but not by PDGF BB. Wroblewski and Edwall (1992) have reported that PDGF AA and PDGF BB homodimers favor rat rib growth plate chondrocyte differentiation and together with IGF-I interact in the regulation of longitudinal bone growth.

A new factor that may play a major role in chondrocyte differentiation has been identified by Hiraki *et al.* (1991a), who have succeeded in purifying from fetal bovine epiphyseal cartilage a 10- to 30-kDa glycosylated protein, named chondromodulin 1 (ChM-1), which stimulates growth of cultured chondrocytes in the presence of FGF. Oligonucleotide primers were derived from the determined N-terminal sequence of the protein and used to amplify bovine epiphyseal cartilage cDNA by polymerase chain reaction and to identify the mRNA for the protein. The expression of this mRNA was highly specific to cartilage because it was not detectable in any other tissue tested. On the basis of the determined, complete nucleotide sequence of the mRNA it was postulated that it encodes a larger precursor that is probably inserted in the cell membrane (Hiraki *et al.*, 1991b). It was also speculated that the mature ChM-1, located at the C terminal of the molecule, is cleaved after the insertion of the precursor in the membrane and that the N-terminal two-thirds of the precursor, which shares a high level of sequence similarity with human and rat pulmonary surfactant apoprotein C [for which reason it was named chondrosurfactant protein (CH-SP)], remains associated with the cell membrane. Further studies are necessary to assess definitively the functions of these proteins.

Agents using adenosine 3′,5′-cyclic monophosphate as second messenger enhance chondrocyte proliferation. In cultured mouse condylar cartilages the addition of parathyroid hormone (PTH) to the culture medium initially stimulated proliferation of chondroprogenitor cells but suppressed subsequent differentiation to chondroblasts (Lewinson *et al.*, 1992). In addition, the hormone induced alterations in hypertrophic chondrocyte differentiation. Parathyroid hormone has a mitogenic effect on chick and rabbit fetal but not postnatal chondrocytes (Koike *et al.*, 1990). In both fetal and postnatal chondrocytes PTH promotes and maintain chondrocytic differentiation. Parathyroid hormone inhibits chondrocyte hypertrophy, and increases alkaline phosphatase activity and onset of mineralization in cultured rabbit growth plate chondrocytes (Kato *et al.*, 1990b). A similar inhibitory effect on chondrocyte hypertrophy is also caused by interleukin 1 (IL-1) (Kato *et al.*, 1993). The inhibition of calcification was observed only when the chondrocytes were exposed to the factors before the onset of calcification. In *in vitro* cultured growth plate chondrocytes the number of 72-kDa receptors for parathyroid hormone on the cell surface was low in resting and proliferating chondrocytes, increased 10-fold in matrix-forming chondrocytes, and thereafter decreased in hypertrophic chondrocytes (Iwamoto et al., 1994).

The parathyroid hormone-related peptide (PTHrP) and PTH have similar biological effects that are mediated through the same PTH receptor. In chick growth plate chondrocytes a dose-dependent stimulation of proliferation by PTHrP was observed (Loveys et al., 1993). At the same time stimulation of proteoglycan synthesis and decreased collagen synthesis and alkaline phosphatase activity were found.

Protein kinase C might be a negative modulator of chondrogenic differentiation during embryonic limb development. Phorbol 12-myristate 13-acetate (PMA), a known activator of protein kinase, inhibits chondrogenesis by mesenchymal cells in vitro (Pacifici and Holtzer, 1977; Lowe et al., 1978). On the other hand, staurosporine, a protein kinase C inhibitor, promotes cartilage differentiation in culture of chick wing bud mesenchymal cells (Kulyk, 1991).

Prostanoids are local hormones derived from arachidonic acid that exert multiple cellular actions through receptors coupled to guanine nucleotide regulatory proteins and that coordinate responses to the circulating hormones promoting prostanoid synthesis (Smith, 1992). Synthesis of and response to prostaglandins (PGs) in articular chondrocytes have been reported by several research groups. Less information is available on PGs in growth plate chondrocytes. Okiji et al. (1993) have examined by immunocytochemistry the localization of prostaglandin 12-synthase in demineralized sections of various calcified tissue-forming cells in rat. Several of these cells, including chondrocytes, were immunoreactive for prostaglandin 12-synthase, suggesting that they were capable of producing PG 12.

Other factors reported to exert an effort on chondrocyte growth and differentiation include EGF, whose receptors were demonstrated on cultured rabbit costal chondrocytes (Kinoshita et al., 1992), HGF (hepatocyte growth factor, a heparin-binding polypeptide mitogen) (Defrances et al., 1992), and natriuretic peptides (Hagiwara et al., 1994).

## J. Angiogenic and Antiangiogenic Activities

Angiogenesis is a complex cascade of events initiated by the production and release of angiogenic factors and/or the switching off of inhibitors of angiogenesis. These events include activation and migration of endothelial cells, degradation and remodeling of extracellular matrix, endothelial cell proliferation, and neovessel formation (Folkman, 1984; Brown and Weiss, 1988). Formation of new vessels occurs in almost all tissues during embryonic development, but in adults it is limited to the uterine wall, in regenerating tissues following wounding, and in pathologies such as neoplasia, arthritis, and proliferative retinopathy (Folkman and Shing, 1992). Cartilage is normally avascular; however, in the growth plate the invasion of the hypertrophic cartilage zone by blood vessels is certainly a crucial step for the replacement of cartilage by bone (Kuettner and Pauli, 1983). As such, growth plate cartilage represents an interesting tissue in the investigation of angiogenesis control (Yabsley and Harris, 1965; Schenk et al., 1968; Brown et al., 1987). Angiogenic inhibitors have been purified from cartilage [cartilage-derived anti-tumor factor (CATF), Takigawa et al., 1988b; cartilage-derived inhibitor (CDI) or tissue inhibitor of metalloprotease (TIMP) (Moses et al., 1990; Moses and Langer, 1991)] and from conditioned medium of scapular chondrocytes [35-kDa chondrocyte-derived inhibitor (ChDI); Moses et al., 1992]. When chondrocytes derived from the permanent cartilaginous sternum of chick embryo were cocultured with endothelial cell aggregates into fibrin or collagen gels, endothelial sprout formation was markedly inhibited (Pepper et al., 1991). Addition of anti-TGF-$\beta$ antibodies significantly reduced the inhibitory effect. Chondrocyte-conditioned medium or exogenously added TGF-$\beta$ had a similar inhibitory activity. In contrast, angiogenic stimulating activity has been found to be associated with mineralized chondrocytes in vitro [endothelial cell-stimulating angiogenesis factor (ESAF) (McFarland et al., 1990; Brown and McFarland, 1992)]. The appearance of the angiogenic activity is in most cases paralleled by the expression of specific gelatinases.

Chick embryo chondrocytes differentiating in culture modulate the expression and release of angiogenesis inhibitors and activators (Descalzi Cancedda et al., 1995). The prechondrogenic undifferentiated cells exhibited a moderate angiogenic activity. When transferred into suspension culture the cells progressively developed a strong inhibitory activity that reached its maximal value at the hypertrophic stage. Interestingly, when the hypertrophic chondrocytes were supplemented with ascorbic acid (i.e., when chondrocytes could organize their own extracellular matrix), the inhibitory activity in the conditioned medium was replaced by a strong angiogenic activity. Angiogenic activity was also exhibited by hypertrophic chondrocytes undergoing maturation to osteoblast-like cells. The maximal angiogenic activity was reached at the time the culture underwent mineralization.

## VII. Oncogenes and Chondrocyte Differentiation

Expression of different oncogenes has been observed at different times during normal chondrocyte differentiation. During in vitro cultivation of mouse mandibular condyles, high, transient c-*fos* expression was found in the entire tissue within 30 min of culture as a result of the mechanical forces applied during dissection (Closs *et al.*, 1990). A second type of c-*fos* expression was instead observed in individual hypertrophic chondrocytes and preceded DNA replication and osteogenic differentiation of these cells. Expression of c-*fos* was also observed by Sandberg *et al.* (1988) by Northern blotting and *in situ* hybridization in the growth plate of developing human long bones. Particularly high levels of c-*fos* expression were detected in chondrocytes bordering the joint space. Indeed, chondrogenic cells and earlier progenitors are transformed by Fos/Jun. In chimeric mice generated with different ES cell clones selected for high exogenous c-*fos* expression, a high frequency of cartilage tumors developed within 3–4 weeks of age. Primary and clonal tumor–derived cell lines were established that expressed high levels of c-*fos*, c-*jun*, and cartilage-specific type II collagen and gave rise to cartilage tumors *in vivo* (Z. Q. Wang *et al.*, 1991, 1993).

The relationship between c-*myc* gene expression and chondrocyte proliferation and maturation was analyzed during endochondral ossification in chick sternum by *in situ* hybridization (Iwamoto *et al.*, 1993c). Regions of proliferating chondrocytes all contained high levels of c-*myc* mRNA, whereas in regions of postmitotic chondrocytes, as hypertrophic cartilage, the mRNA was undetectable. This finding was confirmed by Northern blot analysis and by the observation that *myc* antisense oligonucleotide inhibited proliferation in cultured chondrocytes. Similarly, by immunocytochemistry, Farquharson *et al.* (1992) have shown a restricted localization of the c-Myc protein in the proliferating and differentiating chondrocytes of the growth plate of rat and chick long bones. The expression of c-*myc* was also investigated *in situ* in the chondrocytes of the tibial growth plate of chick with tibial dyschondroplasia, in which chondrocytes are developmentally arrested in the transition stage between proliferation and differentiation (Loveridge *et al.*, 1993). c-*myc* expression was reduced in the transitional chondrocytes, but unaltered levels were observed in the proliferating chondrocytes. Addition of 1,25-dihydroxyvitamin D to the diet, which resulted in a reduced incidence of dyschondroplasia, restored c-Myc production.

The effect of two avian oncogenes on the chondrocyte phenotype and proliferative capacity was investigated (Alemá *et al.*, 1985). Viruses carrying the *src* oncogene suppressed expression of the chondrocyte-specific markers type II collagen and aggrecan. On the other hand, viruses carrying the *myc* oncogene did not interfere with chondrocyte marker expression, but increased the cell proliferative potential.

In some cases, by transfection of chondrocytes with an oncogene DNA or by infection with a retrovirus carrying an activated oncogene, it has been possible to obtain a continuous cell line. Apparently these cell lines are "frozen" in a specific differentiation stage and respond poorly to microenvironment changes. The avian myelocytomatosis virus strain MC29, carrying the oncogene v-*myc*, stimulated quail embryo chondrocytes to proliferate with a progressively reduced doubling time (Gionti *et al.*, 1985). Infected chondrocytes were established in culture as a continuous cell line expressing type II collagen and cartilage proteoglycans but no type X collagen, a marker of hypertrophic chondrocytes. Uninfected control cultures also expressed type X collagen and survived only a few months. Similarly, Quarto *et al.* (1992a) have shown that constitutive expression of v-*myc* keeps quail chondrocytes in stage I (active proliferation and synthesis of type II collagen) and prevents these cells from reconstituting hypertrophic calcifying cartilage when the culture medium is supplemented with ascorbic acid. In agreement with these results, Iwamoto *et al.* (1993c) reported that constitutive overexpression of c-*myc* by retroviral vectors in immature chick chondrocytes maintained the cells in a proliferative state and blocked their maturation into hypertrophic chondrocytes. In these cells the expression of maturation-related genes, such as the gene encoding type X collagen, is prevented. Following infection of primary fetal costal chondrocytes with a recombinant retrovirus (NIH/J2) carrying the *myc* and *raf* oncogenes, a rapidly proliferating clonal line was isolated that maintained a stable chondrocyte phenotype for at least 50 passages (Horton *et al.*, 1988). This line synthesized high levels of cartilage proteoglycan core protein and link protein, but showed reduced type II collagen expression.

By infecting primary embryonic mouse limb bud chondrocytes with a retrovirus carrying the simian virus large T oncogene, a continuous cell line was obtained that still expressed chondrocyte markers, such as type II, IX, and XI collagen as well as cartilage aggrecan and link protein (Mallein-Gerin and Olsen, 1993). On the other hand, by transfection of rabbit articular chondrocytes with a plasmid encoding SV40 early function genes, an immortal cell line (SVRAC) was obtained that displayed an apparently irreversibly dedifferentiated phenotype (Thenet *et al.*, 1992). Infection of chick chondrocytes with the Rous sarcoma virus (RSV), carrying the oncogene v-*src*, resulted in a continuous line presenting reduced serum requirements and able to grow in semisolid medium (Gionti *et al.*, 1989). This line did not express cartilage genes and expressed type I collagen and fibronectin. At variance with RSV-transformed fibroblasts, this line was not tumorigenic following grafting onto the chorioallantoic membrane of embryonated duck eggs.

A clonal cell line with cartilage phenotype and tumorigenicity during a more than 3-year-long culture has been derived from a human chondrosarcoma (Takigawa *et al.*, 1989). After reaching confluence the cells continued to proliferate slowly and formed cartilage nodules. Long-term cultures of the Swarm rat chondrosarcoma chondrocytes have also been established. Under standard adherent conditions a phenotypic instability was encountered. When instead the cells were cultured in agarose, persistence of the chondrocytic phenotype was described (Kucharska *et al.*, 1990). Another chondrogenic cell line (ATDC5) was derived from a teratocarcinoma (Atsumi *et al.*, 1990). Interestingly, when the culture was supplemented with insulin, cells continued to grow even in a postconfluent phase and formed cartilage nodules.

## VIII.  Endochondral Bone Formation Outside the Growth Plate

### A.  Regulation of Fracture Repairs: Callus Formation

In an adult organism, the fracture healing process repeats several stages of endochondral bone formation in the same temporal order (i.e., mesenchyme, cartilage, bone). The temporal and spatial appearance of the different components of these tissues is closely linked to the expression of genes also expressed during developmental bone formation (Sandberg *et al.*, 1993). The location of different collagens in healing fractures of rabbit and rat has been investigated with antibodies directed against the type-specific forms of matrix collagens (Page *et al.*, 1986; Lane *et al.*, 1986). In mechanically stable fractures type I, III, and V collagens rapidly appeared in the primitive mesenchymal callus over the entire periosteal surface. Small nodules of cartilage containing type II and IX collagens were also observed. On the other hand, in mechanically unstable fractures, that is, in areas of motion or anoxia, a chondroid tissue containing type II and IX collagens first appeared. The cartilage was subsequently replaced by endochondral ossification. The presence at this stage of the hypertrophic cartilage-specific type X collagen was demonstrated by Grant *et al.* (1987) in experimental fractures created in chick humerus. An experimental model of fracture healing was used to study production of type I and type II collagen by *in situ* hybridization. The location of the first cells expressing type II collagen, which were adjacent to the cortical bone, suggested that chondrocytes originated from cells derived from the periosteum by differentiation (Sandberg *et al.*, 1989). An accurate investigation of gene expressed in the reparative callus formed after fracture of the rat femur has been published by Jingushi *et al.* (1992). Expression of extracellular matrix protein genes were examined in two callus regions: soft callus (cartilage formation) and hard callus (bone formation). Messenger RNAs for type II collagen and for proteoglycan core protein were maximally expressed in the soft callus during chondrogenesis (day 9). Messenger RNAs for type I collagen and alkaline phosphatase were maximally expressed in the hard callus during endochondral ossification and bone remodeling (day 15) and at 50% maximal value during intramembranous bone formation (day 7). Messenger RNAs for these proteins were detected in the soft callus at low levels during chondrogenesis (day 9) but increased to 80% of maximal levels with chondrocyte hypertrophy and mineralization of the cartilage matrix (day 13). Osteocalcin was expressed in the hard callus during endochondral bone formation and remodeling and during intramembranous bone formation, but never in the bone callus. Osteonectin was expressed in both the hard and soft callus throughout the entire bone repair process. This modulation of gene expression in soft and hard callus as repair progresses suggests a local microenvironmental control.

Apparently the same growth factors that play a role in the control of embryonic endochondral bone formation also play a major role in the control of cartilage and bone formation during fracture repair (Bolander, 1992). Immediately following injury, growth factors, including TGF-$\beta$1 and PDGF, are released into the fracture hematoma by platelets and inflammatory cells. These factors have an influence on the cartilage and intramembranous bone formation in the initial callus. Subsequently chondrocytes and other cells present in the callus become one of the major sources of these and other factors. The expression of TGF-$\beta$1 in the callus from normally healing human fractures has been shown by in situ hybridization (Andrew et al., 1993). Transforming growth factor $\beta$ mRNA was localized in areas of proliferation of mesenchyme, cartilage, and bone. The ability of TGF-$\beta$1 and 2 to stimulate proliferation and differentiation in mesenchymal precursor cells in the periosteum, as it occurs in early fracture healing, was shown by Joyce et al. (1990). Daily injections of both factors into the subperiosteal region of newborn rat femurs resulted in localized chondrogenesis and intramembranous bone formation. The ratio of cartilage to intramembranous bone formation decreased as the dose of TGF-$\beta$ was lowered. After cessation of the injections, endochondral ossification occurred and cartilage was replaced by bone. Moreover, injection of TGF-$\beta$2 stimulated synthesis of TGF-$\beta$1 in chondrocytes and osteoblasts within the newly induced cartilage and bone, suggesting positive autoregulation of TGF-$\beta$. Induction in vivo of bone involving perichondrial cells by a single application of recombinant TGF-$\beta$1 in rabbit ear was reported by Beck et al. (1991).

## B. Calcium Deficiency-Induced Expression of Cartilage Phenotype during Development of Intramembranous Bones

Emergence of a chondrogenic phenotype is observed in typically osteogenic tissue in response to severe systemic calcium deficiency. In calcium-deficient chick embryos (obtained by long-term shell-less cultures), cartilage-specific proteoglycans and type II collagen are produced in the undermineralized calvarial matrix (Jacenko and Tuan, 1986). In view of the reported involvement of TGF-$\beta$ in chondrogenesis the expression of this factor in calvaria of shell-less and control embryos was investigated both at the mRNA and protein level (Sato and Tuan, 1992). Transforming growth factor $\beta$ expression was significantly increased in shell-less calvaria compared to control. The addition of exogenous calcium prevented this increase.

## C. Ectopic Endochondral Bone Formation in Animal Models

Urist (1965) first discovered that, in rabbit, an intramuscular implant of demineralized and lyophilized fragments of bone induces ectopic bone formation on initial, newly formed cartilage. Also, this process recapitulates the pathway of endochondral bone formation. The cascade of events includes activation, migration, and proliferation of mesenchymal progenitor cells; differentiation, hypertrophy, and mineralization of cartilage; angiogenesis and vascular invasion; differentiation, mineralization, and remodelling of bone; and hematopoietic marrow differentiation in the ossicles (Reddi, 1981). Solubilization of the demineralized bone extracellular matrix by chaotropic agents such as guanidine hydrochloride, urea, and sodium dodecyl sulfate (SDS) has led to the indentification of several bone morphogenetic proteins (BMPs). BMPs were in fact originally identified as molecules derived from bone and capable of inducing ectopic cartilage and bone when subcutaneously injected in animal models in most cases after reconstitution with the insoluble residue of the extraction (Sampath et al., 1987; Wang et al., 1988; Luyten et al., 1989; Celeste et al., 1990).

Additional BMPs have been identified by recombinant DNA methodology, cloned, and expressed in vitro (Wozney et al., 1988).

All BMPs so far examined are able to initiate de novo cartilage and bone formation. Recombinant BMP-2, -3, -4, and -7 also induce chondrogenesis and osteogenesis in vivo when they are injected as single proteins (Wozney et al., 1988; Wang et al., 1990; Hammonds et al., 1991).

### D. Chondrocyte Hypertrophy in Osteoarthritic Cartilage

von der Mark *et al.* (1992) have shown the appearance of hypertrophic chondrocytes in osteoarthritic cartilage (OA) by immunostaining for type X collagen with specific antibodies. Freshly isolated OA chondrocytes synthesized mostly type X collagen, whereas normal control chondrocytes synthesized almost exclusively type II collagen. In advanced stages of osteoarthrosis a switch to the deposition of overlapping type I, II, and III collagen was observed (von der Mark *et al.*, 1992). The expression of type X, type I, type II, and type III collagen was compared in normal and degenerated human articular cartilage by *in situ* hybridization (Aigner *et al.*, 1993a,b). At sites of newly formed osteophytic and repair cartilage type X mRNA was strongly expressed and marked areas of endochondral bone formation.

## IX. Disorders in Metabolism of Controlling Agents or in Cell Responses May Lead to Several Deformities Classified as Chondrodysplasias

The chondrodysplasias are inherited disorders of the vertebrate skeleton in which cartilage functions are disturbed. The result is short, deformed bones that grow slowly and frequently exhibit weakened articular surfaces predisposed to osteoarthritis.

Alterations in chondrocyte differentiation have been described in avian tibial dyschondroplasia. Similar abnormalities have been described in pigs, dogs, and horses (Leach and Gay, 1987). In growth plate of affected chick, chondrocytes are arrested in the transitional phase between proliferation and differentiation. The result is an accumulation of unmineralized and avascular cartilage that extends from the growth plate into the metaphysis. Dyschondroplastic chick showed reduced c-*myc* and TGF-$\beta$ expression in the transitional chondrocytes (Loveridge *et al.*, 1993). Following addition of 1,25-dihydroxyvitamin D to the diet, in areas where the lesion was being repaired an increase in c-*myc* and TGF-$\beta$ was observed. Several forms of chondrodysplasias have been described in humans (Rimoin and Lachman, 1990; Spranger, 1992; Horton and Hecht, 1993). In some patients with hypochondroplasia a genetic linkage with a polymorphism of the IGF-I gene was described (Mullis *et al.*, 1991).

Collagens are the extracellular matrix components that are mainly responsible for providing all tissues with structural strength. Nonexpression of cartilage type II collagen was observed in a case of human achondrogenesis (Eyre *et al.*, 1986). The best characterized human chondrodysplasias result from mutations of COL2A1, the gene encoding human type II procollagen. Sixteen different mutations in COL2A1 have been identified to date (Table III). Affected individuals are heterozygous for the mutations, all of which have mapped to the triple-helical domain of the molecule. A majority of the clinical phenotypes falls within the spondyloepiphyseal dysplasia (SED) subclass of chondrodysplasias. The general SED phenotype is dominated by abnormalities of vertebral bodies and epiphyses of long bones with variable involvement of metaphyses; it varies widely in severity, forming a continuum of specific phenotypes. These range from profound dwarfism with death before or at birth [achondrogenesis type II (most severe); hypochondrogenesis (slightly less severe)] to moderately severe dwarfism with early osteoarthritis (SED congenita) to late-onset forms with minimal or no short stature but precocious osteoarthritis of weight-bearing joints (late-onset SED). In addition, patients with the severe to moderately severe SED phenotypes (i.e., SED congenita) exhibit small thorax, cleft palate, facial bone abnormalities, club feet, and frequently ocular problems, such as high-grade myopia and retinal detachment. Several other entities, such as Kniest dysplasia and Strudwick syndrome, fall within this SED continuum. Also, the Stickler syndrome shares many features with the SED phenotype, including eye abnormalities and precocious osteoarthritis.

Another collagen gene, whose mutation has been associated with chondrodysplasia, is the gene encoding type X collagen. Mutations in the type X collagen gene have been found to segregate with Schmid metaphyseal chondrodysplasia, an autosomal dominant disorder of the osseous skeleton (McIntosh et al., 1994). Mutations include a 13-bp deletion, two frameshift mutations, and one missense mutation.

Despite extensive morphological and biochemical analysis of cartilage tissues from patients with chondrodysplasias and osteoarthritis, the mechanisms by which the mutations produce the disease remain poorly understood. Many questions remain unanswered, such as the following: (1) Why do similar mutations produce phenotypes with dramatically different severity? (2) Do abnormal cartilage collagen fibrils result from altered ratios of cartilage collagen molecules or from inclusion of mutant chains into fibrils? (3) Do abnormal fibrils disturb the template functions of growth plate cartilage? (4) How do alterations in the production of cartilage collagens contribute to degeneration of articular cartilages, that is, osteoarthritis?

The availability of technologies that allow designed gene mutations to be introduced into mice has provided means to develop murine models of human CDX and SEDs, and to analyze directly the adverse effects of mutations in a biological and developmental context.

Mutations were designed to disturb particular functions of type II (pro)-

TABLE III

Human Col2a1 Mutations

| Clinical phenotype | Exon | Mutation | Residue | Protein | Authors | Year |
|---|---|---|---|---|---|---|
| Hypochondrogenesis | 33 | G → A | 574 | Gly → Ser | Horton et al. | 1992 |
| Hypochondrogenesis | 43 | G → A | 853 | Gly → Glu | Bogaert et al. | 1992 |
| Hypochondrogenesis | 46 | G → A | 943 | Gly → Ser | Vissing et al. | 1989 |
| SED | 11 | C → T | 75 | Arg → Cys | Williams et al. | 1993 |
| SED | 15 | G → A | 154 | Gly → Arg | Vikkula et al. | 1993 |
| SED | 17 | G → A | 175 | Gly → Arg | Winterpacht et al. | 1994 |
| SED | 19 | G → A | 247 | Gly → Ser | Ritvaniemi et al. | 1994 |
| SED | 20 | Del → skip | 258–273 | 18-aa del | Tiller et al. | 1992 |
| SED | 41 | C → T | 789 | Arg → Cys | Chan et al. | 1991 |
| SED | 44 | 875–877 | 875–877 | 3-aa dupl | Winterpacht et al. | 1994 |
| SED | 48 | Del | 963–999 | 38-aa del | Lee et al. | 1989 |
| SED | 48 | Dupl | 970–984 | 15-aa dupl | Tiller et al. | 1990 |
| SED | 48 | G → A | 997 | Gly → Ser | Chan et al. | 1991 |
| SED | 48 | G → A | 997 | Gly → Ser | Winterpacht et al. | 1993 |
| Kniest | 12 | Del 3'Spl | 91–108 | 28-aa del | Winterpacht et al. | 1993 |
| Kniest | 21 | Skip | 274–279 | 6-aa del | Spranger et al. | In press |
| Kniest | 49 | Del | 1007–12 | 6-aa del | Winterpacht et al. | 1994 |
| Stickler | 7 | C → T | 9 | Arg → stop | Ahmad et al. | 1993 |
| Stickler | 39 | C → T | 732 | Arg → stop | Ahmad et al. | 1991 |
| Stickler | 40 | frameshift | – | stop (ex 42) | Brown et al. | 1992 |
| Stickler | 43 | frameshift | – | stop (ex 44) | Ritvaniemi et al. | 1993 |
| Wagner | 10 | G → A | 67 | Gly → Asp | Kokko et al. | 1993 |
| Familial OA | 21 | G → A | 274 | Gly → Ser | Winterpacht et al. | In press |
| Familial OA | 30 | G → A | 443 | Gly → Ser | Katzenstein et al. | Submitted |
| Familial OA | 31 | C → T | 519 | Arg → Cys | Ala-Kokko et al. | 1990 |
| Familial OA | 31 | C → T | 519 | Arg → Cys | Holderbaum et al. | 1993 |

Modified from M. Metsaranta's doctoral thesis.

collagen (Table IV). From the analysis of these transgenic mice, it is clear that selected mutations of murine Col2a1 behave as dominant negative mutations and can produce mouse phenotypes that resemble human SEDs. The murine phenotypes involve varying degrees of shortening of the bones of the limbs, spine, and face together with structural abnormalities of growth plate and other cartilages and ultrastructural abnormalities of chondrocytes and cartilage collagen fibrils. It appears that the SED phenotype correlates best with a reduction in the abundance of normal cartilage collagen fibrils in cartilage matrix (Garofalo *et al.*, 1991, 1993; Metsäranta *et al.*, 1992; Rintala *et al.*, 1993). Abnormal fibrils can be produced by mutations that disrupt cross-linking sites and also by overproduction of type II collagen in general.

Similar phenotypes were observed by Vandenberg *et al.* (1991) in mice harboring a "mini"-transgene of human COL2A1 designed to disrupt the assembly of type II collagen molecules producing protein suicide. Cheah *et al.* (1993) produced a slightly different chondrodysplasia phenotype when a transgene containing the SV40 T antigen coupled to the Col2a1 promoter was introduced into mice. Expression of the foreign protein was thought to interfere with chondrocyte differentiation. A similar lethal phenotype was generated by Bruggeman *et al.* (1991), who introduced a transgene encoding the diphtheria toxin A chain gene controlled by the rat Col2a1 promoter and enhancer into mice. Two murine chondrodysplasia models have been produced in which the expression of another cartilage

TABLE IV

Col2a1 Mutations in Transgenic Mice

| Mutations | Nucleotide(s) | Exon(s) | Residue(s) | Protein | Ref. |
|-----------|---------------|---------|------------|---------|------|
| Gly-85 | G → T | 11 | 85 | Gly → Cys | Garofalo *et al.* (1991) |
| Lys-87 | A → G | 11 | 87 | Lys → Arg | Garofalo *et al.* (unpublished) |
| N-tel Lys | A → G | 6 | N-tel 122 | Lys → Arg | Metsäranta *et al.* (unpublished) |
| Sil | C → T | 7 | None | No change | Garofalo *et al.* (1993) |
| Del 6 | Del (78 bp) | 6 | N-tel 111–132 | 26-aa del | Katzenstein *et al.* (unpublished) |
| Del 7 | Del (45 bp) | 7 | 1–15 | 15-aa del | Metsäranta *et al.* (1992) |
| Del 27 | Del (54 bp) | 27 | 429–447 | 18-aa del | Garofalo *et al.* (unpublished) |
| Del 15–25 | Del (810 bp) | 15–25 | 142–411 | 270-aa del | Vanhoutte *et al.* (unpublished) |

collagen gene was targeted. In the first, a partially deleted Col9a1 gene/ cDNA under the control of the rat type II collagen promoter was introduced into mice (Nakata et al., 1993). The affected mice were not noticeably small, but exhibited evidence of early osteoarthritis. In the second instance, the mice harbored a partially deleted chick Col10a1 gene designed to produce a shortened a1(X) chain capable of disrupting the assembly of type X collagen molecules (Jacenko et al., 1993a,b). Many affected pups were small, had immunodeficiency, developed severe kyphosis and associated spinal cord compression, and died shortly after birth. The hypertrophic zone of the growth plate was shorter than normal by microscopy, and the mouse was put forth as a possible model for spondylometaphyseal dysplasia. Therefore it is obvious that several transgenic mice with chondrodysplasias have been generated that harbor mutations predicted to alter specific molecular functions. However, the relationship of the murine phenotypes to their human counterparts and consequently their usefulness as models to elucidate disease mechanisms in humans has been somewhat difficult to establish because skeletal development and growth differ between the two species.

Much less information is available on skeletal abnormalities associated with alterations in other extracellular matrix molcules. Nanomelia is a lethal genetic mutation in chickens, characterized by shortened and malformed limbs. Owing to a single base mutation generating a new stop codon in the proteoglycan core protein gene, cultured nanomelic chondrocytes synthesize a truncated aggrecan core protein precursor that is neither processed to a mature proteoglycan, nor translocated and secreted by the cells (Li et al., 1993; Vertel et al., 1993).

## X. Concluding Remarks

In this article we have described differentiation pathway and controlling mechanisms of growth plate chondrocytes during endochondral bone formation. The overall picture emerging from available data obtained while investigating this process both in vivo, in developing embryos, and in vitro, in cultures of differentiating cells, indicates that initial chondrocyte differentiation depends on positional signaling mediated by selected homeobox-containing genes and soluble mediators, such as retinoids and growth factors. In addition, the onset of chondrogenesis and, more importantly, the further development of the chondrocyte differentiation program strongly rely on interactions of the differentiating cells with the microenvironment. Cell–cell and cell–extracellular matrix contacts are an absolute requirement for the correct unfolding of the program. Production of and response to different hormones and growth factors are observed at all times during the process. In several cases, cells producing a factor, or immediately adjacent cells, present on their surface specific receptors for the factor, suggesting that autocrine and paracrine cell stimulations are key elements of the process. In this regard, particularly relevant in promoting onset of chondrogenesis is the role of the TGF-$\beta$ superfamily, and more specifically of the BMP subfamily. Other factors, such as IGFs and perhaps transferrin, most probably play a major role when, preceding additional steps of differentiation, clonal expansion of a cell population that has already reached some differentiation stage occurs by a number of cell divisions. The influence of local microenvironment might also offer an acceptable settlement to the debate as to whether hypertrophic chondrocytes convert to bone cells and live, or remain chondrocytes and die. The expression of the osteogenic potential of hypertrophic chondrocytes is observed *in vitro* only when the cells are exposed to particular microenvironmental culture conditions and is restricted *in vivo* to a site-specific subset of hypertrophic chondrocytes (Fig. 10). Location within the bone rudiment may convey specific differences in the microenvironment to

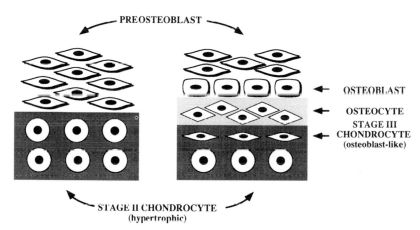

FIG. 10    Bone formation at the chondro–osseous junction of chick long bones. The unfolding of hypertrophic chondrocyte osteogenic potential is restricted *in vivo* to a site-specific subset of cells. Hypertrophic chondrocytes do assume osteoblast-like features following a round of DNA synthesis. At the chondro–osseous junction, acquisition of osteogenic competence occurs nearly simultaneously both in hypertrophic chondrocytes and in committed osteogenic cells. According to this model the initial bone formation must be regarded as a joint endeavor of chondrocyte-derived osteoblast-like cells and of osteoblasts. Therefore osteocytes entrapped in the initial osteoid could derive from both cell populations. It is postulated that the process is triggered by interactions between stage II hypertrophic chondrocytes and preosteoblasts via secretion of paracrine activities and interaction with extracellular matrix molecules deposited by the opposing cell population.

which hypertrophic chondrocytes become exposed during bone development, such as interaction with osteoblast-derived matrix molecules and cellular cross-talk via cell–cell contact or soluble mediators. We would like to suggest that the ultimate fate of hypertrophic chondrocytes may be different at different microanatomical sites. In the subperiosteal and perivascular regions, sites of first bone formation, hypertrophic chondrocytes mature to osteoblast-like cells. At other sites, such as the interior of "degenerating" cartilage, the inherent ability of hypertrophic chondrocytes to convert to an osteoblast-like phenotype may become frustrated and hypertrophic chondrocytes may die, possibly by apoptosis.

## Acknowledgments

This work was supported by grants from Progetti Finalizzati: "Ingegneria Genetica" and "Applicazioni Cliniche della Ricerca Oncologica," CNR (Rome) and by funds from the Associazione Italiana per la Ricerca sul Cancro (Milan). We thank Drs. Rodolfo Quarto and Francesco Minuto for suggestions and helpful discussions, and Silvio Garofalo, who specifically contributed to Section IX. We also thank Ms. Barbara Minuto and Ms. Daniela Giacoppo for editorial and secretarial help.

## References

Adams, S. L., Boettinger, D., Focht, R. J., Holtzer, H., and Pacifici, M. (1982). Regulation of the synthesis of extracellular matrix components on chondroblasts transformed by a temperature-sensitive mutant of Rous sarcoma virus. *Cell (Cambridge, Mass.)* **30,** 373–384.

Adams, S. L., Pallante, K. M., Niu, Z., Leboy, P. S., Golden, E. B., and Pacifici, M. (1991). Rapid induction of type X collagen gene expression in cultured chick vertebral chondrocytes. *Exp. Cell Res.* **193,** 190–197.

Aeschlimann, D., Wetterwald, A., and Paulsson, M. (1993). Expression of tissue transglutaminase in skeletal tissues correlates with events of terminal differentiation of chondrocytes. *J. Cell Biol.* **120,** 1461–1470.

Ahmad, N. N., McDonald-McGinn, D. M., Zackal, E. H., Knowlton, R. G., LaRossa, D., Dimascio, J., and Prockop, D. J. (1993). A second mutation in the type II procollagen gene (COL2A1) causing the Stickler syndrome (arthro-ophtalmopathy) also a premature termination codon. *Am. J. Hum. Genet.* **52,** 39–45.

Ahmad, N. N., Alakokko, L., Knowlton, R. G., Jimenez, S. A., Weaver, E. J., Maguire, J. I., Tasman, W., and Prockop, D. J. (1991). Stop codon in the procollagen-II gene (COL2A1) in a family with the Stickler syndrome (arthro-ophthalmopathy). *Proc. Natl. Acad. Sci. U.S.A.* **88,** 6624–6627.

Ahrens, M., Ankenbauer, T., Schroder, D., Hollnagel, A., Mayer, H., and Gross, G. (1993). Expression of human bone morphogenetic proteins-2 or -4 in murine mesenchymal progenitor C3H10T1/2 cells induces differentiation into distinct mesenchymal cell lineages. *DNA Cell Biol.* **12,** 871–880.

Ahrens, P. B., Solursh, M., and Reiter, R. S. (1977). Stage-related capacity for limb chondrogenesis in cell culture. *Dev. Biol.* **60,** 69–82.

Aigner, T., Bertling, W., Stoss, H., Weseloh, G., and von der Mark, K. (1993a). Independent expression of fibril-forming collagens I, II and III in chondrocytes of human osteoarthritic cartilage. *J. Clin. Invest.* **91**, 829–837.

Aigner, T., Reichenberger, E., Bertling, W., Kirsch, T., Stoss, H., and von der Mark, K. (1993b). Type X collagen expression in osteoarthritic and rheumatoid articular cartilage. *Virchows Arch. B* **63**, 205–211.

Ala-Kokko, L., Baldwin, C. T., Moskowitz, R. W., and Prockop, D. J. (1990). Single base mutation in the type II procollagen gene (COL2A1) as a cause of primary osteoarthritis associated with mild chondrodysplasia. *Proc. Natl. Acad. Sci. U.S.A.* **87**, 6565–6568.

Alemá, S., Tato, F., and Boettinger, D. (1985). *myc* and *src* oncogenes have complementary effects on cell proliferation and expression of specific extracellular matrix components in definitive chondroblasts. *Mol. Cell. Biol.* **5**, 538–544.

Alini, M., Matsui, Y., Dodge, G. R., and Poole, A. R. (1992). The extracellular matrix of cartilage in the growth plate before and during calcification: Changes in composition and degradation of type II collagen. *Calcif. Tissue Int.* **50**, 327–335.

Allen, F., Tickle, C., and Warner, A. (1990). The role of gap junctions in patterning of the chick limb bud. *Development (Cambridge, UK)* **108**, 623–634.

Andrew, J. G., Hoyland, J., Andrew, S. M., Freemont, A. J., and Marsh, D. (1993). Demonstration of TGF-beta 1 mRNA by in situ hybridization in normal human fracture healing. *Calcif. Tissue Int.* **52**, 74–78.

Atsumi, T., Miwa, Y., Kimata, K., and Ikawa, K. (1990). A chondrogenic cell line derived from a differentiating culture of AT805 teratocarcinoma cells. *Cell Differ. Dev.* **30**, 109–116.

Bab, I., Ashton, B. A., Gazit, D., Marx, G., Williamson, M. C., and Owen, M. E. (1986). Kinetics and differentiation of marrow stromal cells in diffusion chambers "in vivo." *J. Cell Sci.* **84P**, 139–151.

Ballock, T. R., Heydemann, A., Wakefield, L. M., Flanders, K. C., Roberts, A. B., and Sporn, M. B. (1993). TGF beta 1 prevents hypertrophy of epiphyseal chondrocytes: Regulation of gene expression for cartilage matrix proteins and metalloproteases. *Dev. Biol.* **158**, 414–429.

Barnard, R., Haynes, K. M., Werther, G. A., and Waters, M. J. (1988). The ontogeny of growth hormone receptors in the rabbit tibia. *Endocrinology (Baltimore)* **122**, 2562–2569.

Basilico, C., and Moscatelli, D. (1992). The FGF family of growth factors and oncogenes. *Adv. Cancer Res.* **59**, 115–159.

Beck, L. S., Deguzman, L., Lee, W. P., Xu, Y., McFatridge, L. A., Gillet, N. A., and Amento, E. P. (1991). In vivo induction of bone by recombinant transforming growth factor beta 1. *J. Bone Miner. Res.* **6**, 961–968.

Bee, J. A., and von der Mark, K. (1990). An analysis of chick limb bud intercellular adhesion underlying the establishment of cartilage aggregates in suspension culture. *J. Cell Sci.* **96**, 527–536.

Bennett, V. D., Weiss, I. M., and Adams, S. L. (1989). Cartilage-specific 5' end of chick alpha 2(I) collagen mRNAs. *J. Biol. Chem.* **15**, 8402–8409.

Bennett, V. D., Pallante, K. M., and Adams, S. L. (1991). The splicing pattern of fibronectin mRNA changes during chondrogenesis resulting in an unusual form of the mRNA in cartilage. *J. Biol. Chem.* **266**, 5918–5924.

Benya, P. D., and Padilla, S. R. (1986). Modulation of the rabbit chondrocyte phenotype by retinoic acid terminates type II collagen synthesis without inducing type I collagen: The modulated phenotype differs from that produced by subculture. *Dev. Biol.* **118**, 296–305.

Benya, P. D., and Padilla, S. R. (1993). Dihydrocytochalasin B enhances transforming growth factor-beta-induced reexpression of the differentiated chondrocyte phenotype without stimulation of collagen synthesis. *Exp. Cell Res.* **204**, 268–277.

Benya, P. D., and Schaffer, J. D. (1982). Dedifferentiated chondrocytes reexpress the differentiated collagen phenotype when cultured in agarose gels. *Cell* (*Cambridge, Mass.*) **30,** 215–224.

Benya, P. D., Brown, P. D., and Padilla, S. R. (1988). Microfilament modification by dihydrocytochalasin B causes retinoic acid-modulated chondrocytes to reexpress the differentiated collagen phenotype without a change in shape. *J. Cell Biol.* **106,** 161–170.

Bernier, S. M., and Goltzman, D. (1993). Regulation of expression of the chondrocytic phenotype in a skeletal cell line (CFK2) in vitro. *J. Bone Miner. Res.* **8,** 475–484.

Bernier, S. M., Desjardins, J., Sullivan, A. K., and Goltzmand, D. (1990). Establishment of an osseous cell line from fetal cat calvaria using an immunocytolytic method of cell selection: Characterization of the cell line and of derived clones. *J. Cell. Physiol.* **145,** 274–285.

Berry, L., and Shuttleworth, C. A. (1989). Expression of the chondrogenic phenotype by mineralizing culures of embryonic chick calvarian bone cells. *Bone Miner.* **7,** 31–45.

Berry, L., Grant, M. E., McClure, J., and Rooney, P. (1992). Bone-marrow-derived chondrogenesis "in vitro." *J. Cell Sci.* **101,** 333–342.

Bianco, P., Fisher, L. W., Young, M. F., Termine, J. D., and Robey, P. G. (1991). Expression of bone sialoprotein (BSP) in developing human tissues. *Calcif. Tissue Int.* **49,** 421–426.

Binderman, I., Greene, R. M., and Pennypacker, J. P. (1979). Calcification of differentiating skeletal mesenchyme in vitro. *Science* **206,** 222–225.

Bogaert, R., Tiller, G. E., Weiss, M. A., Gruber, H. E., Rimoin, D. L., Cohn, D. H., and Eyre, D. R. (1992). An amino acid substitution (Gly853->Glu) in the collagen $\alpha 1(II)$ chain produces hypochondrogenesis. *J. Biol. Chem.* **267,** 22522–22526.

Bohlen, P., Baird, A., Esch, F., Ling, N., and Gospodarowicz, D. (1984). Isolation and partial molecular characterization of pituitary fibroblast growth factor. *Proc. Natl. Acad. Sci. U.S.A.* **81,** 5364–5368.

Bohme, K., Conscience-Egli, M., Tschan, T., Winterhalter, K. H., and Bruckner, P. (1992). Induction of proliferation or hypertrophy of chondrocytes in serum-free culture: The role of insulin-like growth factor-I, insulin, or thyroxine. *J. Cell Biol.* **116,** 1035–1042.

Bohme, K., Winterhalter, K. H., and Bruckner, P. (1994). Terminal differentiation of chondrocytes in culture is a spontaneous process and is arrested by transforming growth factor-$\beta 2$ and basic fibroblast growth factor in synergy. *J. Cell Biol.*

Bolander, M. E. (1992). Regulation of fracture repair by growth factors. *Proc. Soc. Exp. Biol. Med.* **200,** 165–170.

Boskey, A. L. (1992). Mineral-matrix interactions in bone and cartilage. *Clin. Orthop. Relat. Res.* **281,** 244–274.

Boskey, A. L., Stiner, D., Doty, S. B., Binderman, I., and Leboy, P. (1992). Studies of mineralization in tissue culture: Optimal conditions for cartilage calcification. *Bone Miner.* **16,** 1–36.

Bourdon, M. A., and Ruoslahti, E. (1989). Tenascin mediates cell attachment through an RGD-dependent receptor. *J. Cell Biol.* **108,** 1149–1155.

Brookes, S., Smith, R., Casey, G., Dickson, C., and Peters, G. (1989). Sequence organization of the human int-2 gene and its expression in teratocarcinoma cells. *Oncogene* **4,** 429–436.

Brown, C. C., Hembry, R. M., and Reynolds, J. J. (1989). Immunolocalization of metalloprotease and their inhibitor in the rabbit growth plate. *J. Bone J. Surg.* **71,** 580–593.

Brown, P. D., and Benya, P. D. (1988). Alterations in chondrocyte cytoskeletal architecture during phenotypic modulation by retinoic acid and dihydrocytochalasin B-induced reexpression. *J. Cell Biol.* **106,** 171–179.

Brown, R. A., and McFarland, C. D. (1992). Regulation of growth plate cartilage degradation in vitro: Effects of calcification and a low molecular weight angiogenic factor (ESAF). *Bone* **17,** 49–57.

Brown, R. A., and Weiss, J. B. (1988). Neovascularization and its role in the osteoarthritic process. *Ann. Rheum. Dis.* **47**, 881–885.

Brown, R. A., Taylor, C., McLaughlin, B., McFarland, C. D., Weiss, J. B., and Ali, S. Y. (1987). Epiphyseal growth plate cartilage and chondrocytes in mineralizing cultures produce a low molecular mass angiogenic procollagenase activator. *Bone Miner.* **3**, 143–158.

Brown, D. M., Nichols, B. E., Weingeist, T. A., Kimura, A. E., Sheffield, V. C., and Stones, E. M. (1992). Procollagen II gene mutation in Stickler syndrome. *Arch. Ophtalmol.* **110**, 1589–1593.

Brown, R. A., Kayser, M., McLaughlin, B., and Weiss, J. B. (1993). Collagenase and gelatinase production by calcifying growth plate chondrocytes. *Exp. Cell Res.* **208**, 1–9.

Bruckner, P., Horler, I., Mendler, M., Houze, Y., Winterhalter, K. H., Eich-Bender, S. G., and Spycher, M. (1989). Induction and prevention of chondrocyte hypertrophy in culture. *J. Cell Biol.* **109**, 2537–2545.

Bruggeman, L. A., Hou-Xiang, X., Brown, K. S., and Yamada, Y. (1991). Developmental regulation for collagen II gene expression in transgenic mice. *Teratology* **44**, 203–208.

Burch, W. M., Weir, S., and Van Wyk, J. J. (1986). Embryonic chick cartilage produces its own somatomedin-like peptide to stimulate cartilage growth in vitro. *Endocrinology (Baltimore)* **119**, 1370–1376.

Burgeson, R. E., and Hollister, D. W. (1979). Collagen heterogeneity in human cartilage: Identification of several new collagen chains. *Biochem. Biophys. Res. Commun.* **87**, 1124–1131.

Burgess, W. H., Mehlman, T., Marshak, D. R., Fraser, B. A., and Maciag, T. (1986). Structural evidence that endothelial cell growth factor alpha and acidic fibroblast growth factor. *Proc. Natl. Acad. Sci. U.S.A.* **83**, 7216–7220.

Canalis, E., McCarthy, T., and Centrella, M. (1989). The regulation of bone formation by local growth factors. *Bone Miner.* **6**, 27–56.

Capasso, O., Gionti, E., Pontarelli, G., Ambesi-Impiombato, F. S., Nitsch, L., Tajana, G., and Cancedda, R. (1982). The culture of chick embryo chondrocytes and the control of their differentiated functions "in vitro." I. Characterization of the chondrocyte specific phenotypes. *Exp. Cell Res.* **142**, 197–206.

Capasso, O., Tajana, G., and Cancedda, R. (1984). Location of 64K collagen producer chondrocytes in developing chicken embryo tibiae. *Mol. Cell. Biol.* **4**, 1163–1168.

Caplan, A. I. (1991). Mesenchymal stem cells. *J. Orthop. Res.* **9**, 641–650.

Carey, M., Alini, M., Matsui, Y., and Poole, A. R. (1993). Density gradient separation of growth plate chondrocytes. *In Vitro Cell Dev. Biol.* **29**, 117–119.

Carrington, J. L., and Reddi, A. H. (1991). Parallels between development of embryonic and matrix-induced endochondral bone. *BioEssays* **13**, 403–408.

Carrington, J. L., Chen, P., Yanagishita, M., and Reddi, A. H. (1991). Osteogenin (bone morphognetic protein-3) stimulates cartilage formation by chick limb bud cells in vitro. *Dev. Biol.* **146**, 406–415.

Castagnola, P., and Cancedda, R. (1991). Expression of anchorin CII mRNA by cultured chondrocytes. *Cytotechnology* **5**, 41–44.

Castagnola, P., Moro, G., Descalzi Cancedda, F., and Cancedda, R. (1986). Type X collagen synthesis during "in vitro" development of chick embryo tibial chondrocytes. *J. Cell Biol.* **102**, 2310–2317.

Castagnola, P., Torella, G., and Cancedda, R. (1987). Type X collagen synthesis by cultured chondrocytes derived from the permanent cartilaginous region of chick embryo sternum. *Dev. Biol.* **123**, 332–337.

Castagnola, P., Dozin, B., Moro, G., and Cancedda, R. (1988). Changes in the expression of collagen genes show two stages in chondrocyte differentiation "in vitro." *J. Cell Biol.* **106,** 461–467.

Castagnola, P., Bet, P., Quarto, R., Gennari, M., and Cancedda, R. (1991). cDNA cloning and gene expression of chicken osteopontin. *J. Biol. Chem.* **266,** 9944–9949.

Castagnola, P., Tavella, S., Gennari, M., Van De Werken, R., Raffo, P., Raulais, D., Vigny, M., and Cancedda, R. (1995). Retinoic acid-induced heparin binding factor expression by chondrocytes depends on matrix assembly. Submitted for publication.

Celeste, A. J., Iannazzi, J. A., Taylor, R. C., Hewick, R. M., Rosen, V., Wang, E. A., and Wozney, J. M. (1990). Identification of transforming growth factor-β family members present in bone-inductive protein purified from bovine bone. *Proc. Natl. Acad. Sci. U.S.A.* **87,** 9843–9847.

Chan, D., and Cole, W. G. (1991). Low basal transcription of genes for tissue-specific collagens by fibroblast and lymphoblastoid cells. *J. Biol. Chem.* **266,** 12487–12494.

Chan, D., Taylor, T. K. F., and Cole, W. G. (1991). Characterization of an arginine 789 to cysteine substitution in the α1(II) collagen chains of a patient with spondyloepiphyseal dysplasia utilizing low basal transcription of the COL2A1 gene by fibroblasts and analysis of type II collagen from cartilage and cultured chondrocytes. *J. Biol. Chem.* **266,** 12487–12494.

Cheah, K., Levy, A., Kuffner, T., So, C. L., Lovell-Badge, R., Trainor, P. A., and Tam, P. P. L. (1993). Transgenic mouse models of chondrodysplasia in man. *Matrix* **13,** 29.

Chen, P., Carrington, J. L., Hammonds, R. G., and Reddi, A. H. (1991). Stimulation of chondrogenesis in limb bud mesoderm cells by recombinant human bone morphogenetic protein-2b (BMP-2B) and modulation by transforming growth factor beta 1 and beta 2. *Exp. Cell Res.* **195,** 509–515.

Chen, P., Carrington, J. L., Paralkar, V. M., Pierce, G. F., and Reddi, A. H. (1992). Chick limb bud mesodermal cell chondrogenesis: Inhibition by isoforms of platelet-derived growth factor and reversal by recombinant bone morphogenetic protein. *Exp. Cell Res.* **200,** 110–117.

Chuong, C. M., and Edelman, G. M. (1985a). Expression of cell-adhesion molecules in embryonic induction. I. Morphogenesis of nestling feathers. *J. Cell Biol.* **101,** 1009–1026.

Chuong, C. M., and Edelman, G. M. (1985b). Expression of cell-adhesion molecules in embryonic induction. II. Morphogenesis of adult feathers. *J. Cell Biol.* **101,** 1027–1043.

Closs, E. I., Murray, A. B., Schmidt, J., Schon, A., Erfle, V., and Strauss, P. G. (1990). c-*fos* expression precedes osteogenic differentiation of cartilage cells in vitro. *J. Cell Biol.* **111,** 1313–1323.

Cockshutt, A. M., Régnier, F., Vigny, M., Raulais, D., and Chany-Fournier, F. (1993). Retinoic acid-induced heparin-binding factor (RIHB) mRNA and protein are strongly induced in chick embryo chondrocytes treated with retinoic acid. *Exp. Cell Res.* **207,** 430–438.

Coe, M. R., Summers, T. A., Parsons, S. J., Boskey, A. L., and Balian, G. (1992). Matrix mineralization in hypertrophic chondrocyte cultures. Beta glycerophosphate increases type X collagen messenger RNA and the specific activity of pp60$^{c-src}$ kinase. *Bone Miner.* **18,** 91–106.

Coelho, C. N., and Kosher, R. A. (1991). Gap junctional communication during limb cartilage differentiation. *Dev. Biol.* **144,** 47–53.

Coelho, C. N., Upholt, W. B., and Kosher, R. A. (1993a). Ectoderm from various regions of the developing chick limb bud differentially regulates the expression of the chicken homeobox-containing genes GHox-7 and GHox-8 by limb mesenchymal cells. *Dev. Biol.* **156,** 303–306.

Coelho, C. N., Upholt, W. B., and Kosher, R. A. (1993b). The expression pattern of the chicken homeobox-containing gene GHox-7 in developing polydactylous limb buds suggests its involvement in apical ectodermal ridge-directed outgrowth of limb mesoderm and in programmed cell death. *Differentiation (Berlin)* **52**, 129–137.

Collins, M. K. L., and Lopez-Rivas, A. (1993). The control of apoptosis in mammalian cells. *Trends Biochem. Sci.* **18**, 307–309.

Coon, H. G. (1966). Clonal stability and phenotypic expression of chick cartilage cells "in vitro." *Proc. Natl. Acad. Sci. U.S.A.* **55**, 66–73.

Dean, D. D., Muniz, O. E., Berman, I., Pita, J. C., Careno, M. R., Woessner, J. F., and Howell, D. S. (1985). Localization of collagenase in the growth plate of rachitic rats. *J. Clin. Invest.* **76**, 716–722.

Dean, D. D., Muniz, O. E., and Howell, D. S. (1989). Association of collagenase and tissue inhibitor of metalloproteinase (TIMP) with hypertrophic cell enlargement in the growth plate. *Matrix* **9**, 366–375.

Defrances, M. C., Wolf, H. K., Michalopoulos, G. K., and Zarnegar, R. (1992). The presence of hepatocyte growth factor in the developing rat. *Development (Cambridge, UK)* **116**, 387–395.

Delli Bovi, P., Curatola, A. M., Kern, F. G., Greco, A., Ittman, M., and Basilico, C. (1987). An oncogene isolated by transfection of Kaposi's sarcoma DNA encodes a growth factor that is a member of the FGF family. *Cell (Cambridge, Mass.)* **50**, 729–737.

Descalzi Cancedda, F., Manduca, P., Tacchetti, C., Fossa, P., Quarto, R., and Cancedda, R. (1988). Developmentally regulated synthesis of a low molecular weight protein (Ch 21) by differentiating chondrocytes. *J. Cell Biol.* **107**, 2455–2463.

Descalzi Cancedda, F., Asaro, D., Molina, F., Cancedda, R., Caruso, C., Camardella, L., Negri, A., and Ronchi, S. (1990a). The amino terminal sequence of the developmentally regulated Ch 21 protein shows homology with amino terminal sequences of hydrophobic molecule transporters. *Biochem. Biophys. Res. Commun.* **168**, 933–938.

Descalzi Cancedda, F., Dozin, B., Rossi, F., Molina, F., Cancedda, R., Negri, A., and Ronchi, S. (1990b). Ch21 protein, developmentally regulated in chick embryo, belongs to the superfamily of hydrophobic moleules transporters. *J. Biol. Chem.* **265**, 19060–19064.

Descalzi Cancedda, F., Gentili, C., Manduca, P., and Cancedda, R. (1992). Hypertrophic chondrocytes undergo further differentiation in culture. *J. Cell Biol.* **117**, 427–435.

Descalzi Cancedda, F., Melchiori, A., Benelli, R., Gentili, C., Masiello, L., Campanile, G., Cancedda, R., and Albini, A. (1995). Production of angiogenesis inhibitors and stimulators is modulated by cultured growth plate chondrocytes during *in vitro* differentiation: Dependence on extracellular matrix assembly. *Eur. J. Cell Biol.* **66**, in press.

Dessau, W., von der Mark, H., von der Mark, K., and Fisher, S. (1980). Changes in the patterns of collagens and fibronectin during limb bud chondrogenesis. *J. Embryol. Exp. Morphol.* **57**, 51–60.

Detrick, R. J., Dickey, D., and Kinter, C. R. (1990). The effects of N-cadherin misexpression on morphogenesis in *Xenopus* embryos. *Neuron* **4**, 493–506.

Dickson, C., Smith, R., Brookes, S., and Peters, G. (1984). Tumorigenesis by mouse mammary tumor virus: Proviral activation of a cellular gene in the common integration region int-2. *Cell (Cambridge, Mass.)* **37**, 529–536.

Dollé, P., Ruberte, E., Kastner, P., Petkovich, M., Stoner, C. M., Gudas, L. J., and Chambon, P. (1989). Differential expression of genes encoding alfa, beta and gamma retinoic acid receptors and CRABP in the developing limbs of the mouse. *Nature (London)* **342**, 702–705.

Dollé, P., Dierich, A., Le Meur, M., Schimmang, T., Schuhbaur, B., Chambon, P., and Duboule, D. (1993). Disruption of the hoxd-13 gene induces localized heterochrony leading to mice with neotenic limbs. *Cell (Cambridge, Mass.)* **75**, 431–441.

Dorfman, A., Hall, T., Ho, P. L., and Fitch, F. (1979). Clonal antibodies for core protein of chondroitin sulfate proteoglycan. *Proc. Natl. Acad. Sci. U.S.A.* **77,** 3971–3973.

Downie, S. A., and Newman, S. A. (1994). Morphogenetic differences between fore and hind limb precartilage mesenchyme: Relation to mechanisms of skeletal pattern formation. *Dev. Biol.* **162,** 195–208.

Dozin, B., Quarto, R., Rossi, F., and Cancedda, R. (1990). Stabilization of mRNA enhances transcriptional activation of type II collagen gene in differentiating chicken chondrocyte. *J. Biol. Chem.* **265,** 7216–7220.

Dozin, B., Descalzi, F., Briata, L., Hayashi, M., Gentili, C., Hayashi, K., Quarto, R., and Cancedda, R. (1992a). Expression, regulation and tissue distribution of the Ch21 protein during chicken embryogenesis. *J. Biol. Chem.* **267,** 2979–2985.

Dozin, B., Quarto, R., Campanile, G., and Cancedda, R. (1992b). "In vitro" differentiation of mouse embryo chondrocytes: Requirement for ascorbic acid. *Eur. J. Cell Biol.* **58,** 390–394.

Duprez, D., Jeanny, J. C., and Vigny, M. (1993). Localization of RIHB (retinoic acid-induced heparin binding factor) transcript and protein during early chicken embryogenesis and in the developing wing. *Int. J. Dev. Biol.* **37,** 369–380.

Durr, J., Goodman, S., Potocnik, A., von der Mark, H., and von der Mark, K. (1993). Localization of beta 1-integrins in human cartilage and their role in chondrocyte adhesion to collagen and fibronectin. *Exp. Cell Res.* **207,** 235–244.

Ede, D. A. (1983). Cellular condensations and chondrogenesis. In "Cartilage" (B. K. Hall, ed.), Vol. 2, pp. 143–185. Academic Press, New York.

Eichele, G., and Thaller, C. (1987). Characterization of concentration gradients of a morpho-genetically active retinoid in the chick limb bud. *J. Cell Biol.* **105,** 1917–1923.

Ekblom, P., Thesleff, I., Saxén, L., Miettinen, A., and Timpl, R. (1983). Tranferrin as a fetal growth factor: Acquisition of responsiveness related to embryonic induction. *Proc. Natl. Acad. Sci. U.S.A.* **80,** 2651–2655.

Esch, F., Baird, A., Ling, N., Ueno, N., Hill, F., Denoroy, L., Klepper, R., Gospodarowics, D., Bohlen, P., and Guillemin, R. (1985). Primary structure of bovine pituitary basic fibroblast growth factor (FGF) and comparison with the amino-terminal sequence of bovine brain acidic FGF. *Proc. Natl. Acad. Sci. U.S.A.* **82,** 6507–6511.

Eyre, D. R., and Wu, J.-J. (1987). Type XI or $1\alpha$, $2\alpha$, $3\alpha$ collagen. In "Structure and Function of Collagen Types" (R. Mayne and R. E. Burgeson, eds.), pp. 261–281. Academic Press, Orlando, FL.

Eyre, D. R., Upton, M. P., Shapiro, F. D., Wilkinson, R. H., and Vawter, G. F. (1986). Nonexpression of cartilage type II collagen in a case of Langer-Saldino achondrogenesis. *Am. J. Hum. Genet.* **39,** 52–67.

Farquharson, C., Hesketh, J. Q., and Loveridge, N. (1992). The protooncogene c-*myc* is involved in cell differentiation as well as cell proliferation: Studies on growth plate chondrocytes in situ. *J. Cell. Physiol.* **152,** 135–144.

Fife, R. S., Caterson, B., and Myers, S. L. (1985). Identification of link proteins in canine synovial cell cultures and canine articular cartilage. *J. Cell Biol.* **100,** 1050–1055.

Folkman, J. (1984). What is the role of endothelial cells in angiogenesis? *Lab. Invest.* **51,** 601–604.

Folkman, J., and Shing, Y. (1992). Angiogenesis. *J. Biol. Chem.* **267,** 10931–10934.

Franzen, A., Oldberg, A., and Solursh, M. (1989). Possible recruitment of osteoblastic precursor cells from hypertrophic chondrocytes during initial osteogenesis in cartilaginous limbs of young rats. *Matrix* **9,** 261–265.

Frenz, D. A., Jaikaria, N. S., and Newman, S. A. (1989). The mechanism of precartilage mesenchymal condensation: A major role for interaction of the cell surface with the amino-terminal heparin-binding domain of fibronectin. *Dev. Biol.* **136,** 97–103.

Frisch, S. M., and Francis, H. (1994). Disruption of epithelial cell-matrix interactions induces apoptosis. *J. Cell Biol.* **124**, 619–626.

Fujimori, T., Miyatani, S., and Takeichi, M. (1990). Ectopic expression of *N*-cadherin perturbs histogenesis in *Xenopus* embryos. *Development (Cambridge, UK)* **110**, 97–104.

Galotto, M., Campanile, G., Robino, G., Descalzi Cancedda, F., Bianco, P., and Cancedda, R. (1994). Hypertrophic chondrocytes undergo further differentiation to osteoblast-like cells and participate to the initial bone formation in developing chick embryo. *J. Bone Miner. Res.* **9**, 1239–1249.

Galotto, M., Campanile, G., Banfi, A., Trugli, M., and Cancedda, R. (1995). Chondrocyte and osteoblast differentiation stage-specific monoclonal antibodies as a tool to investigate the initial bone formation in developing chick embryo. *Eur. J. Cell Biol.*, in press.

Garcia-Aragon, J., Lobie, P. E., Muscat, G. E., Gobius, K. S., Norstedt, G., and Waters, M. J. (1992). Prenatal expression of the growth hormone (GH) receptor/binding protein in the rat: A role for GH in embryonic and fetal development? *Development (Cambridge, UK)* **114**, 869–876.

Garcia-Martinez, V., Macias, D., Ganan, Y., Garcia-Lobo, J. M., Francia, M. V., Fernandez-Teran, M. A., and Hurle, J. M. (1993). Internucleosomal DNA fragmentation and programmed cell death (apoptosis) in the interdigital tissue of the embryonic chick leg bud. *J. Cell Sci.* **106**, 201–208.

Garofalo, S., Vuorio, E., Metsaranta, M., Rosati, R., Toman, D., Vaughan, J., Lozano, G., Mayne, R., Ellard, J., Horton, W. A., and de Crombrugghe, B. (1991). Reduced amounts of cartilage collagen fibrils and growth plate anomalies in transgenic mice harboring a glycine to cysteine mutation in the mouse proa1(II) collagen gene. *Proc. Natl. Acad. Sci. U.S.A.* **88**, 9648–9652.

Garofalo, S., Matsaranta, M., Ellard, J., Smith, C., Horton, W., and de Crombrugghe, B. (1993). Assembly of cartilage collagen fibrils is disrupted by overexpression of normal type II collagen in transgenic mice. *Proc. Natl. Acad. Sci. U.S.A.* **90**, 3825–3829.

Gatherer, D., Ten Dijke, P., Baird, D. T., and Akhurst, R. J. (1990). Expression of TGF-beta isoforms during first trimester human embryogenesis. *Development (Cambridge, UK)* **110**, 445–460.

Gaunt, S. J., Blum, M., and De Robertis, E. M. (1993). Expression of the mouse goosecoid gene during mi-embryogenesis may mark mesenchymal cell lineages in the developing head, limbs and body wall. *Development (Cambridge, UK)* **117**, 769–778.

Gay, S. W., and Kosher, R. A. (1984). Uniform cartilage differentiation in micromass cultures prepared from a relatively homogeneous population of chondrogenic progenitor cells of the chick limb bud: Effect of prostaglandins. *J. Exp. Zool.* **232**, 317–326.

Geduspan, J. S., and Solursh, M. (1992). A growth-promoting influence from the mesonephros during limb outgrowth. *Dev. Biol.* **151**, 242–250.

Geduspan, J. S., and Solursh, M. (1993). Effects of mesonephros and insulin-like growth factor I on chondrogenesis of limb explants. *Dev. Biol.* **156**, 500–508.

Gelbart, W. M. (1989). The decapentaplegic gene: A TGF-$\beta$ homologue controlling pattern formation in *Drosophila*. *Development (Cambridge, UK)* **107**, 65–74.

Gendron-Maguire, M., Mallo, M., Zhang, M., and Gridley, T. (1993). Hoxa-2 mutant mice exhibit homeotic transformation of skeletal elements derived from cranial neural crest. *Cell (Cambridge, Mass.)* **75**, 1317–1731.

Genge, B. R., Sauer, G. R., Wu, L. N. Y., McLean, F. M., and Wuthier, R. E. (1988). Correlation between phosphatase activity and accumulation of calcium during matrix vesicle-mediated mineralization. *J. Biol. Chem.* **263**, 18513–18519.

Genge, B. R., Wu, L. N. Y., and Adkisson, H. D. (1991). Matrix vesicle annexins exhibit proteolipid-like properties. Selective partioning into lipophilic solvents under acidic conditions. *J. Biol. Chem.* **266**, 10678–10685.

Genge, B. R., Cao, X., Wu, L. N. Y., Buzzi, W. R., Showman, R. W., Arsenault, A. L., Ishikawa, Y., and Wuthier, R. E. (1992). Establishment of the primary structure of the major lipid-dependent $Ca^{2+}$ binding proteins of chicken growth plate cartilage matrix vesicles: Identity within anchorin CII (annexin V) and annexin II. *J. Bone Miner. Res.* **7**, 807–819.

Gentili, C., Bianco, P., Neri, M., Malpeli, M., Campanile, G., Castagnola, P., Cancedda, R., and Descalzi Cancedda, F. (1993). Cell proliferation, extracellular matrix mineralization, and ovotransferrin transient expression during "in vitro" differentiation of chick hypertrophic chondrocytes into osteoblast-like cells. *J. Cell Biol.* **122**, 703–712.

Gentili, C., Doliana, R., Bet, P., Campanile, G., Colombatti, A., Descalzi Cancedda, F., and Cancedda, R. (1994). Ovotransferrin and ovotransferrin receptor expression during chondrogenesis and endochondral bone formation in developing chick embryo. *J. Cell Biol.* **124**, 579–588.

Gerstenfeld, L. C., and Landis, W. (1991). Gene expression and extracellular matrix ultrastructure of a mineralizing chondrocyte cell culture system. *J. Cell Biol.* **112**, 501–513.

Gerstenfeld, L. C., Finer, M. H., and Boedtker, H. (1989). Quantitative analysis of collagen expression in embryonic chick chondrocytes having different developmental fates. *J. Biol. Chem.* **264**, 5112–5120.

Gibson, G. J., Schor, S. L., and Grant, M. E. (1982). Effects of matrix macromolecules on chondrocyte gene expression: Synthesis of a low molecular weight collagen species of cell cultures within collagen gels. *J. Cell Biol.* **93**, 767–774.

Gibson, G. J., Bearman, C. H., and Flint, M. H. (1986). The immunoperoxidase localization of type X collagen in chick cartilage and lung. *Collagen Relat. Res.* **6**, 163–184.

Gilbert, S. F. (1991). Early vertebrate development: Mesoderm and endoderm. *In* "Developmental Biology" (S. F. Gilbert, ed.), pp. 209–215. Sinauer Assoc., Sunderland, MA.

Gimenez-Gallego, G., Rodkey, J., Bennett, C., Rios-Candelore, M., DiSalvo, J., and Thomas, K. (1985). Brain-derived acidic fibroblast growth factor: Complete amino acid sequence and homologies. *Science* **230**, 1385–1388.

Gionti, E., Capasso, O., and Cancedda, R. (1983). The culture of chick embryo chondrocytes and the control of their differentiated functions "in vitro": Transformation by Rous sarcoma virus induces a switch in the collagen type synthesis and enhances fibronectin expression. *J. Biol. Chem.* **258**, 7190–7194.

Gionti, E., Pontarelli, G., and Cancedda, R. (1985). Avian myelocytomatosis virus immortalizes differentiated quail chondrocytes. *Proc. Natl. Acad. Sci. U.S.A.* **82**, 2756–2760.

Gionti, E., Jullien, P., Pontarelli, G., and Sanchez, M. (1989). A continuous line of chicken embryo cells derived from a chondrocyte culture infected with RSV. *Cell Differ. Dev.* **27**, 215–223.

Grandolfo, M., D'Andrea, P., Paoletti, S., Martina, M., Silvestrini, G., Bonucci, E., and Vittur, F. (1993). Culture and differentiation of chondrocytes entrapped in alginate gels. *Calcif. Tissue Int.* **52**, 42–48.

Grant, W. T., Wang, G. J., and Balian, G. (1987). Type X collagen synthesis during endochondral ossification in fracture repair. *J. Biol. Chem.* **262**, 9844–9849.

Green, H., Morkawa, M., and Nixon, T. (1985). A dual effector theory of growth hormone action. *Differentiation (Berlin)* **29**, 195–198.

Grigoriadis, A. E., Heersche, J. N. M., and Aubin, J. E. (1988). Differentiation of muscle, fat, cartilage and bone from progenitor cells present in a bone derived clonal cell population: Effect of dexamethasone. *J. Cell Biol.* **1**, 2139–2151.

Grigoriadis, A. E., Aubin, J. A., and Heersche, J. N. M. (1989). Effects of dexamethasone and vitamin $D_3$ on cartilage differentiation in a clonal chondrogenic cell population. *Endocrinology (Baltimore)* **125**, 2103–2110.

Gunja-Smith, Z., Nagase, H., and Woesser, J. F. (1989). Purification of the neutral proteoglycan-degrading metalloproteinase from human articular cartilage tissue and its identification as stromelysin matrix metalloproteinase-3. *Biochem. J.* **258**, 115–119.

Haack, H., and Gruss, P. (1993). The establishment of murine Hox-1 expression domains during patterning of the limb. *Dev. Biol.* **157**, 410–422.

Habuchi, H., Conrad, H. E., and Glaser, J. H. (1985). Coordinate regulation of collagen and alkaline phosphatase levels in chick embryo chondrocytes. *J. Biol. Chem.* **260**, 13029–13034.

Hagiwara, H., Sakaguchi, H., Itakura, M., Yoshimoto, T., Furuya, M., Tanaka, S., and Hirose, S. (1994). Autocrine regulation of rat chondrocyte proliferation by natriuretic peptide C and its receptor, natriuretic peptide receptor-B. *J. Biol. Chem.* **269**, 10729–10733.

Hale, L. V., Kemick, M. L. S., and Wuthier, R. E. (1986). Effect of vitamin D metabolites on the expression of alkaline phosphatase activity by epiphyseal hypertrophic chondrocytes in primary cell culture. *J. Bone Miner. Res.* **1**, 489–495.

Hamburger, V., and Hamilton, H. L. (1951). A series of normal stages in the development of the chick embryo. *J. Morphol.* **88**, 49–92.

Hammonds, R. G., Swall, R., Dudley, A., Berkemeier, L., Lai, C., Lee, J., Cunningham, N. S., Reddi, A. H., Wood, W. I., and Mason, A. J. (1991). Bone inducing activity of mature BMP-2B produced from a hybrid BMP-2A/2B. *Mol. Endocrinol.* **5**, 149–155.

Hardingham, T. E., and Fosang, A. J. (1992). Proteoglycans: Many forms and many functions. *FASEB J.* **6**, 861–870.

Hascall, V. C. (1981). Proteoglycans: Structure and function. *In* "Biology of Carbohydrates" (V. Ginsburg, ed.), pp. 1–49. Wiley, New York.

Hascall, V. C., Oegema, T. R., and Brown, M. (1976). Isolation and characterization of proteoglycans from chick limb bud chondrocytes grown in vitro. *J. Biol. Chem.* **251**, 3511–3519.

Hayamizu, T. F., Sessions, S. K., Wanek, N., and Bryant, S. V. (1991). Effects of localized application of transforming growth factor beta 1 on developing chick limbs. *Dev. Biol.* **145**, 164–173.

Heine, U., Munoz, E. F., Flanders, K. C., Ellingsworth, L. R., Lam, H. Y., Thompsona, N. L., Roberts, A. B., and Sporn, M. B. (1987). Role of transforming growth factor-beta in the development of the mouse embryo. *J. Cell Biol.* **105**, 2861–2876.

Hiraki, Y., Inoue, H., Kato, Y., and Suzuki, F. (1988). Effect of transforming growth factor-b on cell proliferation and glycosaminoglycan synthesis by rabbit growth-plate chondrocytes in culture. *Biochim. Biophys. Acta* **969**, 91–99.

Hiraki, Y., Tanaka, H., Inoue, H., Kondo, J., Kamizono, A., and Suzuki, F. (1991a). Molecular cloning of a new class of cartilage-specific matrix, chondromodulin-1, which stimulates growth of cultured chondrocytes. *Biochem. Biophys. Res. Commun.* **175**, 971–977.

Hiraki, Y., Inoue, H., Shigeno, C., Sanma, Y., Bentz, H., Rosen, D. M., Asada, A., and Suzuki, F. (1991b). Bone morphogenetic proteins (BMP-2 and BMP-3) promote growth and expression of the differentiated phenotype of rabbit chondrocytes and osteoblastic MC3T3-E1 cells "in vitro." *J. Bone Miner. Res.* **6**, 1373–1385.

Hofmann, C., and Eichele, G. (1994). Retinoids in development. *In* "The Retinoids, Biology, Chemistry and Medicine" (M. B. Sporn, A. B. Roberts, and D. S. Goodman, eds.), pp. 387–441. Raven Press, New York.

Hoffmann, M. F. (1991). Transforming growth factor-beta-related genes in *Drosophila* and vertebrate development. *Curr. Opin. Cell Biol.* **3**, 947–952.

Holderbaum, D., Malemud, C. J., Moskowitz, R. W., and Haqqi, T. M. (1993). Human cartilage from late stage familial osteoarthritis transcribes type II collagen mRNA encoding a cysteine in position 519. *Biochem. Biophys. Res. Commun.* **192**, 1169–1174.

Holtrop, M. E. (1972). The ultrastructure of the epiphyseal plate. II. The hypertrophic chondrocyte. *Calcif. Tissue Res.* **9**, 140–151.

Hoosein, N. M., Brattain, D. E., McKnight, M. K., and Brattain, M. G. (1988). Comparison of the effects of transforming growth factor-beta *N,N*-dimethylformamide, and retinoic acid on transformed and nontransformed fibroblasts. *Exp. Cell Res.* **175**, 125–135.

Horton, W. A., and Hecht, J. T. (1993). Chondrodysplasias. *In* "Extracellular Matrix and Heritable Disorders of Connective Tissue" (P. M. Royce and B. Steinman, eds.), pp. 641–676. Alan R. Liss, New York.

Horton, W. A., Hodd, O. J., Machado, M. A., Ahmed, S., and Griffey, E. S. (1988). Abnormal ossification in thanatophoric dysplasia. *Bone* **9**, 53–61.

Horton, W. A., Machado, M. A., Ellard, J., Campbell, D., Bartley, J., Ramirez, F., Vitale E., and Lee B. (1992). Characterization of a type II collagen (COL2A1) mutation identified in cultured chondrocytes from human hypochondrogenesis. *Proc. Natl. Acad. Sci. U.S.A.* **89**, 4583–4587.

Horton, W. E., and Hassell, J. R. (1986). Independence of cell shape and loss of cartilage matrix production during retinoic acid treatment of cultured chondrocytes. *Dev. Biol.* **115**, 392–397.

Horton, W. E., Yamada, Y., and Hassell, J. R. (1987). Retinoic acid rapidly reduces cartilage matrix synthesis by altering gene transcription in chondrocytes. *Dev. Biol.* **123**, 508–516.

Horton, W. E., Jr., Cleveland, J., Rapp, U., Nemuth, G., Bolander, M., Doege, K., Yamada, Y., and Hassell, J. R. (1988). An established rat cell line expressing chondrocyte properties. *Exp. Cell Res.* **178**, 457–468.

Horwitz, A. L., and Dorfman, A. (1970). The growth of cartilage cells in soft agar and liquid suspension. *J. Cell Biol.* **45**, 434–438.

Hunter, G. K. (1987). An ion-exchange mechanism of cartilage calcification. *Connect. Tissue Res.* **16**, 111–120.

Hunziker, E. B., Schenk, R. K., and Cruz-Orive, L.-M. (1987). Quantitation of chondrocyte performance in growth-plate cartilage during longitudinal bone growth. *J. Bone Jt. Surg., Br. Vol.* **69**, 162–173.

Hunziker, E. B., Wagner, J., and Zapf, J. (1994). Insulin-like growth factor I and growth hormone on development stages of rat growth plate chondrocytes in vivo. *J. Clin. Invest.* **93**, 1078–1086.

Ide, H., and Aono, H. (1988). Retinoic acid promotes proliferation and chondrogenesis in the distal mesodermal cells of chick limb bud. *Dev. Biol.* **130**, 767–773.

Irwin, M. H., Silvers, S. H., and Mayne, R. (1985). Monoclonal antibody against chicken type IX collagen: Preparation, characterization and recognization of the intact form of type IX collagen secreted by chondrocytes. *J. Cell Biol.* **101**, 814–823.

Isaksson, O. G. P., Jansson, J. O., and Gause, I. A. M. (1982). Growth hormone stimulates longitudinal bone growth directly. *Science* **216**, 1237–1239.

Isgaard, J. (1992). Expression and regulation of IGF-I in cartilage and skeletal muscle. *Growth Regul.* **2**, 16–22.

Itagane, Y., Inada, H., Fujita, K., and Isshiki, G. (1991). Interactions between steroid hormones and insulin-like growth factor-I in rabbit chondrocytes. *Endocrinology (Baltimore)* **128**, 1419–1424.

Iwamoto, M., Sato, K., Nakashima, K., Shimazu, A., and Kato, Y. (1989a). Hypertrophy and calcification of rabbit permanent chondrocytes in pelleted cultures: Synthesis of alkaline phosphatase and 1,25-dihydroxycholecalciferol receptor. *Dev. Biol.* **136**, 500–507.

Iwamoto, M., Sato, K., Nakashima, K., Fuchihata, H., Suzuki, F., and Kato, Y. (1989b). Regulation of colony formation of differentiated chondrocytes in soft agar by transforming growth factor beta. *Biochem. Biophys. Res. Commun.* **159**, 1006–1011.

Iwamoto, M., Shimazu, A., Nakashima, K., Suzuki, F., and Kato, Y. (1991). Reduction in basic fibroblast growth factor receptor is coupled with terminal differentiation of chondro-cytes. *J. Biol. Chem.* **266**, 461–467.

Iwamoto, M., Golden, E. B., Adams, S. L., Noji, S., and Pacifici, M. (1993a). Respon-siveness to retinoic acid changes during chondrocyte maturation. *Exp. Cell Res.* **205**, 213–224.

Iwamoto, M., Shapiro, I. M., Yagami, K., Boskey, A. L., Leboy, P. S., Adams, S. L., and Pacifici, M. (1993b). Retinoic acid induces rapid mineralization and expression of mineralization-related genes in chondrocytes. *Exp. Cell Res.* **207**, 413–420.

Iwamoto, M., Yagami, K., Lu Valle, P., Olsen, B. R., Petropoulos, C. J., Ewert, D. L., and Pacifici, M. (1993c). Expression and role of c-*myc* in chondrocytes undergoing endochondral ossification. *J. Biol. Chem.* **268**, 9645–9652.

Iwamoto, M., Jikko, A., Murakami, H., Shimazu, A., Nakashima, K., Iwamoto, M., Takigawa, M., Baba, H., Suzuki, F., and Kato, Y. (1994). Changes in parathy-roid hormone receptors during chondrocyte cytodifferentiation. *J. Biol. Chem.* **269**, 17245–17251.

Iyama, K.-I., Ninomiya, Y., Olsen, B. R., Linsenmayer, T. F., Trelstad, R. L., and Hayashi, M. (1991). Spatiotemporal pattern of type X collagen gene expression and collagen deposi-tion in embryonic chick vertebrae undergoing endochondral ossification. *Anat. Rec.* **229**, 462–472.

Izpisua-Belmonte, J. C., and Duboule, D. (1992). Homeobox genes and pattern formation in the vertebrate limb. *Dev. Biol.* **152**, 26–36.

Izpisua-Belmonte, J. C., Tickle, C., Dollé, P., Wolpert, L., and Duboule, D. (1991). Expres-sion of the homeobox Hox-4 genes and the specification of position in chick wing develop-ment. *Nature (London)* **350**, 585–589.

Izpisua-Belmonte, J. C., Ede, D. A., Tickle, C., and Duboule, D. (1992). The mis-expression of posterior Hox-4 genes in talpid (ta3) mutant wings correlates with the absence of anteroposterior polarity. *Development (Cambridge, UK)*. **114**, 959–963.

Jacenko, O., and Tuan, R. S. (1986). Calcium defficiency induces expression of cartilage-like phenotype in chick embryonic calvaria. *Dev. Biol.* **115**, 215–223.

Jacenko, O., LuValle, P. A., and Olsen, B. R. (1993a). Spondylometaphyseal dysplasia in mice carrying a dominant negative mutation in a matrix protein specific for cartilage-to-bone transition. *Nature (London)* **365**, 56–61.

Jacenko, O., LuValle, P., and Olsen, B. R. (1993b). A truncated α1(X) collagen gene produces spondylometaphyseal dysplasia in transgenic mice. *Matrix* **13**, 39.

Jakowlew, S. B., Dillard, P. J., Winokur, T. S., Flanders, K. C., Sporn, M. B., and Roberts, A. B. (1991). Expression of transforming growth factor-beta s1–4 in chicken embryo chondrocytes and myocytes. *Dev. Biol.* **143**, 135–148.

Jakowlew, S. B., Cubert, J., Danielpour, D., Sporn, M. B., and Roberts, A. B. (1992). Differential regulation of the expression of transforming growth factor beta mRNAs by growth factors and retinoic acid in chicken embryo chondrocytes, myocytes, and fibro-blasts. *J. Cell. Physiol.* **150**, 377–385.

Jiang, T. X., and Chuong, C. M. (1992). Mechanism of skin morphogenesis. I. Analyses with antibodies to adhesion molecules tenascin, N-CAM, and integrin. *Dev. Biol.* **150**, 82–98.

Jiang, T. X., Yi, J. R., Ying, S. Y., and Chuong, C. M. (1993). Activin enhances chondrogen-esis of limb bud cells: Stimulation of precartilaginous mesenchymal condensations and expression of N-CAM. *Dev. Biol.* **155**, 545–557.

Jikko, A., Aoba, T., Murakami, H., Takano, Y., Iwamoto, M., and Kato, Y. (1993). Charac-terization of the mineralization process in cultures of rabbit growth plate chondrocytes. *Dev. Biol.* **156**, 372–380.

Jingushi, S., Joyce, M. E., Flanders, K. C., Hjelmland, L., Roberts, A. B., Sporn, M. B.,

Muniz, O. E., Howell, D. S. Dean, D. D., Ryan, U., and Bolander, M. E. (1990). Distribution of acidic fibroblast growth factor, basic fibroblast growth factor, and transforming growth factor beta 1 in rat growth plate. *In* "Calcium Regulation and Bone Metabolism" (D. V. Cohn, F. H. Glorieux, and T. J. Martin, eds.), pp. 298–303. Elsevier, New York.

Jingushi, S., Joyce, M. E., and Bolander, M. E. (1992). Genetic expression of extracellular matrix proteins correlates with histologic changes during fracture repair. *J. Bone Miner. Res.* **7**, 1045–55.

Joyce, M. E., Roberts, A. B., Sporn, M. B., and Bolander, M. E. (1990). Transforming growth factor-beta and the initiation of chondrogenesis and osteogenesis in the rat femur. *J. Cell Biol.* **110**, 2195–2207.

Kahn, A. J., and Simmons, D. J. (1977). Chondrocyte-to-osteocyte transformation in grafts of perichondrium-free epiphyseal cartilage. *Clin. Orthop. Relat. Res.* **129**, 299–304.

Kataoka, H., and Urist, M. R. (1993). Transplant of bone marrow and muscle-derived connective tissue cultures in diffusion chambers for bioassay of bone morphogenetic protein. *Clin. Orthop. Relat. Res.* **286**, 262–270.

Kato, Y., and Gospodarowicz, D. (1984). Growth requirements of low-density rabbit costal chondrocyte cultures maintained in serum-free medium. *J. Cell. Physiol.* **120**, 354–363.

Kato, Y., and Iwamoto, M. (1990). Fibroblast growth factor is an inhibitor of chondrocyte terminal differentiation. *J. Biol. Chem.* **265**, 5903–5909.

Kato, Y., Hiraki, Y., Inoue, H., Kinoshita, M., Yutani, Y., and Suzuki, F. (1983). Differential and synergistic actions of somatodemedin-like growth factors, fibroblast growth factor and epidermal growth factor in rabbit costal chondrocytes. *Eur. J. Biochem.* **129**, 685–690.

Kato, Y., Iwamoto, M., and Koike, T. (1987). Fibroblast growth factor stimulates colony formation of differentiated chondrocytes in soft agar. *J. Cell. Physiol.* **133**, 491–498.

Kato, Y., Iwamoto, M., Koike, T., Suzuki, F., and Takano, Y. (1988). Terminal differentiation and calcification in rabbit chondrocyte cultures grown in centrifuge tubes: Regulation by transforming growth factors and serum factors. *Proc. Natl. Acad. Sci. U.S.A.* **85**, 9552–9556.

Kato, Y., Shimazu, A., Iwamoto, M., Nakashima, K., Koike, T., Suzuki, F., Nishii, Y., and Sato, K. (1990a). Role of 1,25-dihydroxycholecalciferol in growth plate-cartilage: Inhibition of terminal differentiation of chondrocytes in vitro and in vivo. *Proc. Natl. Acad. Sci. U.S.A.* **87**, 6522–6526.

Kato, Y., Shimazu, A., Nakashima, K., Suzuki, F., Jikko, A., and Iwamoto, M. (1990b). Effects of parathyroid hormone and calcitonin in alkaline phosphatase activity and matrix calcification in rabbit growth plate chondrocyte cultures. *Endocrinology (Baltimore)* **127**, 114–118.

Kato, Y., Nakashima, Z., Iwamoto, M., Murakami, H., Hiranuma, H., Koike, T., Suzuki, F., Fuchihata, H., Ikehara, Y., Noshiro, M., and Jikko, A. (1993). Effects of interleukin-1 on synthesis of alkaline phosphatase, type X collagen, and 1,25-dihydroxyvitamin $D_3$ receptor, and matrix calcification in rabbit chondrocyte cultures. *J. Clin. Invest.* **92**, 2323–2330.

Katzenstein, P., Campbell, D. F., Ellard, J., Machado, M. A., and Horton, W. A. (1995). Gly to ser mutation in COL2A1 in a family with late onset SED and osteoarthritis. Submitted.

Kelley, R. O., and Fallon, J. F. (1978). Identification and distribution of gap junctions in the mesoderm of developing chick limb bud. *J. Embryol. Exp. Morphol.* **46**, 99–110.

Kelley, R. O., and Fallon, J. F. (1983). A freeze fracture and morphometric analysis of gap junctions of limb bud cells: Initial studies on a possible mechanism for morphogenetic signalling during development. *In* "Limb Development and Regeneration" (J. F. Fallon and A. I. Caplan, eds.), Part A, pp. 119–130. Alan R. Liss, New York.

Kimelman, D., and Kirschner, M. (1987). Synergistic induction of mesoderm by FGF and TGF-$\beta$ and the identification of an mRNA coding for FGF in the early *Xenopus* embryo. *Cell (Cambridge, Mass.)* **51**, 869–877.

Kingsley, D. M., Bland, A. E., Grubber, J. M., Marker, P. C., Russell, L. B., Copeland, N. G., and Jenkins, N. A. (1992). The mouse short ear skeletal morphogenesis locus is associated with defects in a bone morphogenetic member of the TGF beta superfamily. *Cell (Cambridge, Mass.)* **71**, 399–410.

Kinoshita, A., Takigawa, M., and Suzuki, F. (1992). Demonstration of receptors for epidermal growth factor on cultured rabbit chondrocytes and regulation of their expression by various growth and differentiation factors. *Biochem. Biophys. Res. Commun.* **183**, 14–20.

Kirsch, T., and von der Mark, K. (1991). Isolation of human type X collagen and immunolocalization in fetal human cartilage. *Eur. J. Biochem.* **196**, 575–580.

Kirsch, T., Swoboda, B., and von der Mark, K. (1992). Ascorbate independent differentiation of human chondrocytes in vitro: Simultaneous expression of types I and X collagen and matrix mineralization. *Differentiation (Berlin)* **52**, 89 100.

Koike, T., Iwamoto, M., Shimazu, A., Nakashima, K., Suzuki, F., and Kato, Y. (1990). Potent mitogenic effects of parathyroid hormone (PTH) on embryonic chick and rabbit chondrocytes. Differential effects of age on growth, proteoglycan, and cyclic AMP responses of chondrocytes to PTH. *J. Clin. Invest.* **85**, 626–631.

Konieczny, S. F., and Emerson, C. P., Jr. (1984). 5-Azacytidine induction of stable mesodermal stem cell lineages from 10T1/2 cells: Evidence for regulatory genes controlling determination. *Cell (Cambridge, Mass.)* **38**, 791–800.

Korkko, J., Ritvanlemi, P., Haataja, L., Kaariainen, H., Kivirikko, K. I., Prockop, D. J., and Ala-Kokko, L. (1993). Mutation in type II procollagen (COL2A1) that substitutes aspartate for glycine $\alpha$1-67 and that causes cataracts and retinal detachment. Evidence for a molecular distinction between the Wagner syndrome and the Stickler syndrome (arthro-ophthalmopathy). *Am. J. Hum. Genet.* **53**, 55–61.

Kosher, R. A., and Solursh, M. (1989). Widespread distribution of type II collagen during embryonic chick development. *Dev. Biol.* **131**, 558–566.

Kosher, R. A., Kulyk, W. M., and Gay, S. W. (1986). Collagen gene expression during limb cartilage differentiation. *J. Cell Biol.* **102**, 1151–1156.

Kucharska, A. M., Kuettner, K. E., and Kimura, J. H. (1990). Biochemical characterization of long-term culture of the Swarm rat chondrosarcoma chondrocytes in agarose. *J. Orthop. Res.* **8**, 781–792.

Kuettner, K. E., and Pauli, B. U. (1983). Vascularity of cartilage. In "Cartilage, Structure Function and Biochemistry" (B. A. Hall, ed.), Vol. 1, pp. 281–312. Academic Press, New York.

Kujawa, M. J., Lennon, D. P., and Caplan, A. (1989). Growth and differentiation of stage 24 limb mesenchyme cells in a serum-free chemically defined medium. *Exp. Cell Res.* **183**, 45–61.

Kulyk, W. M. (1991). Promotion of embryonic limb cartilage differentiation in vitro by staurosporine, a protein kinase C inhibitor. *Dev. Biol.* **146**, 38–48.

Kulyk, W. M., Upholt, W. B., and Kosher, R. A. (1989a). Fibronectin gene expression during limb cartilage differentiation. *Development (Cambridge, UK)* **106**, 449–455.

Kulyk, W. M., Rodgers, B. J., Greer, K., and Kosher, R. A. (1989b). Promotion of embryonic chick limb cartilage differentiation by transforming growth factor-beta. *Dev. Biol.* **135**, 424–430.

Kwan, A. P. L., Dickson, I. R., Freemont, A. J., and Grant, M. E. (1989). Comparative studies of type X collagen expression in normal and rachitic chicken epiphyseal cartilage. *J. Cell Biol.* **109**, 1849–1856.

Lane, J. M., Suda, M., von der Mark, K., and Timpl, R. (1986). Immunofluorescent localization of structural collagen types in endochondral fracture repair. *J. Orthop. Res.* **4,** 318–329.

Langille, R. M., Paulsen, D. F., and Solursh, M. (1989). Differential effects of physiological concentrations of retinoic acid in vitro on chondrogenesis and myogenesis in chick craniofacial mesenchyme. *Differentiation (Berlin)* **40,** 84–92.

Lau, W. F., Tertinegg, I., and Heersche, J. N. (1993). Effects of retinoic acid on cartilage differentiation in a chondrogenic cell line. *Teratology* **47,** 555–563.

Leach, R. M., and Gay, C. V. (1987). Role of epiphyseal cartilage in endochondral bone formation. *J. Nutr.* **117,** 784–790.

Leboy, P. S., Shapiro, I. M., Uschmann, B. D., Oshima, O., and Lin, D. (1988). Gene expression in mineralizing chick epiphyseal cartilage. *J. Biol. Chem.* **263,** 8515–8520.

Leboy, P. S., Vaias, L., Uschmann, B., Golub, E., Adams, S. L., and Pacifici, M. (1989). Ascorbic acid induces alkaline phosphatase, type X collagen, and calcium deposition in cultured chick chondrocytes. *J. Biol. Chem.* **264,** 17281–17286.

Lee, B., Vissing, H., Ramirez, F., Rogers, D., and Rimoin, D. R. (1989). Identification of the molecular defect in a family with spondyloepiphyseal dysplasia. *Science* **244,** 978–980.

Lee, S. J. (1990). Identification of a novel member (GDF-1) of the transforming growth factor-b superfamily. *Mol. Endocrinol.* **4,** 1034–1040.

Lee, S. J., Christakos, S., and Small, M. B. (1993). Apoptosis and signal transduction: Clues to a molecular mechanism. *Curr. Opin. Cell Biol.* **5,** 286–291.

Leonard, C. M., Fuld, H. M., Frenz, D. A., Downie, S. A., Massague, J., and Newman, S. A. (1991). Role of transforming growth factor-beta in chondrogenic pattern formation in the embryonic limb: Stimulation of mesenchymal condensation and fibronectin gene expression by exogenous TGF-beta and evidence for endogeneous TGF-beta-like activity. *Dev. Biol.* **145,** 99–109.

Levitt, D., and Dorfman, A. (1972). The irreversible inhibition of differentiation of limb bud cells by bromodeoxyuridine. *Proc. Natl. Acad. Sci. U.S.A.* **69,** 1253–1257.

Lewinson, D., and Silberman, M. (1992). Chondroclasts and endothelial cells collaborate in the process of cartilage resorption. *Anat. Rec.* **233,** 504–514.

Lewinson, D., Shurtz-Swiirski, R., Shenzer, P., Wingender, E., Mayer, H., and Silberman, M. (1992). Structural changes in condylar cartilage following prolonged exposure to the human parathyroid hormone fragment. *Cell Tissue Res.* **268,** 257–266.

Li, H., Schwartz, N. B., and Vertel, B. M. (1993). cDNA cloning of chick cartilage chondroitin sulfate (aggrecan) core protein and identification of a stop codon in the aggrecan gene associated with the chondrodystrophy, nanomelia. *J. Biol. Chem.* **268,** 23504–11.

Lian, J. B., McKee, M. D., Todd, A. M., and Gertenfeld, L. C. (1993). Induction of bone-related proteins, osteocalcin and osteopontin, and their matrix ultrastructural localization with development of chondrocyte hypertrophy "in vitro." *J. Cell Biol.* **52,** 206–209.

Lindahl, A., Isgaard, J., Nilsson, A., and Isaksson, O. G. P. (1986). Growth hormone potentiates colony formation of epiphyseal chondrocytes in suspension culture. *Endocrinology (Baltimore)* **118,** 1843–1848.

Lindahl, A., Isgaard, J., Carlsson, L., and Isaksson, O. G. P. (1987). Differential effects of growth hormone and insulin-like growth factor I on colony formation of epiphyseal chondrocytes in suspension cultures in rats of different ages. *Endocrinology (Baltimore)* **121,** 1061–1069.

Linsenmayer, T. F., Chen, Q. A., Gibney, E., Gordon, M. K., Marchant, J. K., Mayne, R., and Schmid, T. M. (1991). Collagen types IX and X in the developing chick tibiotarsus: Analysis of mRNA and proteins. *Development (Cambridge, UK)* **111,** 191–196.

Loveridge, N., Farquharson, C., Hesketh, J. E., Jakowlew, S. B., Whitehead, C. C., and Thorp, B. H. (1993). The control of chondrocyte differentiation during endochondral bone growth in vivo: Changes in TGF-beta and the proto-oncogene c-*myc*. *J. Cell Sci.* **105**, 949–956.

Loveys, L. S., Gelb, D., Hurwitz, S. R., Puzas, J. E., and Rosier, R. N. (1993). Effects of parathyroid hormone-related peptide on chick growth plate chondrocytes. *J. Orthop. Res.* **11**, 884–891.

Lowe, M. E., Pacifici, M., and Holtzer, H. (1978). Effects of phorbol-12-myristate-13-acetate on the phenotype program of cultured chondroblasts and fibroblasts. *Cancer Res.* **38**, 2350–2356.

Luyten, F. P., Cunningham, N. S., Muthukumaran, M. S., Hammonds, R. G., Nevins, W. B., Wood, W. I., and Reddi, A. H. (1989). Purification and partial amino acid sequence of osteogenin, a protein initiating bone differentiation. *J. Biol. Chem.* **264**, 13377–13380.

Luyten, F. P., Yu, Y. M., Yanagishita, M., Vukicević, S., Hammonds, R. G., and Reddi, A. H. (1992). Natural bovine osteogenin and recombinant human bone morphogenetic protein-2b are equipotent in the maintenance of proteoglycans in bovine articular cartilage explant cultures. *J. Biol. Chem.* **267**, 3691–3695.

Lyons, K. M., Pelton, R. W., and Hogan, B. L. (1990). Organogenesis and pattern formation in the mouse: RNA distribution patterns suggest a role for bone morphogenetic protein-2A (BMP-2A). *Development (Cambridge, UK)* **109**, 833–844.

Mackem, S., and Mahon, K. A. (1991). Ghox 4.7: A chick homeobox gene expressed primarily in limb buds with limb-type differences in expression. *Development (Cambridge, UK)* **112**, 791–806.

Mackie, E. J., Thesleff, L., and Chiquet-Ehrismann, R. (1987). Tenascin is associated with chondrogenic and osteogenic differentiation "in vivo" and promotes chondrogenesis "in vitro." *J. Cell Biol.* **105**, 2569–2579.

Maden, M., Ong, D. E., Summerbell, D., and Chytil, F. (1988). Spatial distribution of cellular protein binding to retinoic acid in the chick limb bud. *Nature (London)* **335**, 733–735.

Maden, M., Ong, D. E., Summerbell, D., and Chytil, F. (1989). The role of retinoid-binding proteins in the generation of pattern in the developing limb and the nervous system. *Development (Cambridge, UK)* **107**, 109–119.

Mahmood, R., Flanders, K. C., and Morris-Kay, G. M. (1992). Interactions between retinoids and TGF bs in mouse morphogenesis. *Development (Cambridge, UK)* **115**, 67–74.

Mallein-Gerin, F., and Olsen, B. R. (1993). Expression of simian virus 40 large T (tumor) oncogene in mouse chondrocytes induces cell proliferation without loss of the differentiated phenotype. *Proc. Natl. Acad. Sci. U.S.A.* **90**, 3289–3293.

Manduca, P., Castagnola, P., and Cancedda, R. (1988). Dimethylsulphoxide interferes with "in vitro" differentiation of chick embryo endochondral chondrocytes. *Dev. Biol.* **125**, 234–236.

Manduca, P., Descalzi Cancedda, F., and Cancedda, R. (1992). Chondrogenic differentiation in chick embryo osteoblast cultures. *Eur. J. Cell Biol.* **57**, 193–201.

Maor, G., Hochberg, Z., von der Mark, K., Heinegard, D., and Silberman, M. (1989). Human growth hormone enhances chondrogenesis and osteogenesis in a tissue culture system of chondroprogenitor cells. *Endocrinology (Baltimore)* **125**, 1239–1245.

Maor, G., Laron Z., Eshet, R., and Silberman, M. (1993). The early postnatal development of the murine mandibular condyle is regulated by endogenous insulin-like growth factor I. *J. Endocrinol.* **137**, 21–26.

Marics, I., Adelaide, J., Raybaud, F., Mattei, M. G., Coulier, F., Planche, J., de Lapeyriere, O., and Birnbaum, D. (1989). Characterization of the HST-related FGF.6 gene, a new member of the fibroblast growth factor gene family. *Oncogene* **4**, 335–340.

Mark, M. P., Butler, W. T., Prince, C. W., Finkelman, R. D., and Ruch, J. V. (1988).

Developmental expression of 44-kDa bone phosphoprotein (osteopontin) and bone gamma-carboxyglutamic acid (Gla)-containing protein (osteocalcin) in calcifying tissues of rat. *Differentiation (Berlin)* **37**, 123–136.

Massagué, J. (1990). The transforming growth factor-b family. *Annu. Rev. Cell Biol.* **6**, 597–641.

Mayne, R., Vail, M. S., and Miller, E. J. (1975). Analysis of changes in collagen biosynthesis that occur when chick chondrocytes are grown in 5-bromo-2'-deoxyuridine. *Proc. Natl. Acad. Sci. U.S.A.* **72**, 4511–4515.

Mayne, R., Vail, M. S., and Miller, E. J. (1976). The effect of embryo extract on the types of collagen synthesized by cultured chick chondrocytes. *Dev. Biol.* **54**, 230–240.

McFarland, C. D., Brown, R. A., McLaughlin, B., Ali, S. Y., and Weiss, J. B. (1990). Production of endothelial cell stimulating angiogenesis factor (ESAF) by chondrocytes during in vitro cartilage calcification. *Bone Miner.* **11**, 319–333.

McIntosh, I., Abbott, M. H., Warman, M. L., Olsen, B. R., and Francomano, C. A. (1994). Additional mutations of type X collagen confirm Col10A1 as the Schmid metaphyseal chondrodysplasia locus. *Hum. Mol. Genet.* **3**, 303–307.

Meneses, M. F., Krishnan, U. K., and Swee, R. G. (1993). Bizarre parosteal osteochondromatous proliferation of bone (Nora's lesion). *Am. J. Surg. Pathol.* **17**, 691–697.

Metsäranta, M. (1994). Doctoral Thesis. Annales Universitalis Turkunensis.

Metsäranta, M., Young, M. F., Sandberg, M., Termine, J., and Vuorio, E. (1989). Localization of osteonectin expression in human fetal skeletal tissues by in situ hybridization. *Calcif. Tissue Int.* **45**, 146–152.

Metsäranta, M., Garofalo, S., Decker, G., Rintala, M., and de Crombrugghe, B. (1992). Chondrodysplasia in transgenic mice harboring a 15-amino acid deletion in the triple helical domain of proa1(II) collagen chain. *J. Cell Biol.* **118**, 203–212.

Millan, F. A., Denhez, F., Kondaiah, P., and Akhurst, R. J. (1991). Embryonic gene expression patterns of TGF beta 1, beta 2 and beta 3 suggest different developmental functions in vivo. *Development (Cambridge, UK)* **111**, 131–143.

Morgan, B. A., and Tabin, C. J. (1993). The role of homeobox genes in limb development. *Curr. Opin. Genet. Dev.* **3**, 668–674.

Morgan, B. A., Izpisua-Belmonte, J. C., Duboule, D., and Tabin, C. J. (1992). Targeted misexpression of Hox-4.6 in the avian limb bud causes apparent homeotic transformations. *Nature (London)* **358**, 236–239.

Moses, M. A., and Langer, R. (1991). A metalloproteinase inhibitor as an inhibitor of neovascularization. *J. Cell. Biochem.* **47**, 230–235.

Moses, M. A., Sudhalter, J., and Langer, R. (1990). Identification of an inhibitor of neovascularization from cartilage. *Science* **243**, 1408–1410.

Moses, M. A., Sudhalter, J., and Langer, R. (1992). Isolation and characterization of an inhibitor of neovascularization from scapular chondrocytes. *J. Cell Biol.* **119**, 475–482.

Moskalewski, S., and Malejczyk, J. (1989). Bone formation following intrarenal transplantation of isolated murine chondrocytes: Chondrocyte-bone cell transdifferentiation? *Development (Cambridge, UK)* **107**, 473–480.

Muller, P. K., Lemmen, C., Gay, S., Gauss, V., and Kuhn, K. (1977). Immunochemical and biochemical study of collagen synthesis by chondrocytes in culture. *Exp. Cell Res.* **108**, 45–55.

Muller-Glauser, W., Humbel, B., Glatt, M., Strauli, P., Winterhalter, K. H., and Bruckner, P. (1986). On the role of type IX collagen in the extracellular matrix of cartilage: Type IX collagen is localized to intersections of collagen fibrils. *J. Cell Biol.* **102**, 1931–1939.

Mullis, P. E., Patel, M. S., Brickell, P. M., Hindmarsh, P. C., and Brook, C. G. (1991). Growth characteristics and response to growth hormone therapy in patients with hypochondroplasia: Genetic linkage of the insulin-like growth factor I gene at chromosome 12q23 to the disease in a subgroup of these patients. *Clin. Endocrinol. (Oxford)* **34**, 265–274.

Muragaki, Y., Nishimura, I., Henney, A., Ninomiya, Y., and Olsen, B. R. (1990). The alpha 1 (IX) collagen gene gives rise to two different transcripts in both mouse embryonic and human fetal RNA. *Proc. Natl. Acad. Sci. U.S.A.* **87**, 2400–2404.

Nakahara, H., Bruder, S. P., Goldberg, V. M., and Caplan, A. I. (1990). In vivo osteochondrogenic potential of cultured cells derived from the periosteum. *Clin. Orthop. Relat. Res.* **259**, 223–232.

Nakase, T., Nakahara, H., Iwasaki, M., Kimura, T., Kimata, K., Watanabe, K., Caplan, A. I., and Ono, K. (1993). Clonal analysis for developmental potential of chick periosteum-derived cells: Agar gel culture system. *Biochem. Biophys. Res. Commun.* **195**, 1422–1428.

Nakata, K., Nakahara, H., Kimura, T., Kojima, A., Iwasaki, M., Caplan, A. I., and Ono, K. (1992). Collagen gene expression during chondrogenesis from chick periosteum-derived cells. *FEBS Lett.* **16**, 278–282.

Nakata, K., Ono, K., Miyazaki, J.-I., Olsen, B. R., Muragaki, Y., Adachi, E., Yamamura, K., and Kimura, T. (1993). Osteoarthritis associated with mild chondrodysplasia in transgenic mice expressing a1(IX) collagen chains with a central deletion. *Proc. Natl. Acad. Sci. U.S.A.* **90**, 2870–2874.

Neri, M., Descalzi Cancedda, F., and Cancedda, R. (1992). Heat shock response in cultured chick embryo chondrocytes: Osteonectin is a secreted heat shock protein. *Eur. J. Biochem.* **205**, 569–574.

Nilsson, A., Isgaard, A., Lindhahl, A., Dahlström, A., Skottner, A., and Isaksson, O. G. P. (1986). Regulation by growth hormone of number of chondrocytes containing IGF-I in rat growth plate. *Science* **233**, 571–574.

Nishimura, I., Muragaki, Y., and Olsen, B. R. (1989). Tissue-specific forms of type IX collagen-proteoglycan arise from the use of two widely separated promoters. *J. Biol. Chem.* **264**, 20033–20041.

Niswander, L., and Martin, G. R. (1993). FGF-4 regulates expression of EVX-1 in the developing mouse limb. *Development (Cambridge, UK)* **119**, 287–294.

Niswander, L., Tickle, C., Vogel, A., Booth, I., and Martin, A. R. (1993). FGF-4 replaces the apical ectoderm ridge and directs outgrowth and patterning of the limb. *Cell (Cambridge, Mass.)* **75**, 579–587.

Nohno, T., Noji, S., Koyama, E., Ohyama, K., Myokai, F., Kuroiwa, A., Saito, T., and Taniguchi, S. (1991). Involvement of the Chox-4 chicken homeobox genes in determination of anteroposterior axial polarity during limb development *Cell (Cambridge, Mass.)* **64**, 1197–1205.

Nohno, T., Koyama, E., Myokai, F., Taniguchi, S., Ohuchi, H., Saito, T., and Noji, S. (1993). A chicken homeobox gene related to *Drosophila* paired is predominantly expressed in the developing limb. *Dev. Biol.* **158**, 254–264.

Noji, S., Nohno, T., Koyama, E., Muto, K., Ohyama, K., Aoki, Y., Tamura, K., Ohsugi, K., Ide, H., Taniguchi, S., and Saito, T. (1991). Retinoic acid induces polarizing activity but is unlikely to be a morphogen in the chick limb bud. *Nature (London)* **350**, 83–86.

Oberlender, S. A., and Tuan, R. S. (1994). Expression and functional involvement of N-cadherin in embryonic limb chondrogenesis. *Development (Cambridge, UK)* **120**, 177–187.

Oettinger, H. F., and Pacifici, M. (1990). Type X collagen gene expression is transiently up-regulated by retinoic acid treatment in chick chondrocyte cultures. *Exp. Cell Res.* **191**, 292–298.

Ogata, Y., Pratta, M. A., Nagase, H., and Arner, E. C. (1992). Matrix metalloproteinase 9 (92-kDa gelatinase/type IV collagenase) is induced in rabbit articular chondrocytes by cotreatment with interleukin 1b and a protein kinase C activator. *Exp. Cell Res.* **201,** 245–249.

Ohlsson, C., Nilsson, A., Swolin, D., Isaksson, O. G. P., and Lindahl, A. (1993). Establishment of a growth hormone responsive chondrogenic cell line from fetal rat tibia. *Mol. Cell. Endocrinol.* **91,** 167–175.

Ohlsson, C., Isaksson, O., and Lindahl, A. (1994). Clonal analysis of rat tibia growth plate chondrocytes in suspension culture. Differential effects of growth hormone and insulin-like growth factor I. *Growth Regul.* **4,** 1–7.

Okiji, T., Morita, I., Kawashima, N., Kosaka, T., Suda, H., and Murota, S. (1993). Immunohistochemical detection of prostaglandin I2 synthase in various calcified tissue-forming cells in rat. *Arch. Oral Biol.* **38,** 31–36.

Oliver, G., Wright, C. V., Hardwicke, J., and De Robertis, E. M. (1988). A gradient of homeodomain protein in developing forelimbs of *Xenopus* and mouse embryos. *Cell (Cambridge, Mass.)* **55,** 1017–1024.

Oliver, G., Sidell, N., Fiske, W., Heinzmann, C., Mlohandas, T., Sparkes, R. S., and De Robertis, E. M. (1989). Complementary homeo protein gradients in developing limb buds. *Genes Dev.* **3,** 641–650.

Oliver, G., De Robertis, E. M., Wolpert, L., and Tickle, C. (1990). Expression of a homeobox gene in the chick wing bud following application of retinoic acid and grafts of polarizing region tissue. *EMBO J.* **9,** 3093–3099.

Olney, R. C., Smith, R. L., Kee, Y., and Wilson, D. M. (1993). Production and hormonal regulation of insulin-like growth factor binding proteins in bovine chondrocytes. *Endocrinology (Baltimore)* **133,** 563–570.

Osdoby, P., and Caplan, A. I. (1981). First bone formation in the developing chick limb. *Dev. Biol.* **86,** 147–156.

Oshima, O., Leboy, P. S., McDonald, S. A., Tuan, R. S., and Shapiro, I. M. (1989). Developmental expression of genes in chick growth cartilage detected by in situ hybridization. *Calcif. Tissue Int.* **45,** 182–192.

Pacifici, M., and Holtzer, H. (1977). Effects of a tumor-promoting agent on chondrogenesis. *Am. J. Anat.* **150,** 207–212.

Pacifici, M., Cossu, G. M., Molinaro, G., and Tato, F. (1980). Vitamin A inhibits chondrogenesis but not myogenesis. *Exp. Cell Res.* **129,** 469–474.

Pacifici, M., Oshima, O., Fisher, L. W., Young, M. F., Shapiro, I. M., and Leboy, P. S. (1990a). Changes in osteonectin distribution and levels are associated with mineralization of the chicken tibial growth cartilage. *Calcif. Tissue Int.* **47,** 51–61.

Pacifici, M., Golden, E. B., Oshima, O., Shapiro, I. M., Leboy, P. S., and Adams, S. I. (1990b). Hypertrophic chondrocytes. The terminal stage of differentiation in the chondrogenic cell lineage? *Ann. N.Y. Acad. Sci.* **599,** 45–57.

Pacifici, M., Golden, E. B., Adams, S. L., and Shapiro, I. M. (1991a). Cell hypertrophy and type X collagen synthesis in cultured articular chondrocytes. *Exp. Cell Res.* **192,** 266–270.

Pacifici, M., Golden, E. B., Iwamoto, M., and Adams, S. L. (1991b). Retinoic acid treatment induces type X collagen gene expression in cultured chick chondrocytes. *Exp. Cell Res.* **195,** 38–46.

Pacifici, M., Iwamoto, M., Golden, E. B., Leatherman, J. L., Lee, Y.-S., and Choung, C.-M. (1993). Tenascin is associated with articular cartilage development. *Dev. Dyn.* **198,** 123–134.

Page, M., Hogg, J., and Ashhurst, D. E. (1986). The effects of mechanical stability on the macromolecules of the connective tissue matrices produced during fracture healing. I. The collagens. *Histochem. J.* **18,** 251–265.

Panganiban, G. E. F., Rashka, K. E., Neitzel, M. D., and Hoffmann, F. M. (1990). Biochemical characterization of the *Drosophila* dpp protein, a member of the transforming growth factor beta family of growth factors. *Mol. Cell. Biol.* **10,** 2669–2677.

Partanen, A. M., Thesleff, I., and Ekblom, P. (1984). Transferrin is required for early tooth morphogenesis. *Differentiation (Berlin)* **27,** 59–66.

Paulsen, D. F., Langille, R. M., Dress, V., and Solursh, M. (1988). Selective stimulation of "in vitro" limb-bud chondrogenesis by retinoic acid. *Differentiation (Berlin)* **39,** 123–130.

Pelton, R. W., Dickinson, M. E., Moses, H. L., and Hogan, B. L. (1990). In situ hybridization analysis of TGF beta 3 RNA expression during mouse development: Comparative studies with TGF beta 1 and beta 2. *Development (Cambridge, UK)* **110,** 609–620.

Pepper, M. S., Montesano, R., Vassalli, J.-D., and Orci, L. (1991). Chondrocytes inhibit endothelial sprout formation in vitro: Evidence for involvement of a transforming growth factor-beta. *J. Cell. Physiol.* **146,** 170–179.

Poole, A. R. (1991). The growth plate: Cellular physiology, cartilage assembly, and mineralization. *In* "Cartilage: Molecular Aspects" (B. Hall and S. Newman, eds.), pp. 179–211. CRC Press, Boca Raton, FL.

Poole, A. R., and Pidoux, I. (1986). Proteoglycans in health and disease: Structure and functions. *Biochem. J.* **236,** 1–14.

Poole, A. R., and Pidoux, I. (1989). Immunoelectron microscopic studies of type X collagen in endochondral ossification. *J. Cell Biol.* **109,** 2547–2554.

Prockop, D. J., Kivirrikko, K. I., Tuderman, L., and Guzman, N. A. (1979). The biosynthesis of collagen and its disorders. *N. Engl. J. Med.* **301,** 13–17.

Quarto, R., Dozin, B., Tacchetti, C., Campanile, G., Malfatto, C., and Cancedda, R. (1990). "In vitro" development of hypertrophic chondrocytes starting from selected clones of dedifferentiated cells. *J. Cell Biol.* **110,** 1379–1386.

Quarto, R., Dozin, B., Tacchetti, C., Robino, G., Zenke, M., Campanile, G., and Cancedda, R. (1992a). Constitutive *myc* expression impairs hypertrophy and calcification in cartilage. *Dev. Biol.* **149,** 168–176.

Quarto, R., Campanile, G., Cancedda, R., and Dozin, B. (1992b). Thyroid hormone, insulin and glucocorticoids are sufficient to support chondrocyte differentiation to hypertrophy: A serum free analysis. *J. Cell Biol.* **119,** 989–995.

Quarto, R., Dozin, B., Bonaldo, P., Cancedda, R., and Colombatti, A. (1993). Type VI collagen expression is upregulated in the early events of chondrocyte differentiation. *Development (Cambridge, UK)* **117,** 245–251.

Ramdi, H., Legay, C., and Lievremont, M. (1993). Influence of matricial molecules on growth and differentiation of entrapped chondrocytes. *Exp. Cell Res.* **207,** 449–454.

Reddi, A. H. (1981). Cell biology and biochemistry of endochondral bone development. *Cell Res.* **1,** 209–226.

Reddi, A. H. (1992). Regulation of cartilage and bone differentiation by bone morphogenetic proteins. *Curr. Opin. Cell Biol.* **4,** 850–855.

Reese, C. A., Wiedemann, H., Kuhn, K., and Mayne, R. (1982). Characterization of a highly soluble collagenous molecule isolated from chicken hyaline cartilage. *Biochemistry* **21,** 826–830.

Reginato, A. M., Lash, J. W., and Jiminez, S. A. (1986). Biosynthetic expression of type X collagen in embryonic chick sternum cartilage during development. *J. Biol. Chem.* **261,** 2897–2904.

Richman, J. M., and Diewert, V. M. (1988). The fate of Merckel's cartilage chondrocytes in ocular culture. *Dev. Biol.* **129,** 48–60.

Rimoin, D. L., and Lachman, R. S. (1990). The chondrodysplasias. *In* "Principles and Practice of Medical Genetics" (A. E. H. Emery and D. L. Rimoin, eds.), pp. 895–992. Churchill-Livingstone, Edinburgh.

Rintala, M., Metsäranta, M., Garofalo, S., de Crombrugghe, B., Vuorio, E. I., and Rönning,

O. (1993). Abnormal craniofacial development and cartilage structure in transgenic mice harbouring a Gly → Cys mutation in the cartilage specific type II collagen gene. *J. Craniofacial Genet. Dev. Biol.* **13**, 137–146.

Ritvenaniemi, P., Ala-Kokko, L., Korkko, J., Haataja, L., Kaarialnen, H., Kivirikko, K. I., and Prockop, D. J. (1993). Identification of gly to asp acid mutation in type II procollagen gene in a family with the Stickler syndrome by denaturing gradient gel electrophoresis. *Genomics* **17**, 218–221.

Roach, H. I. (1992). Trans-differentiation of hypertrophic chondrocytes into cells capable of producing a mineralized bone matrix. *Bone Miner.* **19**, 1–20.

Roach, H. I., and Shearer, J. R. (1989). Cartilage resorption and endochondral bone formation during the development of long bones in chick embryos. *Bone Miner.* **6**, 289–309.

Robinson, D., Efrat, M., Mendes, D. G., Halperin, N., and Nevo, Z. (1993). Implants composed of carbon fiber mesh and bone-marrow-derived, chondrocyte enriched cultures for joint surface reconstruction. *Bull. Hosp. Jt. Dis. Orthop. Inst.* **53**, 75–82.

Rogina, B., Coelho, C. N., Kosher, R. A., and Upholt, W. B. (1992). The pattern of expression of the chicken homolog of HOX11 in the developing limb suggests a possible role in the ectodermal inhibition of chondrogenesis. *Dev. Dyn.* **193**, 92–101.

Ruberte, E., Friederich, V., Morriss-Kay, G., and Chambon, P. (1992). Differential distribution pattern of CRABP I and CRABP II transcripts during mouse embryogenesis. *Development (Cambridge, UK)* **115**, 973–978.

Rubin, J. S., Osada, H., Finch, P. W., Taylor, W. G., Rudikoff, S., and Aaronson, S. A. (1989). Purification and characterization of a newly identified growth factor specific for epithelial cells. *Proc. Natl. Acad. Sci. U.S.A.* **86**, 802–806.

Ruiz, I., Altaba, A., and Melton, D. A. (1990). Axial patterning and the establishment of polarity in the frog embryo. *Trends Genet.* **6**, 57–64.

Russell, S. M., and Spencer, E. M. (1985). Local injections of human or rat growth hormone or of purified human somatomedin-C stimulate unilateral tibial epiphyseal growth in hypophysectomized rats. *Endocrinology (Baltimore)* **116**, 2563–2567.

Ryan, M. C., and Sandell, L. J. (1990). Differential expression of a cysteine-rich domain in the amino-terminal propeptide of type II (cartilage) procollagen by alternative splicing of mRNA. *J. Biol. Chem.* **265**, 10334–10339.

Sakamoto, H., Mori, M., Taira, M., Yoshida, T., Matsukawa, S., Shimizu, K., Sekiguchi, M., Terada, M., and Sugimura, T. (1986). Transforming gene from human stomach cancers and a noncancerous portion of stomach mucosa. *Proc. Natl. Acad. Sci. U.S.A.* **83**, 3997–4001.

Säkselä, O., Moscatelli, D., Sommer, A., and Rifkin, D. B. (1988). Endothelial cell-derived heparan sulfate binds basic fibroblast growth factor and protects it from proteolytic degradation. *J. Cell Biol.* **107**, 743–751.

Sampath, T. H., Muthukumaran, N., and Reddi, A. H. (1987). Isolation of osteogenin, an extracellular matrix-associated bone-inductive protein, by heparin affinity chromatography. *Proc. Natl. Acad. Sci. U.S.A.* **84**, 7109–7113.

Sandberg, M. M., and Vuorio, E. I. (1987). Localization of types I, II, and III collagen mRNAs in developing human skeletal tissues by in situ hybridization. *J. Cell Biol.* **104**, 1077–1084.

Sandberg, M. M., Vuorio, T., Hirvonen, H., Alitalo, K., and Vuorio, E. I. (1988). Enhanced expression of TGF-beta and c-fos mRNAs in the growth plates of developing human long bones. *Development (Cambridge, UK)* **102**, 461–470.

Sandberg, M. M., Aro, H. T., Multimaeki, P., Aho, H., and Vuorio, E. I. (1989). In situ localization of collagen production by chondrocytes and osteoblasts in fracture callus. *J. Bone Jt. Surg., Am. Vol.* **71**, 69–77.

Sandberg, M. M., Aro, H. T., and Vuorio, E. I. (1993). Gene expression during bone repair. *Clin. Orthop. Relat. Res.* **289**, 292–312.

Sandell, L. J., Morris, N., Robbins, J. R., and Goldring, M. B. (1991). Alternatively spliced type II procollagen mRNAs define distinct populations of cells during vertebral development: Differential expression of the amino-propeptide. *J. Cell Biol.* **114**, 1307–1319.

Sara, V. R., and Carlsson-Skwirut, C. (1988). The biosynthesis and regulation of fetal insulin-like growth factors. *In* "The Endocrine Control of the Fetus" (W. Kuntzel and A. Jensen, eds.), pp. 223–235. Springer-Verlag, Berlin.

Sato, K., Iwamoto, M., Nakashima, K., Suzuki, F., and Kato, Y. (1990). $1\alpha,25$-dihydroxyvitamin $D_3$ stimulates colony formation of chick embryo chondrocytes in soft agar. *Exp. Cell Res.* **187**, 335–338.

Sato, T., and Tuan, R. S. (1992). Effect of systemic calcium deficiency on the expression of transforming growth factor-beta in chick embryonic calvaria. *Dev. Dyn.* **193**, 300–313.

Schenk, R. J., Wiener, J., and Spiro, D. (1968). Fine structural aspects of vascular invasion of the tibial epiphyseal plate of growing rats. *Acta Anat.* **68**, 1–17.

Schiltz, J. R., Mayne, R., and Holtzer, H. (1973). The synthesis of collagen and glycosaminoglycans by dedifferentiated chondroblasts in culture. *Cell Differ.* **1**, 97–108.

Schlechter, N. L., Russell, S. M., Spencer, E. M., and Nicoll, C. S. (1986). Evidence suggesting that the direct growth-promoting effect of growth hormone on cartilage in vivo is mediated by local production of somatomedin. *Proc. Natl. Acad. Sci. U.S.A.* **83**, 7932–7934.

Schmid, T. M., and Conrad, H. E. (1982). A unique low molecular weight collagen secreted by cultured chick embryo chondrocytes. *J. Biol. Chem.* **257**, 12444–12450.

Schmid, T. M., and Linsenmayer, T. F. (1985a). Developmental acquisition of type X collagen in the embryonic chick tibiotarsus. *Dev. Biol.* **107**, 373–381.

Schmid, T. M., and Linsenmayer, T. F. (1985b). Immunohistochemical localization of short chain cartilage collagen (type X) in avian tissues. *J. Cell Biol.* **100**, 598–605.

Schmid, T. M., Bonen, D. K., Luchene, L., and Linsenmayer, T. F. (1991). Late events in chondrocyte differentiation: Hypertrophy, type X collagen synthesis and matrix calcification. *In Vivo* **5**, 533–540.

Schofield, J. N., and Wolpert, L. (1990). Effect of TGF beta 1, TGF beta 2, and bFGF on chick cartilage and muscle cell differentiation. *Exp. Cell Res.* **191**, 144–148.

Schofield, J. N., Rowe, A., and Brickell, P. M. (1992). Position-dependence of retinoic acid receptor-beta gene expression in the chick limb bud. *Dev. Biol.* **152**, 344–353.

Schwartz, Z., Nasatzky, E., Ornoy, A., Brooks, B. P., Soskolne, W. A., and Boyan, B. D. (1994). Gender-specific, maturation-dependent effects of testosterone on chondrocytes in culture. *Endocrinology (Baltimore)* **134**, 1640–1647.

Scott, M. P. (1993). A rational nomenclature for vertebrate homeobox (HOX) genes. *Nucleic Acids Res.* **21**, 1687–1688.

Shapiro, S. S., and Poon, J. P. (1976). Effect of retinoic acid on chondrocyte glycosaminoglycan biosynthesis. *Arch. Biochem. Biophys.* **174**, 74–81.

Silberman, M., Lewinson, D., Gonen, H., Lizarbe, M. A., and von der Mark, K. (1983). In vitro transformation of chondroprogenitor cells into osteoblasts and the formation of new membrane bone. *Anat. Rec.* **206**, 373–383.

Simeone, A., Acampora, D., Arcioni, L., Andrews, P. W., Boncinelli, E., and Mavilio, F. (1990). Sequential activation of HOX 2 homeobox genes by retinoic acid in human embryonal carcinoma cells. *Nature (London)* **346**, 763–766.

Smith, S. M., and Eichele, G. (1991). Temporal and regional differences in the expression pattern of distinct retinoic acid receptor beta transcripts in the chick embryo. *Development (Cambridge, UK)* **111**, 245–252.

Smith, W. L. (1992). Prostanoid biosynthesis and mechanisms of action. *Am. J. Physiol.* **263**, 181–191.

Solursh, M. (1983). Cell-cell interactions and chondrogenesis. *In* "Cartilage" (B. K. Hall, ed.), Vol. 2, pp. 121–141. Academic Press, New York.

Solursh, M., and Reiter, R. S. (1980). Evidence for histogenic interactions during "in vitro" limb chondrogenesis. *Dev. Biol.* **78,** 141–150.

Solursh, M., Ahrens, P. B., and Reiter, R. S. (1978). A tissue culture analysis of the steps in limb chondrogenesis. *In Vitro* **14,** 51–61.

Solursh, M., Jensen, K. L., Zanetti, N. C., Linsenmayer, T. F., and Reiter, R. S. (1984). Extracellular matrix mediates epithelial effects on chondrogenesis in vitro. *Dev. Biol.* **105,** 451–457.

Solursh, M., Jensen, K. L., Reiter, R. S., Schmid, T. M., and Linsenmayer, T. F. (1986). Environmental regulation of type X collagen production by cultures of limb mesenchyme, mesectoderm and sternal chondrocytes. *Dev. Biol.* **117,** 90–101.

Spranger, J., Menger, H., Mundlos, S., Winterpacht, A., and Zabel, B. Kniest dysplasia is caused by dominant collagen II gene (COL2A1) mutations: parental somatic mosaicism manifesting as Stickler phenotype and mild spondyloepiphyseal dysplasia. *Pediatr. Radiol.* (in press).

Sporn, M. B., and Roberts, A. B. (1990). "Peptide Growth Factors and Their Receptors," Vols. I and II. Springer-Verlag, New York.

Sporn, M. B., and Roberts, A. B. (1992). Transforming growth factor-beta: recent progress and new challenges. *J. Cell Biol.* **119,** 1017–1021.

Spranger, J. (1992). International classification of osteochondrodysplasias. *Eur. J. Pediatr.* **151,** 407–415.

Stirpe, N. S., and Goetinck, P. F. (1989). Gene regulation during cartilage differentiation: Temporal and spatial expression of link protein and cartilage matrix protein in the developing limb. *Development (Cambridge, UK)* **107,** 23–33.

Stocum, D. L., Davis, R. M., Leger, M., and Conrad, H. E. (1979). Development of the tibiotarsus in the chick embryo: Biosynthetic activities of histologically distinct regions. *J. Embryol. Exp. Morphol.* **54,** 155–170.

Storm, E. E., Huynh, T. V., Copeland, N. G., Jenkins, N. A., Kingsley, D. M., and Lee, S.-J. (1994). Limb alterations in brachypodism mice due to mutations in a new member of the TGFb-superfamily. *Nature (London)* **368,** 639–643.

Strauss, P. G., Closs, E. I., Schmidt, J., and Erfle, V. (1990). Gene expression during osteogenic differentiation of mandibular condyles "in vitro." *J. Cell Biol.* **110,** 1369–1378.

Sullivan, R., and Klagsburn, M. (1985). Purification of cartilage-derived growth factor by heparin affinity chromatography. *J. Biol. Chem.* **260,** 2399–2403.

Summerbell, D., Lewis, J., and Wolpert, L. (1973). Positional information in chick limb morphogenesis. *Nature (London)* **244,** 492–496.

Suzuki, F. (1994). Regulation of cartilage differentiation and metabolism. *Bone Miner. Res.* **8,** 115–142.

Suzuki, F., Takse, T., Takigawa, M., Uchida, A., and Shimomura, Y. (1981). Simulation of the initial stage of endochondral ossification: In vitro sequential culture of growth-cartilage cells and bone marrow cells. *Proc. Natl. Acad. Sci. U.S.A.* **78,** 2368–2372.

Sylvia, V. L., Swartz, Z., Schuman, L., Morgan, R. T., Mackey, S., Gomez, R., and Boyan, B. D. (1993). Maturation-dependent regulation of protein kinase-C activity by vitamin-D(3) metabolites in chondrocyte cultures. *J. Cell. Physiol.* **157,** 271–278.

Szabo, P., Muratoglu, S., Bachrati, C., Malpeli, M., Neri, M., Dozin, B., Deak, F., Cancedda, R., and Kiss, I. (1995). Expression of the cartilage matrix protein gene at various stages of chondrocyte differentiation. Submitted for publication.

Tabin, C. J. (1991). Retinoids, homeoboxes, and growth factors: Toward molecular models for limb development. *Cell (Cambridge, Mass.)* **66,** 199–217.

Tacchetti, C., Quarto, R., Nitsch, L., Hartmann, D. J., and Cancedda, R. (1987). "In vitro" morphogenesis of chick embryo hypertrophic cartilage. *J. Cell Biol.* **105,** 999–1006.

Tacchetti, C., Quarto, R., Campanile, G., and Cancedda, R. (1989). Calcification of "in vitro" developed hypertrophic cartilage. *Dev. Biol.* **132,** 442–447.

Tacchetti, C., Tavella, S., Dozin, B., Quarto, R., Robino, G., and Cancedda, R. (1992). Cell condensation in chondrogenic differentiation. *Exp. Cell Res.* **200,** 26–33.

Takigawa, M., Enomoto, M., Shirai, E., Nishii, Y., and Suzuki, F. (1988a). Differentiation effects of 1α,25-dihydroxycholecalciferol and 24R,25-dihydroxycholecalciferol on the proliferation and the differentiated phenotype of rabbit costal chondrocytes in culture. *Endocrinology (Baltimore)* **122,** 831–839.

Takigawa, M., Shirai, E., Enomoto, M., Hiraki, Y., Suzuki, F., Shiio, T., and Yugari, Y. (1988b). Cartilage derived anti tumor factor (CATF): Partial purification and correlation of inhibitory activity against tumor growth with anti-angiogenic activity. *J. Bone Miner. Metab.* **6,** 83–92.

Takigawa, M., Tajima, K., Pan, H. O., Enomoto, M., Kinoshita, A., Suzuki, F., Takano, Y., and Mori, Y. (1989). Establishment of a clonal human chondrosarcoma cell line with cartilage phenotypes. *Cancer Res.* **49,** 3996–4002.

Tamarin, A., Crawley, A., Lee, J., and Tickle, C. (1984). Analysis of upper beak defects in chicken embryos following treatment with retinoic acid. *J. Embryol. Exp. Morphol.* **84,** 105–123.

Taniguchi, Y., Tanaka, T., Gotoh, K., Satoh, R., and Inazu, M. (1993). Transforming growth factor beta 1-induced cellular heterogeneity in the periosteum of rat parietal bones. *Calcif. Tissue Int.* **53,** 122–126.

Tavella, S., Raffo, P., Tacchetti, C., Cancedda, R., and Castagnola, P. (1994). N-cam and N-cadherin expression during *in vitro* chondrogenesis. *Exp. Cell Res.* **215,** 354–362.

Thaller, C., and Eichele, G. (1987). Identification and spatial distribution of retinoids in the developing chick limb bud. *Nature (London)* **327,** 625–628.

Thaller, C., and Eichele, G. (1990). Isolation of 3,4-didehydroretinoic acid, a novel morphogenetic signal in the chick wing bud. *Nature (London)* **345,** 815–819.

Thaller, C., Hofmann, C., and Eichele, G. (1993). 9-*cis*-Retinoic acid, a potent inducer of digit pattern duplications in the chick wing bud. *Development (Cambridge, UK)* **118,** 957–965.

Thenet, S., Benya, P. D., Demignot, S., Feunteun, J., and Adolphe, M. (1992). SV40-immortalization of rabbit articular chondrocytes: Alteration of differentiated functions. *J. Cell. Physiol.* **150,** 158–167.

Thesingh, C. W., Groot, C. G., and Wassenaar, A. M. (1991). Transdifferentiation of hypertrophic chondrocytes in murine metatarsal bones, induced by cocultured cerebrum. *Bone Miner.* **12,** 25–40.

Thesleff, I., and Ekblom, P. (1984). Role of transferrin in branching morphogenesis, growth and differentiation of the embryonic kidney. *J. Embryol. Exp. Morphol.* **82,** 147–161.

Thomas, J. T., Boot-Handford, R. P., and Grant, M. E. (1990). Modulation of type X collagen gene expression by calcium beta-glycerophosphate and levamisole: Implication for a possible role for type X collagen in endochondral bone formation. *J. Cell Sci.* **95,** 639–648.

Thompson, A. Y., Piez, K. A., and Seyedin, S. M. (1985). Chondrogenesis in agarose gel culture. A model for chondrogenic induction, proliferation and differentiation. *Exp. Cell Res.* **157,** 483–494.

Thomsen, G., Woolf, T., Whitman, M., Sokol, S., Vaughan, J., Vale, W., and Melton, D. A. (1990). Activins are expressed early in *Xenopus* embryogenesis and can induce axial mesoderm and anterior structures. *Cell (Cambridge, Mass.)* **63,** 485–493.

Thomson, B. M., Bennett, J., Dean, V., Triffitt, J., Meikle, M. C., and Loveridge, N. (1993). Preliminary characterization of porcine bone marrow stromal cells: Skeletogenic potential, colony-forming activity, and response to dexamethasone, transforming growth factor b, and basic fibroblast growth factor. *J. Bone Miner. Res.* **8,** 1173–1183.

Thorogood, P. V., and Hinchliffe, J. R. (1975). An analysis of the condensation process during chondrogenesis in the embryonic chick hind limb. *J. Embryol. Exp. Morphol.* **33,** 581–606.

Thorp, B. H., Anderson, I., and Jakowlew, S. B. (1992). Transforming growth factor-beta 1, -beta 2, and -beta 3 in cartilage and bone cells during endochondroal ossification in the chick. *Development (Cambridge, UK)* **114,** 907–911.

Tickle, C. (1980). The polarizing region and limb development. *In* "Development in Mammals" (M. H. Johnson, ed.), Vol. 4, pp. 101–136. Elsevier/North-Holland, Amsterdam.

Tickle, C., Summerbell, D., and Wolpert, L. (1975). Positional signalling and specification of digits in chick limb morphogenesis. *Nature (London)* **254,** 199–202.

Tickle, C., Alberts, B., Wolpert, L., and Lee, J. (1982). Local application of retinoic acid to the limb bud mimics the action of the polarizing region. *Nature (London)* **296,** 564–565.

Tickle, C., Lee, J., and Eichele, G. (1985). A quantitative analysis of the effect of all-*trans*-retinoic acid on the pattern of chick wing development. *Dev. Biol.* **109,** 82–95.

Tiller, G. E., Rimoin, D. L., Murray, L. W., and Cohn, D. H. (1990). Tandem duplication within at type II collagen gene (COL2AI) exon in an individual with spondyloepiphyseal dysplasia. *Proc. Natl. Acad. Sci. U.S.A.* **87,** 3889–3893.

Tiller, G. E., Weis, M. A., Eyre, D. R., Rimoin, D. L., and Cohn, D. H. (1992). An RNA splicing mutation (G + 5 IVS 20) in the gene for type II collagen (COL2A1) produces spondyloepiphyseal dysplasia congenita (SEDC). *Am. J. Hum. Genet.* **51,** A37.

Tomasek, J. J., Mazurkiewicz, J. E., and Newman, S. A. (1982). Nonuniform distribution of fibronectin during avian limb development. *Dev. Biol.* **90,** 118–126.

Tschan, T., Hoerler, I., Houze, Y., Winterhalter, K. H., Richter, C., and Bruckner, P. (1990). Resting chondrocytes in culture survive without growth factors, but are sensitive to toxic oxygen metabolites. *J. Cell Biol.* **111,** 257–260.

Tschan, T., Bohme, K., Conscience, E. M., Zenke, G., Winterhalter, K. H., and Bruckner, P. (1993). Autocrine or paracrine transforming growth factor-beta modulates the phenotype of chick embryo sternal chondrocytes in serum-free agarose culture. *J. Biol. Chem.* **5,** 5156–5161.

Tsonis, P. A. (1991). 1,25-Dihydroxyvitamin $D_3$ stimulates chondrogenesis of the chick limb bud mesenchymal cells. *Dev. Biol.* **143,** 130–134.

Tuan, R. S., and Lynch, M. H. (1983). Effect of experimentally induced calcium deficiency on the developmental expression of collagen types in chick embryonic skeleton. *Dev. Biol.* **100,** 374–386.

Urios, P., Duprez, D., Le Caer, J. P., Courtois, Y., Vigny, M., and Laurent, M. (1991). Molecular cloning of RIHB, a heparin binding protein regulated by retinoic acid. *Biochem. Biophys. Res. Commun.* **175,** 617–624.

Urist, M. R. (1965). Bone formation by autoinduction. *Science* **150,** 893–899.

Vandenberg, P., Khillan, J. S., Prockop, D. J., Lelminen, H., and Kontusaari, S. (1991). Expression of a partially deleted gene of human type II procollagen (COL2A1) in transgenic mice produces a chondrodysplasia. *Proc. Natl. Acad. Sci. U.S.A.* **88,** 7640–7644.

van de Werken, R., Gennari, M., Bet, P., Tavella, S., Molina, F., Cancedda, R., and Castagnola, P. (1993). Modulation of tensin and vimentin expression in chick embryo developing cartilage and cultured differentiating chondrocytes. *Eur. J. Biochem.* **217,** 787–790.

Vasan, N., and Lash, J. W. (1975). Chondrocyte metabolism as affected by vitamin A. *Calcif Tissue Res.* **19,** 99–107.

Vertel, B. M., Walters, L. M., Grier, B., Maine, N., and Goetinck, P. F. (1993). Nanomelic chondrocytes synthesize, but fail to translocate, a truncated aggrecan precursor. *J. Cell Sci.* **104,** 939–948.

Vetter, U., Zapf, J., Heit, W., Helbing, G., Heinze, E., Froesch, E. R., and Teller, W. M. (1986). Human fetal and adult chondrocytes. Effect of insulin-like growth factors I and II, insulin, and growth hormone on clonal growth. *J. Clin. Invest.* **77,** 1903–1908.

Vigny, M., Raulais, D., Puzenat, N., Duprez, D., Hartmann, M. P., Jeanny, J. C., and Curtois, Y. (1989). Identification of a new haprin-binding protein localized within chick basement membranes. *Eur. J. Biochem.* **186**, 733–740.

Vikkula, M., Ritvaniemi, P., Vuorio, A. F., Kaitilia, I., Ala-Kokko, L., and Peltonen, L. (1993). A mutation in the aminoterminal triple helix of type II collagen causing severe osteochondrodysplasia. (*GENOMICS*) **16**, 282–285.

Vissing, H., D'Alessio, M., Lee, B., Ramirez, F., Godfrey, M., and Hollister, D. W. (1989). Glycine, to serine substitution in the triple helical domain of proα1(II) collagen results in a lethal perinatal form of short-limbed dwarfism. *J. Biol. Chem.* **264**, 18266–18267.

von der Mark, K., and von der Mark, H. (1977). The role of three genetically distinct collagen types in endochondral ossification and calcification of cartilage. *J. Bone Jt. Surg.* **59**, 458–464.

von der Mark, K., von der Mark, H., and Gay, S. (1976). Study of differential collagen synthesis during development of the chick embryo by immunofluorescence. *Dev. Biol.* **53**, 153–170

von der Mark, K., van Menxel, M., and Weideman, H. (1982). Isolation and characterization of new collagens from chick cartilage. *Eur. J. Biochem.* **124**, 57–62.

von der Mark, K., Kirsch, T., Nerlich, A. G., Kua, A., Weseloh, G., Glickert, K., and Stuess, H. (1992). Type X collagen synthesis in human osteoarthritic cartilage: Indication of chondrocyte hypertrophy. *Arthritis Rheum.* **35**, 806–811.

Wanek, N., Gardiner, D. M., Muneoka, K., and Bryant, S. V. (1991). Conversion by retinoic acid of anterior cells into ZPA cells in the chick wing bud. *Nature (London)* **350**, 81–83.

Wang, E. A., Rosen, V., Cordes, P., Hewick, R. M., Kritz, M. J., Luzenberg, P. *et al.* (1988). Purification and characterization of the distinct bone-inducing factors. *Proc. Natl. Acad. Sci. U.S.A.* **85**, 9484–9488.

Wang, E. A., Rosen, V., D'Alessandro, J. S., Baudy, M., Cordes, P., Harada, T., Isreal, D. I., Hewick, R. M., Kerns, K. M., Lapan, P., *et al.* (1990). Recombinant human bone morphogenetic protein induces bone formation. *Proc. Natl. Acad. Sci. U.S.A.* **87**, 2220–2224.

Wang, E. A., Israel, D. I., Kelly, S., and Luxenberg, D. P. (1993). Bone morphogenetic protein-2 causes commitment and differentiation in C3H10T1/2 and 3T3 cells. *Growth Factors* **9**, 57–71.

Wang, Z. Q., Grigoriadis, A. E., Mohle-Steinlein, U., and Wagner, E. F. (1991). A novel target cell for c-fos-induced oncogenesis: Development of chondrogenic tumors. *EMBO J.* **10**, 2437–2450.

Wang, Z. Q., Grigoriadis, A. E., and Wagner, E. F. (1993). Stable murine chondrogenic cell lines derived from c-*fos*-induced cartilage tumors. *J. Bone Miner. Res.* **8**, 839–847.

West, C. M., Lanza, R., Rosenbloom, J., Lowe, M., Holtzer, H., and Avdalović, N. (1979). Fibronectin alters the phenotypic properties of cultured chick embryo chondroblasts. *Cell (Cambridge, Mass.)* **17**, 491–501.

Wharton, K. A., Thomsen, G. H., and Gelbart, W. M. (1991). *Drosophila* 60A gene, a new TGF beta family member, is closely related to human bone morphogenetic proteins. *Proc. Natl. Acad. Sci. U.S.A.* **88**, 9214–9218.

Widelitz, R. B., Jiang, T. X., Murray, B. A., and Chuong, C. M. (1993). Adhesion molecules in skeletogenesis: II. Neural cell adhesion molecules mediate precartilaginous mesenchymal condensations and enhance chondrogenesis. *J. Cell. Physiol.* **156**, 399–411.

Williams, C. J., Considine, E., Knowlton, R. G., Reginato, A. J., Neuman, G., Harrison, D. A., Bixton, P. G., Jimenez, S. A., and Prockop, D. J. (1993). Spondyloepiphyseal dysplasia and precocious osteoarthritis in a family with an Arg75 → Cys mutation in the procollagen type II gene (COL2A1). *Hum. Genet.* **92**, 499–505.

Winterpacht, A., Hilbert, M., Schwarze, U., Mundios, S., Spranger, J., and Zabel, B. U.

(1993). Kniest and Stickler dysplasia phenotypes caused by collagen type II gene (COL2A1) defect. *Nature Genetics* **3**, 323–326.

Winterpacht, A., Hilbert, M., Scharze, U., Mundlos, S., Spranger, J., and Zabel, B. U. Familial early osteoarthrosis with mild chondrodysplasia due to a type II procollagen gene (COL2A1) point mutation resulting in a serine for glycine 274 substitution. (in press).

Wozney, J. M. (1992). The bone morphogenetic protein family and osteogenesis. *Mol. Reprod. Dev.* **32**, 160–167.

Wozney, J. M., Rosen, V., Celeste, A. J., Mitsock, C. L. M., Whitters, M. J., Kriz, R. W., Hewick, R. M., and Wang, E. A. (1988). Novel regulators of bone formation: Molecular clones and activities. *Science* **242**, 1528–1534.

Wozney, J. M., Rosen, V., Byrne, M., Celeste, A. J., Moutsatsos, I., and Wang, E. A. (1990). Growth factors influencing bone development. *J. Cell Sci.* **13**, 149–156.

Wroblewski, J., and Edwall, C. (1992). PDGF BB stimulates proliferation and differentiation in cultured chondrocytes from rat rib growth plate. *Cell Biol. Int. Rep.* **16**, 133–144.

Wu, L. N., Genge, B. R., and Wuthier, R. E. (1991a). Association between proteoglycans and matrix vesicles in the extracellular matrix of growth plate cartilage. *J. Biol. Chem.* **266**, 1187–1194.

Wu, L. N., Genge, B. R., Lloyd, G. C., and Wuthier, R. E. (1991b). Collagen-binding proteins in collagenase-released matrix vesicles from cartilage. Interaction between matrix vesicle proteins and different types of collagen. *J. Biol. Chem.* **266**, 1195–1203.

Yabsley, R. H., and Harris, W. R. (1965). The effect of shaft fractures and periosteal stripping on the vascular supply to epiphyseal plates. *J. Bone Jt. Surg.* **47**, 551–566.

Yasui, N., Ono, K., Konoomi, H., and Nagai, Y. (1984). Transitions in collagen types during endochondral ossification in human growth cartilage. *Clin. Orthop. Relat. Res.* **183**, 215–218.

Yasui, N., Benya, P. D., and Nimni, M. E. (1986). Coordinate regulation of type IX and type II collagen synthesis during growth of chick chondrocytes in retinoic acid or S-bromo-2'-deoxyuridine. *J. Biol. Chem.* **261**, 7997–8001.

Yokouchi, Y., Ohsugi, K., Sasaki, H., and Kuroiwa, A. (1991a). Chicken homeobox gene Msx-1: Structure, expression in limb buds and effect of retinoic acid. *Development (Cambridge, UK)* **113**, 431–444.

Yokouchi, Y., Sasaki, H., and Kuroiwa, A. (1991b). Homeobox gene expression correlated with the bifurcation process of limb cartilage development. *Nature (London)* **353**, 443–445.

Yoshimura, M., Jimenez, S. A., and Kaji, A. (1981). Effects of viral transformation on synthesis and secretion of collagen and fibronectin-like molecules by embryonic chick chondrocytes in culture. *J. Biol. Chem.* **256**, 9111–9117.

Yoshioka, C., and Yagi, T. (1988). Electron microscopic observations on the fate of hypertrophic chondrocytes in condylar cartilage of rat mandible. *J. Craniofacial Genet. Dev. Biol.* **8**, 253–264.

Young, H. E., Ceballos, E. M., Smith, J. C., Mancini, M. L., Wright, R. P., Ragan, B. L., Bushell, I., and Lucas, P. A. (1993). Pluripotent mesenchymal stem cells reside within avian connective tissue matrices. *In Vitro Cell Dev. Biol.* **29**, 723–736.

Zanetti, N. C., and Solursh, M. (1984). Induction of chondrogenesis in limb mesenchymal cultures by disruption of the actin cytoskeleton. *J. Cell Biol.* **99**, 115–123.

Zhan, X., Bates, B., Hu, X. G., and Goldfarb, M. (1988). The human FGF-5 oncogene encodes a novel protein related to fibroblast growth factors. *Mol. Cell. Biol.* **8**, 3487–3495.

Zhao, C. Q., Zhou, X., Eberspaecher, H., Solursh, M., and de Combrugghe, B. (1993). Cartilage homeoprotein 1, a homeoprotein selectively expressed in chondrocytes. *Proc. Natl. Acad. Sci. U.S.A.* **90**, 8633–8637.

# Index